Studies in Mechanobiology, Tissue Engineering and Biomaterials

Volume 6

Series Editor
Amit Gefen, Ramat Aviv, Israel

Further volumes of this series can be found on our homepage:
http://www.springer.com/series/8415

Aldo R. Boccaccini · Sian E. Harding
Editors

Myocardial Tissue Engineering

Springer

Prof. Aldo R. Boccaccini
Institute of Biomaterials
Department of Materials Science
  and Engineering
University of Erlangen-Nuremberg
91058 Erlangen
Germany
e-mail: aldo.boccaccini@ww.
    uni-erlangen.de

Prof. Sian E. Harding
Translational Medicine Building
Imperial College London
Hammersmith Campus
Du Cane Road
London W12 0NN
UK
e-mail: sian.harding@imperial.ac.uk

ISSN 1868-2006
ISBN 978-3-642-18055-2
DOI 10.1007/978-3-642-18056-9
Springer Heidelberg Dordrecht London New York

e-ISSN 1868-2014
e-ISBN 978-3-642-18056-9

Library of Congress Control Number: 2011936229

© Springer-Verlag Berlin Heidelberg 2011

This work is subject to copyright. All rights are reserved, whether the whole or part of the material is concerned, specifically the rights of translation, reprinting, reuse of illustrations, recitation, broadcasting, reproduction on microfilm or in any other way, and storage in data banks. Duplication of this publication or parts thereof is permitted only under the provisions of the German Copyright Law of September 9, 1965, in its current version, and permission for use must always be obtained from Springer. Violations are liable to prosecution under the German Copyright Law.

The use of general descriptive names, registered names, trademarks, etc. in this publication does not imply, even in the absence of a specific statement, that such names are exempt from the relevant protective laws and regulations and therefore free for general use.

Cover design: WMXDesign GmbH, Heidelberg

Printed on acid-free paper

Springer is part of Springer Science+Business Media (www.springer.com)

Preface

Cardiovascular diseases (CVD) are the leading cause of death in the industrialised world [1] and now have risen to a similar position for developing countries. Myocardial infarction is the main cause of death in patients with CVD, although damage to heart muscle can also occur from infection, drugs, alcohol, chemotherapeutic agents or because of congenital conditions. While much progress has been made in preventing acute deaths from myocardial infarction, in the patients who survive the initial acute event there is a common progression to heart failure. This condition has a prognosis that is poorer than many cancers and even transplantation does not fully restore life expectancy.

The adult heart cannot adequately repair the damaged tissue, as mature contracting cardiomyocytes have severely limited ability to divide [2], and the stem/progenitor population within the adult heart apparently do not provide the scale of replacement needed after substantial damage [3]. Thus the result of myocardial infarction is the formation of scar tissue with different contractile, mechanical and electrical properties to that of normal myocardium, and a heart which is unable to deliver sufficient blood to meet the body's metabolic requirements [4].

Both materials and cells of various kinds have been developed in an attempt to reverse this decline, with varying degrees of success. Myocardial tissue engineering (MTE) represents the combination of these strategies to solve problems of delivery, retention and support of introduced cells, as well as to harness additional therapeutic properties of the materials themselves. This book contains 11 chapters written by leading experts in MTE, giving a complete analysis of the field and presenting the latest advances from the research and applications points of view. The chapters cover all relevant aspects of MTE strategies, including cell sources, specific TE techniques and biomaterials used.

Many different cell types have been suggested for cell therapy in the framework of MTE, including, skeletal myoblasts, fetal cardiomyocytes, embryonic or induced pluripotent stem cell-derived cardiomyocytes, and resident adult stem/progenitor cells, all having advantages and disadvantages. Cell replacement therapy approaches are being proposed as convenient alternatives for regeneration or repair of the damaged myocardium and strategies are discussed in this book.

The book covers also a complete range of biomaterials, examining different aspects of their application in MTE, such as biocompatibility with cardiac cells, mechanical capability and compatibility with the mechanical properties of the native myocardium as well as degradation behaviour in vivo and in vitro. Although a great deal of research is being carried out in the field, this book also addresses many questions that still remain unanswered and highlights those areas in which further research efforts are required. Some of the chapters give also an insight into clinical trials and possible novel cell sources for cell therapy in MTE.

A commonly proposed method of introducing new cells into the damaged myocardium is by injection in suspension into either the circulating blood or directly into the myocardium or around the infarct. Cell delivery by direct injection can be inefficient with substantial cell loss, while intracoronary infusion may be contraindicated because of the larger size of pre-differentiated cardiomyocytes. This situation has prompted search for alternative delivery techniques for the cells whereby biomaterials play a key role in the success of several myocardial tissue regeneration strategies. In this context engineered three-dimensional scaffolds, in the form of mesh, patch or foam, cultured with relevant cells, provide a basic structure to be implanted into the infarcted region of the heart. The approach involves stitching an engineered cardiac construct, made from cardiomyocytes seeded into a suitable scaffold or matrix, directly onto the infarct region. This process ensures delivery of cells directly to the desired site, i.e. damaged myocardium. The use of structural biomaterial patches which will act not only as delivery vector for cells but also as mechanical support to the injured myocardium is being investigated worldwide and several examples of the research being carried out are presented in different chapters of this book. The expectation is that while supporting the ventricle, the wall stress will be reduced and potential scar expansion and further deterioration will be prevented. Either differentiated cardiomyocytes or cardiomyocyte progenitors could be grown in thin layer on the biomaterial patch, which would then be attached with cells in contact with the infarct area [5]. A linked hypothesis is that the residual vascularisation of the infarct region, coupled with the high resistance of embryonic cells to hypoxia, should preserve the cells until their own vascular networks are established. A number of engineered biomaterial patches are being proposed and investigated, indicating that a great variety of synthetic and natural polymers, and their combinations, is being investigated for applications in MTE approaches, as also discussed in dedicated chapters of this book.

As a summary, the book includes an introductory chapter to the field of MTE (Tissue Engineering for Cardiac Regeneration, Gaetani et al.), and two chapters describing new strategies in MTE (Inherently Bio-Active Scaffolds: Intelligent Constructs to Model the Stem Cell Niche, Di Nardo et al., Strategies for Myocardial Tissue Engineering: The Beat Goes On, Akhayari et al.). Next, a group of chapters focus on cell-based approaches (Creating Unique Cell Microenvironments for the Engineering of a Functional Cardiac Patch, Dvir et al., Intramyocardial Stem Cell Transplantation Without Tissue Engineered Constructs: The Current Clinical Situation, Donnodorf et al., Tissue Engineered Myocardium,

Zimmerman). The application of injectable bioamaterials for MTE is described in Injectable Materials for Myocardial Tissue Engineering (Singelyn et al.) while Tissue Engineering Approaches for Myocardial Bandage: Focus on Hydrogel Constructs (Giraud et al.) presents the application of hydrogel constructs in MTE. Two chapters focus on advanced polymeric biomaterials for MTE: Engineering of Multifunctional Scaffolds for Myocardial Repair Through Nanofunctionalization and Microfabrication of Novel Polymeric Biomaterials, (Rossellini et al.), focusing on new strategies for nanofunctionalisation and microfabrication of MTE constructs, and Electrospun Nanocomposites and Stem Cells in Cardiac Tissue Engineering, (Genovese et al.), describing electrospinning based scaffolds. Finally, Heart Valve Tissue Engineering (Chester et al.) presents the engineering of heart valves.

We are grateful to all authors and referees who have contributed with their efforts and timely support to make this book possible.

References

1. Lloyd-Jones, D., Adams, R.J., Brown, T.M. et al.: Heart disease and stroke statistics—2010 Update: a report from the american heart association. Circulation. **121**, 46–215 (2010)
2. Anversa, P., Leri, A., Kajstura, J., Nadal-Ginard, B.: Myocyte growth and cardiac repair. J. Mol. Cell. Cardiol. **34**, 91–105 (2002)
3. Dimmeler, S., Zeiher, A.M., Schneider, M.D.: Unchain my heart: the scientific foundations of cardiac repair. J. Clin. Investig. **115**, 572–583 (2005)
4. Vunjak-Novakovic, G., Tandon, N., Godier, A. et al.: Challenges in cardiac tissue engineering. Tissue. Eng. (Part B). **16**, 169–187 (2010)
5. Stuckey, D.J., Ishii, H., Chen, Q.Z.: Magnetic resonance imaging evaluation of remodeling by cardiac elastomeric tissue scaffold biomaterials in a rat model of myocardial infarction. Tissue. Eng. (Part A). **A16**, 3395–3402 (2010)

Prof. Aldo R. Boccaccini
Prof. Sian E. Harding

Contents

Tissue Engineering for Cardiac Regeneration. 1
Roberto Gaetani, Pieter A. F. Doevendans, Elisa Messina
and Joost P. G. Sluijter

**Inherently Bio-Active Scaffolds: Intelligent Constructs
to Model the Stem Cell Niche** 29
Paolo Di Nardo, Marilena Minieri, Annalisa Tirella, Giancarlo Forte
and Arti Ahluwalia

Strategies for Myocardial Tissue Engineering: The Beat Goes On. . . . 49
Payam Akhyari, Mareike Barth, Hug Aubin and Artur Lichtenberg

**Creating Unique Cell Microenvironments for the Engineering
of a Functional Cardiac Patch** 81
Tal Dvir, Jonathan Leor and Smadar Cohen

**Intramyocardial Stem Cell Transplantation Without Tissue
Engineered Constructs: The Current Clinical Situation** 95
Peter Donndorf and Gustav Steinhoff

Tissue Engineered Myocardium. 111
Wolfram-Hubertus Zimmermann

Injectable Materials for Myocardial Tissue Engineering 133
Jennifer M. Singelyn and Karen L. Christman

**Tissue Engineering Approaches for Myocardial Bandage:
Focus on Hydrogel Constructs** 165
Marie Noëlle Giraud and Hendrik Tevaearai

Engineering of Multifunctional Scaffolds for Myocardial Repair Through Nanofunctionalization and Microfabrication of Novel Polymeric Biomaterials 187
Elisabetta Rosellini, Caterina Cristallini, Niccoletta Barbani and Paolo Giusti

Electrospun Nanocomposites and Stem Cells in Cardiac Tissue Engineering 215
Jorge A. Genovese, Cristiano Spadaccio, Alberto Rainer and Elvio Covino

Heart Valve Tissue Engineering............................ 243
Adrian H. Chester, Magdi H. Yacoub and Patricia M. Taylor

Author Index 267

Tissue Engineering for Cardiac Regeneration

Roberto Gaetani, Pieter A. F. Doevendans, Elisa Messina and Joost P. G. Sluijter

Abstract Tissue engineering is an interdisciplinary field that involves engineering, chemistry, biology and medicine and is emerging in the last decade as a possible approach to regenerate an injured organ by using cells, matrix, biological active molecules and physiologic stimuli. In the cardiovascular field, cardiac tissue engineering (CTE) is suggested as an alternative approach for direct cell transplantation and aims to regenerate an injured myocardial ventricular wall or to repair congenital defects. This chapter will focus on the strategies developed for CTE and in particular on materials and cells used and the advantages that CTE could offers compared to conventional cell therapy.

Abbreviations

BMCs	Bone marrow cells
CHF	Congestive heart failure
CPCs	Cardiac progenitor cells
CTE	Cardiac tissue engineering
EC	Endocardial
ECM	Extracellular matrix

R. Gaetani (✉) · P. A. F. Doevendans
Department of Cardiology, HLCU, University Medical Center Utrecht,
Utrecht, The Netherlands
e-mail: R.Gaetani@umcutrecht.nl

R. Gaetani · E. Messina
Department of Experimental Medicine, Cenci-Bolognetti Foundation, Pasteur Institute,
University La Sapienza, Rome, Italy

J. P. G. Sluijter
Interuniversity Cardiology Institute of the Netherlands (ICIN), Utrecht, The Netherlands

EDV	End-diastolic volume
EDAWth	End-diastolic anterior wall thickness
EF	Ejection fraction
EPCs	Endothelial progenitor cells
FS	Fractional shortening
G-CSF	Granulocyte-colony stimulating factor
IC	Intracoronary
IV	Intravenous
IM	Intramyocardial
LV-EDV	Left ventricular end-diastolic volum
LVID	Left ventricular internal diameter
MI	Myocardial infarction
RGD	Arginine–glycine–aspartic acid

1 Introduction

Congestive heart failure (CHF) and coronary artery disease are the leading cause of morbidity and mortality in western society. Every year, numerous patients are diagnosed with congestive heart failure. In the past decade, medical, interventional and surgical therapies to treat cardiovascular diseases have advanced significantly. Although these therapies have improved the quality of life of many patients, limited strategies are available to prevent the development of heart failure, which is related to the loss of cardiomyocytes compromising myocardial contractile function [1, 2]. Heart transplantation is currently the last chance for end-stage heart failure patients, but is hampered by severe shortage of donor organs and also by rejection of the transplanted heart.

The possibility of using stem cell-based therapies for people suffering from acute myocardial infarction (MI) or living with CHF has encouraged basic, pre-clinical and clinical research to address this issue. Cell-based therapies are a promising alternative, given the basic assumption that left ventricular dysfunction is largely due to the loss of a critical number of cardiomyocytes and therefore it could be potentially reversed by implantation of new contractile cells into the post-infarction scar.

During the last few years, cellular therapy for the diseased heart has shown encouraging results on cardiac function in animal models of heart ischemia using a variety of different cells, even without a clear cardiovascular differentiation of the transplanted cells [3]. Several years after the first human clinical applications using non-cardiac stem cells as a therapy for acute myocardial infarction [4], heart failure or refractory angina, it should be recognized that the results are mixed, with benefits ranging from absent to transient, but at most marginal [3].

Preclinical studies demonstrated promising results and also short-term effects in patients were observed, however in most trials long-term follow-up has shown that conventional pharmacological therapy often has the same late outcome as cellular therapy [5–8]. Cell therapy decreased the death rate of the endogenous myocytes, improved neoangiogenesis or positively affected ventricular remodelling in short-term follow-up, probably by secretion of paracrine factors. The small improvements in ventricular function though, especially in the long-term follow-up, is probably due to the inability of the cells to actually form new cardiomyocytes. Another important issue concerning cell therapy is that this approach suffers from limitations related to variable cell retention, survival, and significant cell death or apoptosis soon after implantation in the diseased myocardium [9]. Furthermore, in ischemic heart disease both cardiomyocytes and the extracellular matrix (ECM) are destroyed or modified. Therefore, it could be important to combine procedures aiming at regenerating both myocardial cells and the ECM.

Cardiac tissue engineering (CTE) offers the possibility to combine cells and ECM giving a mechanical support to the diseased myocardium and the proper environment for the transplanted cells.

The focus of this chapter will be on CTE strategies that are currently being developed for heart regeneration.

2 Cardiac Tissue Engineering for Heart Regeneration

Tissue engineered constructs provide the promise to restore cardiac function after surgical repair. The insertion of patches or conduits is frequently required to correct congenital heart defects in infants or acquired and congenital heart diseases in adults [10, 11]. The currently available synthetic graft materials are inert and become stiff, calcify and may require re-operation because of disruption or thrombosis. For infants, the synthetic materials do not grow as the child becomes older and none of the materials begin to contract in synchrony with the heart [10].

With the increased knowledge in (stem) cell biology, CTE is developing as a strategy to combine scaffold material and cells in order to provide a regenerative approach, thereby aiming to restore the myocardial tissue. The matrix should provide a mechanical support to ventricular chamber integrity in order to limit ventricular wall dilatation, provide a favorable environment to the transplanted cells and enhance cell survival, proliferation and differentiation. The ideal matrix should be (1) biodegradable, (2) not induce any immune response by the host, (3) support electro-mechanical properties of the heart and be replaced with newly synthesized ECM. Two main approaches of CTE are thus far performed: in vitro and in vivo CTE (see Fig. 1).

- In vitro CTE: focuses on seeding cells onto pre-formed porous scaffolds. The cells are cultivated under precise culture conditions in order to enhance survival, proliferation and differentiation (see Fig. 1a). The formed patch is then applied onto the epicardial surface of the heart in order to provide mechanical support to the damaged heart and enhance cardiac performance. Secretion of paracrine factors from the cells is likely to be relevant too. The matrix and cultivation

(a)
In vitro CTE:
cells are seeded or mixed with matrix and cultivated in vitro

(b)
In vivo CTE:
cells are mixed with the matrix and injected into the infarcted myocardium

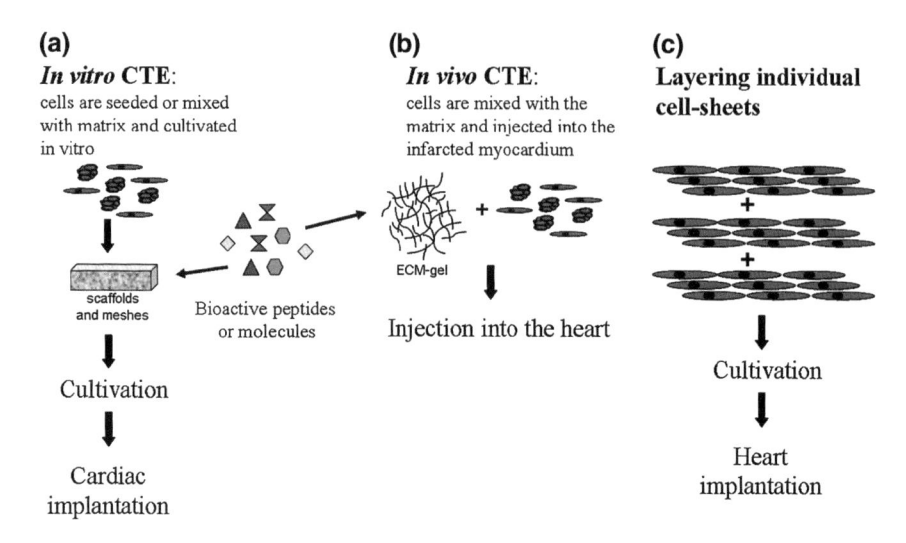

(c)
Layering individual cell-sheets

Fig. 1 CTE strategies

parameters play an important role to guide and organize seeded cells in order to control and obtain a myocardial patch that is as similar to the native heart tissue as possible. The advantage of the in vitro approach is the possibility to control construct shape, size, cell differentiation rate and organization of the seeded cells. The main limitation of this approach is the need for nutrient diffusion that limits the thickness of the constructs. Therefore vascularization is required in addition to mature contractile myocardial cells (see Tables 1 or 2). The use of a perfusion-bioreactor can improve viability of the cultivated cells [12, 13] and/or enhance differentiation by applying mechanical stress [13, 14].

- In vivo or in situ CTE: has emerged in the last years and involves injection of a mixture of biomaterials and cells. With this approach, cells and biomaterial are mixed and delivered by injection into the ventricular wall (see Fig. 1b). This will improve survival and retention of the transplanted cells and provide a mechanical support to the damaged ventricle at the same time (see Tables 3 and 4). The advantage is that regeneration and replacement of the damaged tissue is done in the natural milieu of the damaged myocardium. This approach is easy and feasible, but cell growth and differentiation cannot be controlled as tightly as in the in vitro model.

3 Survival and Engraftment of Transplanted Cell: Strategies to Improve Myocardial Regeneration

The first limitation of direct cell injection is the considerable cell death that occurs in the first seven days after infarction due to inflammation and ischemia. Zhang et al. found that high levels of cardiomyocyte death occured whitin 4 days after

Table 1 Experimental design of in vivo CTE studies

Scaffold	Cell type	Animal model	Application time	Evaluation time	Groups	Ref.
Fibrin glue	Myoblast (5×10^6)	Rat	1 w post MI	5 w post MI	BSA; cells; matrix; cells + matrix	[48, 49]
Alginate or fibrin glue	No cells	Rat	5 w post MI	10 w post MI	BSA; alginate; fibrin glue;	[50]
Fibrin glue	BMMCs (5×10^6)	Rat	3 w post MI	8 w post MI	Medium; cells; cells + matrix	[51]
Collagen/Matrigel/fibrin	No cells	Rat	1 w post MI	6 w post MI	Medium; collagen; Matrigel; fibrin;	[63]
Collagen	No cells	Rat	1 w post MI	7 w post MI	Saline; matrix	[64]
Alginate	Neonatal CM (1×10^6)	Rat	2 w post MI/ 10 w post MI	10 w post MI/ 18 w post MI	Saline; cells; alginate/saline; matrix	[72]
Alginate	No cells	Rat	1 w post MI	5 w post MI	Saline; alginate; alginate/VEGF; alginate/PDGF; alginate/VEGF/PDGF	[73]
Alginate + RGD	No cells	Rat	5 w post MI	10 w post MI	BSA; alginate; alginate + RGD	[75]
Alginate + RGD	No cells	Rat	1 w post MI	9 w post MI	BSA; alginate; alginate + RGD	[74]
Self assembling peptide nanofibers (NF)-PDGF	No cells	Rat	After MI	2 w post MI	Sham; PDGF-BB (50 ug and 100 ug, single injection); PDGF-BB (50 ug and 100 ug)	[33]
Self assembling NF-IGF-1	Rat CPCs (100000)	Rat	After MI	4 w post MI	PBS; CPCs; NF-IGF-1; CPCs + NF-IGF-1;	[34]

Abbreviation CM: Cardimyocytes; BMMCs: Bone marrow mononuclear cells; HUCBCs: Human umbilical cord blood cells; CPCs: Cardiac progenitor cells; MI: Myocardial infarction; w: week

Table 2 Benefits of in vivo CTE application on cardiac function

Scaffold/cells	Improvement in cardiac function							Vasculogenesis	Ref.
	W.T.	E.F.	S.V.	LVSD	LVDD	F.S.	FAC		
Fibrin glue/myoblast	↑#§	↑#§	↑#§	n. d.	n. d.	n. d.	n. d.	↑°⊙	[48, 49]
Alginate (A) or fibrin glue(FG)	FG: ↑§ A: ↑§	n. d.	FG: →§ A: ↑§	FG: →§ A: ↑§	FG: →§ A: ↑§	n. d.	n. d.	FG: ↑§ A: ↑§	[50]
Fibrin glue/BMMCs	n. d.	n. d.	n. d.	n. d.	n. d.	n. d.	n. d.	↑#°	[51]
Collagen/matrigel/fibrin	n. d.	n. d.	n. d.	n. d.	n. d.	n. d.	n. d.	↑§	[63]
Collagen	↑§	↑§	↑§	±§	±§	n. d.	n. d.	n. d.	[64]
Alginate/neonatal CM	2w: ↑§ 10w: ±§	n. d.	2 w: ±§ 10w: ↑§	2 w: ↑§ 10 w: ±§	2 w: ↑§ 10 w: ±§	↑§	2 w: ±§ 10 w: ↑§	n. d.	[72]
Alginate + VEGF/PDGF	n. d.	↑	n. d.	n. d.	↑	n. d.	n. d.	↑Alg. VEGF or Alg. VEGF/PDGF versus CTR	[73]
Alginate + RGD	↑§	n. d.	n. d.	↑§	↑§	↑§	n. d.	↑§ ± Alg. RGD versus Alg.	[75]
Alginate + RGD	↑§	n. d.	↑Alg. only	↑Alg. only	↑Alg. only	n. d.	↑Alg. only	→	[74]
NF-PDGF	n. d.	n. d.	n. d.	↑PN-PDGF	↑PN-PDGF	↑NF-100 versus other groups	n. d.	→	[33]
NF-IGF-1	↑#§	↑#*°§	n. d.	n. d.	n. d.	n. d.	n. d.	↑#*°§	[34]

Abbreviations W.T: Wall thickness; E.F: Ejection fraction; S.V: Stroke volume; LVSD: Left ventricular systolic dimension; LVDD: Left ventricular diastolic dimension; F.S: Fractional shortening; V.R: Ventricular remodelling; F.A.C: fractional area change
↑: Statistically significant improvement; ±: Not significant increase; →: No improvement
Cells — matrix versus control; * Cells + matrix versus matrix; ° Cells + matrix versus cells; § Matrix versus cells; ⊙ Cells versus matrix

Table 3 Experimental design of in vitro CTE studies

Scaffold	Cell type	Animal model	Application time	Evaluation time	Groups	Ref.
Collagen	No cells	Rat	After MI	6 w post MI	Sham; MI; sham + scaffold; MI + scaffold; MI + G-CSF; MI + scaffold + G-CSF	[58]
Collagen	HUCBCs	Mouse	After MI	6 w post MI	Sham; cells; matrix; matrix + cells	[59]
Alginate	Fetal heart cells (3×10^5)	Rat	1 w post MI	10 w post MI	Sham; cells; matrix; matrix + cells	[69]
Alginate (in vivo pre-vascularization)	Fetal CM (2.5×10^6)	Rat	1 w post MI	5 w post MI	Sham; matrix (in vivo bioreactor); matrix + cells (in vitro); matrix + cells (in vivo);	[71]
EHT	Neonatal heart cells (2.5×10^6 cells/ EHT)	Rats	2 w post MI	6 w post MI	Matrix; EHT	[79]
Cell-sheet engineering	Adipose tissue-MSCs D.F.	Rat	4 w post MI	4 w post MI	MSCs-based sheet D.F.-based sheet	[88]
Cell-sheet engineering	Neonatal CM/ECs	Rat	2 w post MI	6 w post MI	Sham; CMs (2.4×10^6); CMs + ECs (2×10^5); CMs + ECs (4×10^5); CMs + ECs (8×10^5).	[89]
Urinary bladder matrix (UBM)	No cells MSCs[95]	Dog	Right ventricular defect/after surgery	8 w after surgery	Sham; UBM, and Dacron.	[92, 95]
Small intestinal submucosa (SIS)	MSCs	Rabbit	4 w post MI	8 w post MI	Sham; MI; SIS group; MSCs-seeded SIS group;	[96]

Abbreviation CM: Cardimyocytes; BMMCs: Bone marrow mononuclear cells; HUCBCs: Human umbilical cord blood cells; CPCs: Cardiac progenitor cells; MI: Myocardial infarction; w: week

Table 4 Benefits of in vitro CTE application on cardiac function

Scaffold/cells	Improvement in cardiac function							Vasculogenesis	Ref.
	W.T.	E.F.	S.V.	LVSD	LVDD	V.R.	FAC/FS		
Collagen	n. d	n. d.	n. d.	n. d.	n. d.	↑§	n. d.	↑§ ↑↑matrix + G-CSF versus matrix	[58]
Collagen	↑#*°	↑#*	n. d.	n. d.	↑#*°	n. d.	n. d.	n. d.	[59]
Alginate/fetal heart cells	↑#	n. d.	↑#	↑#	↑#	n. d.	n. d.	↑#	[69]
Alginate fetal C.M.	↑#§	n. d.	↑#§	↑#§	↑#§	n. d.	↑#§	↑#§	[71]
EHT/neonatal heart cells	n. d	n. d.	n. d.	n. d.	↑*	n. d.	↑*	n. d.	[79]
Cell-sheet engineering aMSCs/D.F	↑MSCs versus DF and sham	n. d.	n. d.	n. d.	↑MSCs versus DF and sham	↑MSCs versus DF and sham	↑MSCs versus DF and sham	↑MSCs versus DF and sham	[88]
Cell-sheet engineering Neonatal CM./ECs	↑#	n. d.	n. d.	n. d.	n. d.	n. d.	↑ ECs (4×10^5/ 8×10^5)	↑CMs/ECs versus CM	[89]
UBM	↑§, UBM versus Dracon	n. d.	n. d.	n. d.	n. d.	n. d.	n. d.	↑UBM versus Dracon	[92, 95]
SIS/MSCs	↑#*§	↑#*§	n. d.	↑#*§	↑#§	n. d.	n. d.	↑#*§	[96]

Abbreviations W.T: Wall thickness; E.F: Ejection fraction; S.V: Stroke volume; LVSD: Left ventricular systolic dimension; LVDD: Left ventricular diastolic dimension; F.S: Fractional shortening; V.R: Ventricular remodelling; F.A.C: fractional area change
↑: Statistically significant improvement; ±: Not significant increase; →: No improvement
Cells + matrix versus control; * Cells + matrix versus matrix; ° Cells + matrix versus cells; § Matrix versus control; ^ Matrix versus cells; ⊙ Cells versus matrix

implanting into injured hearts and suggested ischemia as the main cause [15]. Similar data were reported in various animal models, thereby using different cell types and routes of administration. Generally more than 50% of cells die within the first days of implantantion [9], and only 5–10% of surviving cells could be reported in a pig [16, 17] and a rat model [18].

The ischemic environment induces cell death and needs vascularization since survival of transplanted cardiomyocytes is directly proportional to tissue perfusion [19]. Injected cells show signs of irreversible ischemic injury after 1 day [15]. The lack of matrix can also induce cell death and is mediated by the Anoikis signaling pathway. This is particularly important during collection and preparation of the cells for transplantation [20].

Different strategies were developed in order to enhance cell survival upon transplantation and include the use of bioactive molecules to prevent cell death. Different studies demonstrated that cell engraftment can be enhanced by cytoprotective molecules. This includes transgenic overexpression of Akt, BCL-2 [21, 22] or hypoxia inducible factor-1α (HIF-1α) [23]. In addition in vitro preconditionig [24, 25] or simultaneous administration with IGF-1 [26–28], VEGF [29–31] or free radical scavenger [32] are all strategies that demonstrated to increase the benefits of cell therapy.

However, co-administration of bioactive molecules results in a temporary benefit related to the time of administration. Alternatively, genetic modification of transplantable cells, in order to enhance survival, is hampered for possible tumorigenic side effects.

To this end, CTE represents a possible solution to increase cell survival and cell retention in the heart. Different bioactive molecules can be linked to a specific matrix in order to provide a controlled release of the molecules. Coupling PDGF-BB and IGF-1 to self-assembling peptide nanofibers demonstrated a prolonged delivery of the growth factors and resulted in an improved cardiac function comapared to a control group that received only a single injection [28, 33]. Interestingly, recently it was demonstrated that injection of rat cardiac progenitor cells (CPCs) together with IGF-1 modified peptide nanofinbers (NF-IGF-1) enhanced heart function. The combination preserved LV function and increased myocyte regeneration as compared to saline, cells or NF-IGF-1 only [34].

Although we are able to improve cell survival by increasing resistance to death stimuli, another important problem is related to cell retention in the heart. Different studies demonstrated that a large number of cells are trapped in other organs if injected systemically or are launched from the site of myocardial injection [35, 36].

When cells are injected into an arrested heart by thiopental injection, the cell retention is 4 times higher than injection into a beating heart. This indicates that cardiac contraction and/or coronary perfusion are important in the early washout of cells [37]. In addition, an increase in cell retention is obtained if one applies fibrin glue on the site of injection [37]. Survival and retention can also be improved by injecting cells together with a matrix such as liquid collagen [38, 39] or fibrin glue [40] resulting in improved cardiac function and remodeling.

In summary, it is evident that the first step to optimize the repair of a damaged myocardium is to increase cell retention and survival. This aspect is particularly important if we consider the absolute number of cardiomyocytes that are lost after MI and that one of the main limitations of cellular therapy is the number of cells that will be needed. CTE offers different approaches for this problem and each strategy needs to be studied in relation to the time of administration, disease state (acute versus chronic MI) and cells used.

4 The Importance of Extracellular Matrix

The view that the ECM is a passive component of the heart and only provides support for cells is changed in the last decade. ECM plays an important role in tissue homeostasis and regulates cell proliferation, differentiation and migration. During embryonic development ECM patterns vary in a specific manner and ECM components such as collagen, fibronectin or laminin are temporally expressed in different stages [41]. These findings could also be translated in vitro where 3D cultures altered embryonic stem cell differentiation as compared to 2D, by inducing a more mature phenotype [42]. Accordingly, it was demonstrated that increasing laminin and fibronectin concentrations in a collagen based matrix drives embryonic stem cell differentiation to cardiac or endothelial lineages, respectively [43]. ECM is also involved in cardiomyocyte and fibroblast function in response to the myocardial physical forces that activate specific intracellular signalling pathways [44, 45].

To summarize, it is evident that in order to obtain cardiac regeneration and restore ventricular function by CTE, the fundamental choice is an appropriate cell source and matrix. The matrix should support growth, proliferation and differentiation of the cells and should have physical and mechanical properties that support the natural structure of the heart. On the other end, the scaffolds should enhance ingrowth and vascularisation of the host cells and electro-mechanical integration with the ventricular tissue. Different CTE strategies can be performed and the choice of the matrix must be made according to the type of application and cells used.

5 Natural ECM for CTE Application

5.1 Fibrin Scaffold

Fibrin is a natural biopolymer composed of monomers of fibrinogen, a soluble plasma glycoprotein, and is involved in blood coagulation, platelet activation, signal transduction and inflammatory responses. Fibrinogen is composed of three polypeptide chains; two of them, Aα and Bβ, are cleaved by the protease thrombin thereby exposing their polymerization sites [46]. The cleaved monomers will then self-associate to form a fibrin netwok that is further stabilized by the transglutaminase factor XIIIa and cross-linking of γ chains to form a stable polymer [46]. The use of fibrin in the clinic is already well established as a haemostatic agent and used as a sealant in surgical procedures or for haemophiliac patients [47].

The main advantages using fibrin in scaffolds is the possibilty to produce autologous fibrin matrix from patients peripheral blood, thereby reducing the costs of the matrix and possible viral infections [47]. Moreover, other advantages are the adhesion properties of the materials and the possibilty of an easy and homogeneous cell distribution. In the last decade the use of fibrin, by means of hydrogel or glue, as potential matrix for cardiac tissue regeneration was investigated.

Fibrin glue is normally obtained after fractionation of pooled plasma. The two main components are a cryoprecipitate, concentrated fibrinogen and aprotinin, and a combination of thrombin and Calcium Chloride ($CaCl_2$). Using this system, it is possible to separate cells and matrix until the experimental setup needs polymerization such as formation of the bio-complex in vivo immediately after an experimental MI or in the post-infarction scar in a chronic model. Delivering in vivo with a Duploject applicator, allows simultaneous mixing of cells into the matrix and delivery in vivo [46].

Christman et al. transplanted myoblast, fibrin glue, or myoblast in fibrin glue in rats at 1 week post-MI. In the control group, the authors reported a significant deterioration in fractional shortening (%FS) and a significant decrease in wall thickness after 4 weeks of MI. In the treatment groups, the preserved %FS and infarct wall thickness, but no significant differences could be observed between treated groups [48]. By histological analysis, it was shown that cells were engrafted after 4 weeks and formed a viable graft. Interestingly, when only myoblasts were injected they were present mainly in the border zone, but when cells were injected within the fibrin scaffold they survived also in the infarct zone, indicating a positive effect of the matrix in the migratory properties of the cells. In another paper these authors reported that, 24 h post injection, no differences of engrafted cells could be observed between cells transplanted with or without fibrin glue. However, after 4 weeks the percentage of engrafted cells was significantly higher with the use of fibrin glue (9.7 $\pm$ 4.2% versus 4.3 $\pm$ 1.5%), indicating that injection of cells with fibrin glue improves survival or proliferation of the injected cells, but not cell retention per se [49]. Although no differences in the infarct size were observed between groups, both fibrin glue treatments displayed a small significant reduction of the scar size as compared with cell injection only, probably due to improved neovascularization in the infarcted area [49]. When fibrin glue is injected directly into the scar in chronic heart failure (e.g., 5 weeks after MI), the treatment improved FS and increased wall thickness and left ventricular internal diameter (LVID) already at 2 days post-injection. Interestingly, 5 weeks after injection (10 weeks after MI), deterioration of FS was observed in both groups, while wall thickness was preserved in the fibrin-injected group [50]. This study also confirmed previous reported results regarding the angiogenic properties of the fibrin glue.

In a similar approach, bone marrow mononuclear cells were transplanted 3 weeks after MI into the infarct region. Eight weeks after injection, fibrin–matrix injected cells were found in the infarct region and improved tissue regeneration by an increased micro-vessel density and a significant larger internal diameter of new capillaries [51]. Similar results were obtained by injecting bone marrow-derived mesenchymal stem cells in fibrin glue as compared to treatment with vehicle, cells or fibrin only [52]. Here, authors suggested that an increase in the levels of VEGF [52] in the infarct scar was a possible mechanism for the increased neovascularization observed.

In an attempt to generate vascularized 3D cardiac tissue, Birla et al. mixed neonatal cardiomyocytes and fibrin gel in a small silicone tube, made a longitudinal slit and placed the tube around the femoral artery of a recipient rat in order to generate an intrinsic supply. Three weeks after implementation, the explanted construct was full of vascularized tissue with "cardiac-like" contractile properties and responding to epinephrine and Ca^{2+} ion stimulation [53]. Recently a similar approach was performed with primary cardiac cells, isolated from neonatal rat hearts. After 3 weeks, the implanted tissue presented a high number of viable cells with aligned fibers, neovascularization, collagen production and cardiac contractile properties [54].

One of the advantages of injectable biopolymers is the possibility to deliver the hydrogel percutaneously via catheters, thereby avoiding large surgical procedures and the associated risk and costs. Martens et al. studied different catheter delivery methods with variable concentrations of fibrinogen and thrombin and demonstrated that this approach was feasible for future clinical applicability [40]. In addition, they demonstrated that, by using human MSC, cardiac-specific retention is increased after transplantation in a rat model of acute MI (39.7% ± 7.0% versus 22.6 ± 3.5%) when cells are transplanted with fibrin compared to saline. This resulted in a decrease in the presence of MSCs in liver and kidney, but not in the lungs [40]. Additionally, fibrin hydrogel can be coupled to biologically active peptides in a relatively simple reaction thereby improving the properties and the advantage of the matrix. Different bioactive domains have been synthesized with transglutaminase sequences such as laminin, collagen and the arginine–glycine–aspartic acid (RGD)-containing peptides. The incorporation of the heparin binding domain allows the possibility for a slow delivery of heparin binding growth factors by a cell-mediated degradation [47]. These possible modifications are also important to enhance the mechanical and biological properties of the matrix, thereby increasing cell survival, proliferation and differentiation of cultivated cells.

5.2 Collagen Scaffold

Collagen is one of the main ECM compounds in all organs. Extracellular cardiac connective tissue is composed mainly of collagen with smaller amounts of laminin, fibronectin and elastin. The ECM is important to provide structural and functional integrity and contributes to coordinated mechanical support of heart function. Two main types of collagen are present in the heart, type I and type III, covering almost 85% of total collagen. Following ischemia the balance of these two types of collagen fibers is shifted. The relative amount of collagen I decreases while in total there is increased collagen deposition or pathological fibrosis. Fibrosis contributes to adverse myocardial remodeling. This emphasizes the important role of collagen for the normal activity of the heart.

Preformed porous dry collagen is already used in clinical setting for 30 years as a haemostatic agent to repair tissue injuries and prevent hemorrhages [55]. Kofidis et al. utilized preformed collagen matrix for the first time to cultivate neonatal rat cardiomyocytes and presented a model of in vitro contractile cardiac tissue. Despite the fact that the central scaffold area was 2 mm thick, good viability and good homogenous cell distribution was reported. The construct, having cardiomyocyte bundles, displayed a synchronous and macroscopical visible conctraction up to 12 weeks in culture. Contractile forces could be regulated with administration of epinephrine [56, 57]. When these constructs were transplanted into the epicardial suface of infarcted rat hearts, they induced neovascularization and reduced left ventricular dilatation up to 3 weeks. Interestingly, co-administration with granulocyte-colony stimulating factor (G-CSF) lead to an increase in vascularization, probably due to an increased stem cell mobilization and homing to the infarcted myocardium [58]. Similar results were found when human umbilical cord derived stem cells were seeded in these collagen matrices and transplanted into 7 days old infarcts in a mouse. After 45 days, the cell–matrix combination displayed a lower End-diastolic volume (EDV) and improved ejection fraction (EF) as compared to monotreatments [59].

The clinical relevance of this matrix was also tested in a small non randomized clinical trial with patients having a chronic scar, an EF < 35% and indication of concomitant single off-pump coronary artery bypass graft surgery (OP-CABG). Ten patients received direct injection of bone marrow mononuclear cells and ten patients received cells seeded into a scaffold of $7 \times 5 \times 0.6$ cm and sutured over the infarct and peri-infarct zones. After 1 year follow up (10 ± 3.5 months), no adverse events were reported, suggesting feasibility and safety of the treatment. In addition, an increased EF, improved deceleration time, decreased left ventricular end-diastolic volume (LV-EDV), and reduced scar area thickness could be observed. Although promising, further studies are warranted in order to better evaluate cardiac function improvement through the scaffold because of the concomitant CABG mediated revascularization [60, 61].

Further modifications of this pre-formed collagen matrix was studied by Schussler et al. They coupled an RGD peptide to the collagen material in order to increase viability, contractile performance and differentiation of seeded cardiomyocytes [62]. When compared to non-modified collagen, RGD modified scaffold increased viability of transplanted cardiomyocytes up to 1 month in vitro. In addition, the cells were better aligned and elongated, with the presence of cross striation, and contractile performance was increased threefold as compared with the non-modified scaffolds.

Collagen can also be used in a liquid form thereby having the advantage that cells are more homogeneous distributed than in the preformed scaffold. The easy gelation at physiological temperature and the possibility to combine with other hydrogels (e.g., EHT model) are beneficial features of the material. In addition to in vitro preparation for tissue engineering. Huang et al. injected liquid collagen in a rat model of reperfusion injury. They demonstrated a higher number of capillaries 5 weeks after transplantation when compared to saline groups, but similar to fibrin or matrigel usage [63]. In a similar study, injection of collagen improved cardiac stroke volume and EF. However, in contrast with the previous study, no matrix infiltrating cells or induction of vascularization were observed in the histological analysis [64].

In summary, the easy use of collagen matrix makes it a good candidate for CTE considering the beneficial influence on cell attachment, proliferation and functional maturation of the cultivated/injected cells. Moreover, it mimics the native environment because collagen is one of the main ECM components of the heart.

5.3 Hyaluronic Acid

Another component of the natural ECM is hyaluronic acid (HA), a high molecular weight linear polysaccharide. It regulates different physiologic processes by interacting with other ECM components, such as proteoglycans, thereby maintaining hydration and supporting the biomechanical properties of the tissue. On the other end, HA can bind cellular surfaces and can regulate cell attachment, migration, proliferation, inflammation and wound healing responses [65]. When cardiomyocytes were cultivated on HA micro-fluidic patterns, they elongated and aligned along the HA pattern and formed contractile organoids with a diameter of ~ 16 μm [66]. In addition, Ventura et al. developed a mixture ester of hyaluronan with butyric and retinoic acid and reported the potential of this matrix to enhance cardiogenic differentiation of human mesenchymal stem cells [67]. When the matrix is transplanted into infarcted heart, cardiac function is improved compared to PBS injection [68].

Only a few applications of HA in tissue engineering were tested, mostly in combination with other matrices, either synthetic or natural.

5.4 Alginate

Alginate is a polysaccharide derived from the cell wall of brown algae and is a polymer composed of mannuronic and glucuronic acid. One of the main advantages of this material is the easy gelation in the presence of Ca^{2+} or other divalent cations, and, together with a high biocompatibility, represents a good material for cell immobilization or drug encapsulation.

Leor et al. seeded fetal cardiomyocytes on a sodium alginate matrix, which showed good viability in vitro and formed a beating cardiomyocyte structure. When transplanted in vivo, cells within the graft were able to differentiate further into mature cardiomyocytes, the grafts were also vascularized and infiltrated by inflammatory cells [69]. For the next step Amir et al. cultivated fetal cardiomyocytes in a preformed highly porous alginate scaffold [70]. After 4 days in culture the patch was transplanted into the peritoneal cavity of a rat and the animal was used as a bioreactor. One week later, the transplanted patch was vascularized with host cells and applied on the injured cardiac rat heart. Four weeks after transplantation, the sutured patches were still present on the epicardial surface, but without expression of cardiac markers and infiltration of myofibroblasts and macrophages. Echocardiographic analysis revealed an improved ventricular function with synchronized contractile activity [70]. Cardiomyocytes were also seeded into an alginate patch with growth factor reduced matrigel and a mixture of pro-angiogenic and pro-survival factors [71]. This showed the presence of bundles of cardiac muscle after the vascularisation process and integration with the host myocardium 4 weeks after transplantation. Moreover, they could observe scar thickening, reduced detrimental remodeling and improved cardiac function [71].

The application of alginate as a gel directly injected into the heart was also reported. When alginate was injected directly in the myocardial wall of a 7 days old MI, the alginate matrix significantly increased scar thickness, systolic and diastolic wall thickening, and cardiac function as compared to saline or cardiomyocyte injection only. Similar beneficial effects in scar thickness and remodeling were obtained also when alginate is injected in a 2 months old infarct [72].

Hao et al. used a VEGF-A or PDGF-BB modified alginate scaffold and injected this into a MI model [73]. The controlled release of the growth factors (alone or in combination) from the matrix was confirmed experimentally in vitro. Although the addition of growth factors induced vessel formation, increased smooth muscle cell infiltration, no differences in left ventricle diastolic-diameter and E.F could be observed [73]. RGD peptide was also coupled to alginate scaffold and when implanted into a MI model, different results in term of in vivo vascularisation and cardiac function were reported by two different studies when compared with unmodified alginate [74, 75].

5.5 *Engineered Heart Tissue*

One of the first well established 3D-myocardial in vitro models was performed almost 10 years ago by Zimmerman et al. and called engineered heart tissue (EHT) [76]. EHTs consist of a mixture of liquid collagen, growth factor-reduced matrigel and neonatal rat cardiomyocytes [76, 77]. The mixture is then casted into a circular mold, statically cultivated for 7 days and subjected to phasic mechanical stretch for another 7 days. The resulting engineered tissue is composed of multicellular clusters and longitudinally oriented cell bundles with features similar to that of native differentiated myocardium, although having smaller sized cardiomyocytes compared to adult tissue. Histological and ultra-structural analysis showed that the tissue is composed mainly of cardiomyocytes with the presence of smooth muscle cells, primitive capillaries, fibroblasts and macrophages. After 1 week in culture, full development of the cardiomyocytes was achieved and resulted in a high degree of sarcomeric organization with electrophysiological and contractile properties similar to that of native heart muscle [77]. Interestingly, when transplanted into the heart the constructs retained their contractile properties but, despite a good engraftment, the lack of functional cardiac improvement and the problems of immunogenicity have led to further improvements of this approach [78]. Three modifications of the culture conditions were done in order to increase therapeutic benefit of the transplanted tissue [79]: (1) the construct was cultivated under auxotonic load, elevated oxygen concentration; (2) the culture medium was supplemented with insuline in order to increase graft sizes, (3) 5 circular EHTs were stacked together to form a contractile multi-loop of longitudinally orientated muscle strands. These new EHT-constructs integrated fully in the host myocardium 4 weeks after in vivo implantation, having a diameter of about 450 µm, and formed new capillaries connected with recipient vessels. Engraftment and electrical coupling of the EHT patches is demonstrated by the propagation of electrical responses in remote myocardium and functional improvements of myocardial function [79]. In order to reduce immunogenicity of the EHT and establish culture condition more close to clinical translation, EHT could be generated with a mixture of heart derived cells (cardiomyocytes, smooth muscle cells, fibroblast and endothelial cells), the use of growth factors to avoid serum contamination and the use of triiodothyronine and insulin to facilitate withdrawal of Matrigel [80]. In addition, pouch-like EHTs could be developed and placed over the entire ventricular chamber, from apex to base of a rat heart [81].

This EHT model was also produced with embryonic stem (ES) cell-derived cardiomyocytes with in vitro characteristics similar to the previously published studies [82]. After transplantation subcutaneously into immunodeficent mouse no formation of teratoma was reported.

6 Building a Cardiac Tissue without ECM Components: Cell-Sheet Based Engineering

An interesting approach has been developed by cell sheet tissue engineering. To make sheets, cells are cultured on temperature-responsive culture dishes, made of poly-*N*-isopropylacrylamide. This material is hydrophobic and promotes cell attachment and proliferation in the appropriate medium at 37°C. By decreasing the temperature to 32°C, the polymer becomes hydrophilic and cell sheets detach from the plate. Shimizu et al. seeded primary-isolated cardiomyocytes on these sheets [83] and demonstrated that these cell layers could be collected. Furthermore when implanted subcutaneously in rats, the transplanted tissue displayed a good cardiomyocyte organization with the presence of sarcomeric structures and spontaneous contractile activity [84]. Individual cell-sheets were collected and layered on top of each other, the sheets fused, started beating synchronously and became vascularized in vivo within 4 weeks [84, 85]. Moreover when placed around a dissected abdominal aorta, this myocardial tube increased in thickness and displayed a mature sarcomeric structure [86] probably due to an hypertrophic response of the cardiomyocyte to the mechanical stress of the host blood flow [14]. When the patch is transplanted onto infarcted rat hearts, donor cardiomyocytes formed connections via gap junctions and intercalated disks with the host tissue [87]. In addition, when adipose tissue-derived MSCs were cultured similarly and transplanted in a chronic MI model, the transplanted sheets increased in size compared to native tissue and contributed to improved cardiac function. The engrafted sheets mainly consist of cells expressing the mesenchymal markers and only a small fraction expressed endothelial, cardiac or smooth muscle cell markers [88]. When increasing numbers of endothelial cells were added, this increased vasculature formation and microvessel density, both in vitro and in vivo. Interestingly, 4 weeks after implantation this was also related to an improved fractional shortening, end-diastolic anterior wall thickness, increased secretion of VEGF, bFGF, and HGF, and less formation of fibrosis [89].

One of the main advantages of the thermo-responsive cell-sheet approach is the lack of any additional ECM matrix components, thereby reducing the problem of immunogenicity and bio-compatibility of the host. However different issues need to be resolved such as the cell source and the limited number of sheets that can be layered to avoid blocking of nutrition and oxygen diffusion and cell death. The possibilities to use endothelial progenitor cells, in combination with other cell types, and the good vascularization of the grafts in vivo are important features of this approach and can improve the applicability.

7 Tissue Derived Extracellular Matrix

The use of native tissue and isolating the ECM from cardiac tissue is another intriguing opportunity for the further development of tissue engineering. The main advantage of this approach is the presence of all the different matrix components and the native tissue architecture. Different tissue-specific protocols for the isolation of the ECM have been developed in the last years, thereby using different de-cellularization agents to preserve secondary matrix organization.

Badylak et al. used a porcine derived a-cellular matrix to repair a right ventricular defect in a pig model. The authors used a patch of porcine small intestinal submucosa (SIS) or urinary bladder matrix (UBM), demonstrating that the implanted patch was completely replaced by native tissue with different cell types, vascularization and reduced detrimental tissue remodeling [90]. The same group implanted UBM and expanded polytetrafluoroethlyene (ePTFE) to the left ventricular wall in a pig model of MI [91]. The matrix was still visible at 1 month but bioresorbed after 3 months. At 3 months the UBM group showed presence of myofibroblast and regions of cells that expressed both cardiac sarcomeric and α-smooth muscle actin, indicating a positive effect in the endogenous—mediated cardiac regeneration [91].

When UBM was patched in a canine model with right ventricular defects, it resulted in an increase in global systolic and diastolic function, higher number of cardiomyocytes in the scar region and improved vascularization after 8 weeks [92, 93]. In addition, the improvement in global function was linearly correlated with increased number of cardiomyocytes in the UBM patch-transplanted region [94].

Potapova et al. pre-differentiated hMSCs to the cardiogenic lineage and seeded differentiated and undifferentiated cells in the native isolated matrix. After a period ranging from 3 to 8 days in culture, the seeded patch was placed around the right ventricle of a canine model. Interestingly, when cells are not pre-differentiated in vitro, the recovery of regional function is similar to previously reported results with only the matrix. However, the group that received pre-differentiated cells an improved regional global function and high numbers of striated cells were reported [95].

MSCs were also cultivated in small intestinal submucosa (SIS) patches for 5–7 days and than transplanted on the left ventricular wall of a rabbit model of MI. After 4 weeks the use of matrix improved cardiac function in both groups compared to control. The use of MSCs in this study is also related to a small increase in cardiac function (not in all parameters analyzed) and higher increase in vasculogenesis, as compared to the groups treated with matrix only [96].

Another intriguing approach is the use of a-cellular matrix as a gel that can be injected into the heart. Singelyn et al. recently reported the isolation of ECM from pig ventricle as a gel [97]. This cardiac ECM is a viscous liquid solution on ice or at room temperature and will become a gel at 37°C [97, 98]. Preliminary studies demonstrated that the gel, when placed in a clinically compatible catheter, can be pushed through with minimal resistance, indicating the clinical applicability of the materials [97]. Recently, it was reported that the isolation of pericardial ECM [99] and SIS [100] is possible as a gel and can be used for CTE.

The importance of the native tissue architecture is reported by Ott et al. [101]. In this study they generated the complete bioartificial heart. The authors de-cellularized a whole rat heart, thereby preserving its fiber orientation of collagen I and III, laminin and fibronectin. Subsequently, the heart was repopulated by intramural injection of cardiac derived cells and by infusion of rat aortic endothelial cells. When the recellularized heart is cultivated in a bioreactor for 8 days, providing a simulated systolic and diastolic medium flow, the construct displayed electrical activity and responded to drugs [101]. This seminal paper should be considered a milestone in tissue engineering of whole organs, by decellularizing a whole heart, keeping ECM architecture intact, and re-cellularization to obtain potentially a functional pumping organ. However, although intriguing, challenging and amazing, the clinical applicability of this model is limited. Maybe it is possible to rebuild cardiac tissue partly or use the model for drug toxicity screening.

8 Synthetic Extracellular Matrix for Cardiac Tissue Engineering

Synthetic ECM includes different classes of biodegradable materials such as polyester, polylactones and polyuretanes. The most widely known members of the polyester family used for synthetic TE applications are poly-glycolic acid (PGA) [102–104], poly-lactic acid (PLA) [105] and their copolymers poly(lactic-glycolic) acid (PLGA) [106]. The main advantage is that porosity, density, structure and composition can be controlled and designed as needed for different cell types and applications. For this reason their use is reported in different organ systems. Combining different amounts of the materials makes a controlled degradation upon implantation possible [107]. Unfortunately, an immune reaction was reported after implantation [108] and accumulation of the acid degradation metabolite can be toxic for cells [109].

An interesting class of biocompatible synthetic materials is polyurethanes (PU) based polymers. The peculiarity of this material is that mechanical characteristics and biological degradability can be regulated by different polymerization processes and the specific structure of the two main components; dioles and isocyanates [11]. PU film, patterned with laminin lanes, can support the growth of spatially organized neonatal cardiomyocytes [110]. Accordingly, Siepe et al. cultivated myoblasts on highly porous PU based scaffolds and demonstrated its applicability for CTE [111, 112]. The implanted cell/PU-scaffold improved ventricular function in a similar way as myoblast cell transplantation. On the other end, the implantation of the matrix without cells failed to prevent cardiac deterioration [112]. Cultivation of cardiomyocytes into nanofibrous polycaprolactone meshes was also demonstrated [113, 114]. However, the use of this synthetic material in TE is limited compared to the other described materials.

Synthetic materials lack the biological properties that are typical for natural materials that could potentially influence growth and differentiation of cells in a more physiologic way. For these reasons different combinations were tested by combining natural and synthetic materials. The advantage to use a synthetic material is that structure and physical properties can be designed prior to synthesis and in accordance with the type of application and cultivated cells. In most hybrid applications the use of a synthetic polymer is needed to design the structure of the matrix and to provide specific mechanical properties and biodegradability. On the other end natural scaffolds can enhance survival, proliferation and differentiation of the seeded cells. The relative simple technique to couple biological peptides in order to prevent cell death, enhance proliferation, or differentiation of cells is also another advantage in the use of synthetic materials. To this purpose different combinations of synthetic and natural materials were tested for cell culture and tissue engineering. Alginate was combined with PLGA microspheres, carrying alginate lysate, in order to enhance and control the biodegradation rate [115]. Other applications combined collagen material with polycaprolactone for growth of fibroblasts and keratinocytes [116]. In the same way, polycaprolactone was combined with chitosane [117], fibrin [118], fibronectin [118] and gelatin [119]. Another model was reported by Chen et al. where they described a hybrid heart patch formed of elastomer, poly(glycerol sebacate) and demonstrated that embryonic stem cells grow and differentiate in here and used for myocardial TE [120].

In summary, synthetic materials represent another class of polymers that can be used for tissue engineering. Very few really convincing applications were performed thus far, however, the easy way to control mechanical properties and biodegradation rates are a great advantage in the use of these materials, especially if combined with a natural material and/or with bioactive peptides. Further studies are needed in relation to the specific cell type and of application type in order to increase biological properties and establish good culture conditions in terms of proliferation, viability and differentiation. Finally, an important aspect that needs to be evaluated is the possible immune reaction or possible toxic products that can be produced during biodegradation in vivo.

9 Conclusion

CTE is an evolving field that is receiving increased attention as a realistic approach to repair damaged myocardium. Cardiac cell therapy was extensively studied in the last decade and different clinical trials were performed with several cell types. Despite the confusing results, feasibility and safety of cellular therapy was demonstrated, although the benefits, in terms of long-term improvement of cardiac function, are still doubtful.

CTE offers the advantage to protect transplanted cells by providing a good environment and by protecting cells from death signals, probably the two main problems of direct cell injection. In addition, transplantation of cells together with a matrix provides also a mechanical support to the infarcted ventricular wall and can prevent post-infarct ventricular dilatation.

The choice of the appropriate cell source and matrix play a fundamental role for a successful TE application. For cardiac use, cardiomyocytes were largely used as a possible cell source for heart regeneration, however, their use is not clinically applicable and other cells sources need to be considered. Cardiac progenitor cells seem to be a promising cell type due to their intrinsic capacity to proliferate, thereby providing enough cell numbers, and for their natural commitment and differentiation potential to the myocardial lineage [3].

CTE is developing and changing together with new knowledge in the stem cell field. The matrix is not only a passive supporting material for cultivation and transplantation of cells, but has to be an active component of the biological scaffold that can drive growth and differentiation. The possibility to modify the material with bioactive peptides, growth factors, cytokines, or other bioactive peptides was demonstrated to improve both in vitro and in vivo efficacies of the scaffold. The possibility to combine different types of matrices can further improve culture conditions and mimic the natural environment of the heart as much as possible. Matrix modifications or combinations can have benefits not only for the transplanted cells, but also for the injured organ. The possibility to prevent endogenous cell death, to stimulate endogenous vasculogenesis or to provide specific physical–mechanical characteristics are all studied and demonstrated advantages of TE applications.

In summary, TE is an evolving field with a realistic possibility to be a therapeutic approach to repair damaged myocardium. However, many studies need to be performed to improve knowledge on (cardiac) progenitor cells usage. In addition, inreased guideness of growth and differentiation, including vascularization of the tissue and evaluation of concerns regarding biological safety of the application need to be considered.

References

1. Udelson, J.E., Patten, R.D., Konstam, M.A.: New concepts in post-infarction ventricular remodeling. Rev. Cardiovasc. Med. **4**(Suppl 3), S3–S12 (2003)
2. Anversa, P., Li, P., Zhang, X., et al.: Ischaemic myocardial injury and ventricular remodelling. Cardiovasc. Res. **27**(2), 145–157 (1993)
3. Gaetani, R., Barile, L., Forte, E., et al.: New perspectives to repair a broken heart. Cardiovasc. Hematol. Agents. Med. Chem. **7**(2), 91–107 (2009)
4. Menasche, P., Hagege, A.A., Vilquin, J.T., et al.: Autologous skeletal myoblast transplantation for severe postinfarction left ventricular dysfunction. J. Am. Coll. Cardiol. **41**(7), 1078–1083 (2003)
5. Assmus, B., Schachinger, V., Teupe, C., et al.: Transplantation of progenitor cells and regeneration enhancement in acute myocardial infarction (TOPCARE-AMI). Circulation **106**(24), 3009–3017 (2002)
6. Schachinger, V., Assmus, B., Britten, M.B., et al.: Transplantation of progenitor cells and regeneration enhancement in acute myocardial infarction: final one-year results of the TOPCARE-AMI trial. J. Am. Coll. Cardiol. **44**(8), 1690–1699 (2004)
7. Wollert, K.C., Meyer, G.P., Lotz, J., et al.: Intracoronary autologous bone-marrow cell transfer after myocardial infarction: the BOOST randomised controlled clinical trial. Lancet **364**(9429), 141–148 (2004)
8. Meyer, G.P., Wollert, K.C., Lotz, J., et al.: Intracoronary bone marrow cell transfer after myocardial infarction: eighteen months' follow-up data from the randomized, controlled BOOST (BOne marrOw transfer to enhance ST-elevation infarct regeneration) trial. Circulation **113**(10), 1287–1294 (2006)
9. Robey, T.E., Saiget, M.K., Reinecke, H., et al.: Systems approaches to preventing transplanted cell death in cardiac repair. J. Mol. Cell Cardiol. **45**(4), 567–581 (2008)
10. Zimmermann, W.H., Cesnjevar, R.: Cardiac tissue engineering: implications for pediatric heart surgery. Pediatr. Cardiol. **30**(5), 716–723 (2009)
11. Akhyari, P., Kamiya, H., Haverich, A., et al.: Myocardial tissue engineering: the extracellular matrix. Eur. J. Cardiothorac. Surg. **34**(2), 229–241 (2008)

12. Radisic, M., Yang, L., Boublik, J., et al.: Medium perfusion enables engineering of compact and contractile cardiac tissue. Am. J. Physiol. Heart Circ. Physiol. **286**(2), H507–H516 (2004)
13. Brown, M.A., Iyer, R.K., Radisic, M.: Pulsatile perfusion bioreactor for cardiac tissue engineering. Biotechnol. Prog. **24**(4), 907–920 (2008)
14. Fink, C., Ergun, S., Kralisch, D., et al.: Chronic stretch of engineered heart tissue induces hypertrophy and functional improvement. Faseb J. **14**(5), 669–679 (2000)
15. Zhang, M., Methot, D., Poppa, V., et al.: Cardiomyocyte grafting for cardiac repair: graft cell death and anti-death strategies. J. Mol. Cell Cardiol. **33**(5), 907–921 (2001)
16. Hudson, W., Collins, M.C., deFreitas, D., et al.: Beating and arrested intramyocardial injections are associated with significant mechanical loss: implications for cardiac cell transplantation. J. Surg. Res. **142**(2), 263–267 (2007)
17. Freyman, T., Polin, G., Osman, H., et al.: A quantitative, randomized study evaluating three methods of mesenchymal stem cell delivery following myocardial infarction. Eur. Heart J. **27**(9), 1114–1122 (2006)
18. Hayashi, M., Li, T.S., Ito, H., et al.: Comparison of intramyocardial and intravenous routes of delivering bone marrow cells for the treatment of ischemic heart disease: an experimental study. Cell Transpl. **13**(6), 639–647 (2004)
19. Reinecke, H., Zhang, M., Bartosek, T., et al.: Survival, integration, and differentiation of cardiomyocyte grafts: a study in normal and injured rat hearts. Circulation **100**(2), 193–202 (1999)
20. Zvibel, I., Smets, F., Soriano, H.: Anoikis: roadblock to cell transplantation? Cell Transpl. **11**(7), 621–630 (2002)
21. Nakamura, Y., Yasuda, T., Weisel, R.D., et al.: Enhanced cell transplantation: preventing apoptosis increases cell survival and ventricular function. Am. J. Physiol. Heart Circ. Physiol. **291**(2), H939–H947 (2006)
22. Li, W., Ma, N., Ong, L.L., et al.: Bcl-2 engineered MSCs inhibited apoptosis and improved heart function. Stem Cells **25**(8), 2118–2127 (2007)
23. Azarnoush, K., Maurel, A., Sebbah, L., et al.: Enhancement of the functional benefits of skeletal myoblast transplantation by means of coadministration of hypoxia-inducible factor 1alpha. J. Thorac. Cardiovasc. Surg. **130**(1), 173–179 (2005)
24. Maurel, A., Azarnoush, K., Sabbah, L., et al.: Can cold or heat shock improve skeletal myoblast engraftment in infarcted myocardium? Transplantation **80**(5), 660–665 (2005)
25. Jayakumar, J., Suzuki, K., Sammut, I.A., et al.: Heat shock protein 70 gene transfection protects mitochondrial and ventricular function against ischemia-reperfusion injury. Circulation **104**(12 Suppl 1), I303–I307 (2001)
26. Liu, T.B., Fedak, P.W., Weisel, R.D., et al.: Enhanced IGF-1 expression improves smooth muscle cell engraftment after cell transplantation. Am. J. Physiol. Heart Circ. Physiol. **287**(6), H2840–H2849 (2004)
27. Laflamme, M.A., Chen, K.Y., Naumova, A.V., et al.: Cardiomyocytes derived from human embryonic stem cells in pro-survival factors enhance function of infarcted rat hearts. Nat. Biotechnol. **25**(9), 1015–1024 (2007)
28. Davis, M.E., Hsieh, P.C., Takahashi, T., et al.: Local myocardial insulin-like growth factor 1 (IGF-1) delivery with biotinylated peptide nanofibers improves cell therapy for myocardial infarction. Proc. Natl. Acad. Sci. USA **103**(21), 8155–8160 (2006)
29. Retuerto, M.A., Schalch, P., Patejunas, G., et al.: Angiogenic pretreatment improves the efficacy of cellular cardiomyoplasty performed with fetal cardiomyocyte implantation. J. Thorac. Cardiovasc. Surg. **127**(4), 1041–1049 (2004). discussion 1049–1051
30. Wang, Y., Haider, H.K., Ahmad, N., et al.: Combining pharmacological mobilization with intramyocardial delivery of bone marrow cells over-expressing VEGF is more effective for cardiac repair. J. Mol. Cell Cardiol. **40**(5), 736–745 (2006)
31. Matsumoto, R., Omura, T., Yoshiyama, M., et al.: Vascular endothelial growth factor-expressing mesenchymal stem cell transplantation for the treatment of acute myocardial infarction. Arterioscler. Thromb. Vasc. Biol. **25**(6), 1168–1173 (2005)

32. Suzuki, K., Murtuza, B., Beauchamp, J.R., et al.: Dynamics and mediators of acute graft attrition after myoblast transplantation to the heart. Faseb J. **18**(10), 1153–1155 (2004)
33. Hsieh, P.C., Davis, M.E., Gannon, J., et al.: Controlled delivery of PDGF-BB for myocardial protection using injectable self-assembling peptide nanofibers. J. Clin. Invest. **116**(1), 237–248 (2006)
34. Padin-Iruegas, M.E., Misao, Y., Davis, M.E., et al.: Cardiac progenitor cells and biotinylated insulin-like growth factor-1 nanofibers improve endogenous and exogenous myocardial regeneration after infarction. Circulation **120**(10), 876–887 (2009)
35. Muller-Ehmsen, J., Krausgrill, B., Burst, V., et al.: Effective engraftment but poor mid-term persistence of mononuclear and mesenchymal bone marrow cells in acute and chronic rat myocardial infarction. J. Mol. Cell Cardiol. **41**(5), 876–884 (2006)
36. Hou, D., Youssef, E.A., Brinton, T.J., et al.: Radiolabeled cell distribution after intramyocardial, intracoronary, and interstitial retrograde coronary venous delivery: implications for current clinical trials. Circulation **112**(Suppl 9), I150–I156 (2005)
37. Terrovitis, J., Lautamaki, R., Bonios, M., et al.: Noninvasive quantification and optimization of acute cell retention by in vivo positron emission tomography after intramyocardial cardiac-derived stem cell delivery. J. Am. Coll. Cardiol. **54**(17), 1619–1626 (2009)
38. Kutschka, I., Chen, I.Y., Kofidis, T., et al.: Collagen matrices enhance survival of transplanted cardiomyoblasts and contribute to functional improvement of ischemic rat hearts. Circulation **114**(Suppl 1), I167–I173 (2006)
39. Kutschka, I., Chen, I.Y., Kofidis, T., et al.: In vivo optical bioluminescence imaging of collagen-supported cardiac cell grafts. J. Heart Lung Transpl. **26**(3), 273–280 (2007)
40. Martens, T.P., Godier, A.F., Parks, J.J., et al.: Percutaneous cell delivery into the heart using hydrogels polymerizing in situ. Cell Transpl. **18**(3), 297–304 (2009)
41. Corda, S., Samuel, J.L., Rappaport, L.: Extracellular matrix and growth factors during heart growth. Heart Fail. Rev. **5**(2), 119–130 (2000)
42. Akins, R.E., Rockwood, D., Robinson, K.G., et al.: Three-dimensional culture alters primary cardiac cell phenotype. Tissue Eng. Part A **16**(2), 629–641 (2010)
43. Battista, S., Guarnieri, D., Borselli, C., et al.: The effect of matrix composition of 3D constructs on embryonic stem cell differentiation. Biomaterials **26**(31), 6194–6207 (2005)
44. Fomovsky, G.M., Thomopoulos, S., Holmes, J.W.: Contribution of extracellular matrix to the mechanical properties of the heart. J. Mol. Cell Cardiol. **62**, 1331–1338 (2009)
45. Bowers, S.L., Banerjee, I., Baudino, T.A.: The extracellular matrix: At the center of it all. J. Mol. Cell Cardiol. **48**(3), 474–482 (2010)
46. Mosesson, M.W.: Fibrinogen and fibrin structure and functions. J. Thromb. Haemost. **3**(8), 1894–1904 (2005)
47. Ahmed, T.A., Dare, E.V., Hincke, M.: Fibrin: A versatile scaffold for tissue engineering applications. Tissue Eng. Part B. Rev. **14**, 199–215 (2008)
48. Christman, K.L., Fok, H.H., Sievers, R.E., et al.: Fibrin glue alone and skeletal myoblasts in a fibrin scaffold preserve cardiac function after myocardial infarction. Tissue Eng. **10**(3–4), 403–409 (2004)
49. Christman, K.L., Vardanian, A.J., Fang, Q., et al.: Injectable fibrin scaffold improves cell transplant survival, reduces infarct expansion, and induces neovasculature formation in ischemic myocardium. J. Am. Coll. Cardiol. **44**(3), 654–660 (2004)
50. Yu, J., Christman, K.L., Chin, E., et al.: Restoration of left ventricular geometry and improvement of left ventricular function in a rodent model of chronic ischemic cardiomyopathy. J. Thorac. Cardiovasc. Surg. **137**(1), 180–187 (2009)
51. Ryu, J.H., Kim, I.K., Cho, S.W., et al.: Implantation of bone marrow mononuclear cells using injectable fibrin matrix enhances neovascularization in infarcted myocardium. Biomaterials **26**(3), 319–326 (2005)
52. Huang, N.F., Lam, A., Fang, Q., et al.: Bone marrow-derived mesenchymal stem cells in fibrin augment angiogenesis in the chronically infarcted myocardium. Regen Med. **4**(4), 527–538 (2009)

53. Birla, R.K., Borschel, G.H., Dennis, R.G., et al.: Myocardial engineering in vivo: formation and characterization of contractile, vascularized three-dimensional cardiac tissue. Tissue Eng. **11**(5–6), 803–813 (2005)
54. Birla, R.K., Dhawan, V., Dow, D.E., et al.: Cardiac cells implanted into a cylindrical, vascularized chamber in vivo: pressure generation and morphology. Biotechnol. Lett. **31**(2), 191–201 (2009)
55. Silverstein, M.E., Keown, K., Owen, J.A., et al.: Collagen fibers as a fleece hemostatic agent. J. Trauma. **20**(8), 688–694 (1980)
56. Kofidis, T., Akhyari, P., Wachsmann, B., et al.: A novel bioartificial myocardial tissue and its prospective use in cardiac surgery. Eur. J. Cardiothorac. Surg. **22**(2), 238–243 (2002)
57. Kofidis, T., Akhyari, P., Wachsmann, B., et al.: Clinically established hemostatic scaffold (tissue fleece) as biomatrix in tissue- and organ-engineering research. Tissue Eng. **9**(3), 517–523 (2003)
58. Gaballa, M.A., Sunkomat, J.N., Thai, H., et al.: Grafting an acellular 3-dimensional collagen scaffold onto a non-transmural infarcted myocardium induces neo-angiogenesis and reduces cardiac remodeling. J. Heart Lung Transpl. **25**(8), 946–954 (2006)
59. Cortes-Morichetti, M., Frati, G., Schussler, O., et al.: Association between a cell-seeded collagen matrix and cellular cardiomyoplasty for myocardial support and regeneration. Tissue Eng. **13**(11), 2681–2687 (2007)
60. Chachques, J.C., Trainini, J.C., Lago, N., et al.: Myocardial assistance by grafting a new bioartificial upgraded myocardium (MAGNUM clinical trial): one year follow-up. Cell Transpl. **16**(9), 927–934 (2007)
61. Chachques, J.C., Trainini, J.C., Lago, N., et al.: Myocardial assistance by grafting a new bioartificial upgraded myocardium (MAGNUM trial): clinical feasibility study. Ann. Thorac. Surg. **85**(3), 901–908 (2008)
62. Schussler, O., Coirault, C., Louis-Tisserand, M., et al.: Use of arginine-glycine-aspartic acid adhesion peptides coupled with a new collagen scaffold to engineer a myocardium-like tissue graft. Nat. Clin. Pract. Cardiovasc. Med. **6**(3), 240–249 (2009)
63. Huang, N.F., Yu, J., Sievers, R., et al.: Injectable biopolymers enhance angiogenesis after myocardial infarction. Tissue Eng. **11**(11–12), 1860–1866 (2005)
64. Dai, W., Wold, L.E., Dow, J.S., et al.: Thickening of the infarcted wall by collagen injection improves left ventricular function in rats: a novel approach to preserve cardiac function after myocardial infarction. J. Am. Coll. Cardiol. **46**(4), 714–719 (2005)
65. Stern, R., Asari, A.A., Sugahara, K.N.: Hyaluronan fragments: an information-rich system. Eur. J. Cell Biol. **85**(8), 699–715 (2006)
66. Khademhosseini, A., Eng, G., Yeh, J., et al.: Microfluidic patterning for fabrication of contractile cardiac organoids. Biomed. Microdevices **9**(2), 149–157 (2007)
67. Ventura, C., Cantoni, S., Bianchi, F., et al.: Hyaluronan mixed esters of butyric and retinoic Acid drive cardiac and endothelial fate in term placenta human mesenchymal stem cells and enhance cardiac repair in infarcted rat hearts. J. Biol. Chem. **282**(19), 14243–14252 (2007)
68. Lionetti, V., Cantoni, S., Cavallini, C., et al.: Hyaluronan mixed esters of butyric and retinoic acid affording myocardial survival and repair without stem cell transplantation. J. Biol. Chem. **285**(13), 9949–9961 (2010)
69. Leor, J., Aboulafia-Etzion, S., Dar, A., et al.: Bioengineered cardiac grafts: a new approach to repair the infarcted myocardium? Circulation **102**(19 Suppl 3), III56–III61 (2000)
70. Amir, G., Miller, L., Shachar, M., et al.: Evaluation of a peritoneal-generated cardiac patch in a rat model of heterotopic heart transplantation. Cell Transpl. **18**(3), 275–282 (2009)
71. Dvir, T., Kedem, A., Ruvinov, E., et al.: Prevascularization of cardiac patch on the omentum improves its therapeutic outcome. Proc. Natl. Acad. Sci. U S A. **106**(35), 14990–14995 (2009)
72. Landa, N., Miller, L., Feinberg, M.S., et al.: Effect of injectable alginate implant on cardiac remodeling and function after recent and old infarcts in rat. Circulation **117**(11), 1388–1396 (2008)

73. Hao, X., Silva, E.A., Mansson-Broberg, A., et al.: Angiogenic effects of sequential release of VEGF-A165 and PDGF-BB with alginate hydrogels after myocardial infarction. Cardiovasc. Res. **75**(1), 178–185 (2007)
74. Tsur-Gang, O., Ruvinov, E., Landa, N., et al.: The effects of peptide-based modification of alginate on left ventricular remodeling and function after myocardial infarction. Biomaterials **30**(2), 189–195 (2009)
75. Yu, J., Gu, Y., Du, K.T., et al.: The effect of injected RGD modified alginate on angiogenesis and left ventricular function in a chronic rat infarct model. Biomaterials **30**(5), 751–756 (2009)
76. Zimmermann, W.H., Fink, C., Kralisch, D., et al.: Three-dimensional engineered heart tissue from neonatal rat cardiac myocytes. Biotechnol. Bioeng. **68**(1), 106–114 (2000)
77. Zimmermann, W.H., Schneiderbanger, K., Schubert, P., et al.: Tissue engineering of a differentiated cardiac muscle construct. Circ. Res. **90**(2), 223–230 (2002)
78. Zimmermann, W.H., Didie, M., Wasmeier, G.H., et al.: Cardiac grafting of engineered heart tissue in syngenic rats. Circulation **106**(12 Suppl 1), I151–I157 (2002)
79. Zimmermann, W.H., Melnychenko, I., Wasmeier, G., et al.: Engineered heart tissue grafts improve systolic and diastolic function in infarcted rat hearts. Nat. Med. **12**(4), 452–458 (2006)
80. Naito, H., Melnychenko, I., Didie, M., et al.: Optimizing engineered heart tissue for therapeutic applications as surrogate heart muscle. Circulation **114**(Suppl 1), I72–I78 (2006)
81. Yildirim, Y., Naito, H., Didie, M., et al.: Development of a biological ventricular assist device: preliminary data from a small animal model. Circulation **116**(Suppl 11), I16–I23 (2007)
82. Guo, X.M., Zhao, Y.S., Chang, H.X., et al.: Creation of engineered cardiac tissue in vitro from mouse embryonic stem cells. Circulation **113**(18), 2229–2237 (2006)
83. Shimizu, T., Yamato, M., Kikuchi, A., et al.: Two-dimensional manipulation of cardiac myocyte sheets utilizing temperature-responsive culture dishes augments the pulsatile amplitude. Tissue Eng. **7**(2), 141–151 (2001)
84. Shimizu, T., Yamato, M., Isoi, Y., et al.: Fabrication of pulsatile cardiac tissue grafts using a novel 3-dimensional cell sheet manipulation technique and temperature-responsive cell culture surfaces. Circ. Res. **90**(3), e40 (2002)
85. Shimizu, T., Sekine, H., Isoi, Y., et al.: Long-term survival and growth of pulsatile myocardial tissue grafts engineered by the layering of cardiomyocyte sheets. Tissue Eng. **12**(3), 499–507 (2006)
86. Sekine, H., Shimizu, T., Yang, J., et al.: Pulsatile myocardial tubes fabricated with cell sheet engineering. Circulation **114**(Suppl 1), I87–I93 (2006)
87. Sekine, H., Shimizu, T., Kosaka, S., et al.: Cardiomyocyte bridging between hearts and bioengineered myocardial tissues with mesenchymal transition of mesothelial cells. J. Heart Lung Transpl. **25**(3), 324–332 (2006)
88. Miyahara, Y., Nagaya, N., Kataoka, M., et al.: Monolayered mesenchymal stem cells repair scarred myocardium after myocardial infarction. Nat. Med. **12**(4), 459–465 (2006)
89. Sekine, H., Shimizu, T., Hobo, K., et al.: Endothelial cell coculture within tissue-engineered cardiomyocyte sheets enhances neovascularization and improves cardiac function of ischemic hearts. Circulation **118**(Suppl 14), S145–S152 (2008)
90. Badylak, S., Obermiller, J., Geddes, L., et al.: Extracellular matrix for myocardial repair. Heart Surg. Forum. **6**(2), E20–E26 (2003)
91. Robinson, K.A., Li, J., Mathison, M., et al.: Extracellular matrix scaffold for cardiac repair. Circulation **112**(Suppl 9), I135–I143 (2005)
92. Kochupura, P.V., Azeloglu, E.U., Kelly, D.J., et al.: Tissue-engineered myocardial patch derived from extracellular matrix provides regional mechanical function. Circulation **112**(Suppl 9), I144–I149 (2005)
93. Badylak, S.F., Kochupura, P.V., Cohen, I.S., et al.: The use of extracellular matrix as an inductive scaffold for the partial replacement of functional myocardium. Cell Transpl. (15 Suppl 1), S29–S40 (2006)

94. Kelly, D.J., Rosen, A.B., Schuldt, A.J., et al.: Increased myocyte content and mechanical function within a tissue-engineered myocardial patch following implantation. Tissue Eng. Part A **15**(8), 2189–2201 (2009)
95. Potapova, I.A., Doronin, S.V., Kelly, D.J., et al.: Enhanced recovery of mechanical function in the canine heart by seeding an extracellular matrix patch with mesenchymal stem cells committed to a cardiac lineage. Am. J. Physiol. Heart Circ. Physiol. **295**(6), H2257–H2263 (2008)
96. Tan, M.Y., Zhi, W., Wei, R.Q., et al.: Repair of infarcted myocardium using mesenchymal stem cell seeded small intestinal submucosa in rabbits. Biomaterials **30**(19), 3234–3240 (2009)
97. Singelyn, J.M., DeQuach, J.A., Seif-Naraghi, S.B., et al.: Naturally derived myocardial matrix as an injectable scaffold for cardiac tissue engineering. Biomaterials **30**(29), 5409–5416 (2009)
98. Eitan, Y., Sarig, U., Dahan, N., et al.: Acellular cardiac extracellular matrix as a scaffold for tissue engineering: In vitro cell support, remodeling and biocompatibility. Tissue. Eng. Part C Methods **16**(4), 671–683 (2010)
99. Seif-Naraghi, S.B., Salvatore, M.A., Schup-Magoffin, P.J., et al.: Design and characterization of an injectable pericardial matrix gel: a potentially autologous scaffold for cardiac tissue engineering. Tissue Eng. Part A **16**(6), 2017–2027 (2010)
100. Crapo, P.M., Wang, Y.: Small intestinal submucosa gel as a potential scaffolding material for cardiac tissue engineering. Acta. Biomater. **6**(6), 2091–2096 (2010)
101. Ott, H.C., Matthiesen, T.S., Goh, S.K., et al.: Perfusion-decellularized matrix: using nature's platform to engineer a bioartificial heart. Nat. Med. **14**(2), 213–221 (2008)
102. Bursac, N., Papadaki, M., Cohen, R.J., et al.: Cardiac muscle tissue engineering: toward an in vitro model for electrophysiological studies. Am. J. Physiol. **277**(2 Pt 2), H433–H444 (1999)
103. Papadaki, M., Bursac, N., Langer, R., et al.: Tissue engineering of functional cardiac muscle: molecular, structural, and electrophysiological studies. Am. J. Physiol. Heart Circ. Physiol. **280**(1), H168–H178 (2001)
104. Carrier, R.L., Papadaki, M., Rupnick, M., et al.: Cardiac tissue engineering: cell seeding, cultivation parameters, and tissue construct characterization. Biotechnol. Bioeng. **64**(5), 580–589 (1999)
105. Lo, H., Kadiyala, S., Guggino, S.E., et al.: Poly(L-lactic acid) foams with cell seeding and controlled-release capacity. J. Biomed. Mater. Res. **30**(4), 475–484 (1996)
106. McDevitt, T.C., Angello, J.C., Whitney, M.L., et al.: In vitro generation of differentiated cardiac myofibers on micropatterned laminin surfaces. J. Biomed. Mater. Res. **60**(3), 472–479 (2002)
107. Zong, X., Bien, H., Chung, C.Y., et al.: Electrospun fine-textured scaffolds for heart tissue constructs. Biomaterials **26**(26), 5330–5338 (2005)
108. Bostman, O.M.: Intense granulomatous inflammatory lesions associated with absorbable internal fixation devices made of polyglycolide in ankle fractures. Clin. Orthop. Relat. Res. **278**, 193–199 (1992)
109. Taylor, M.S., Daniels, A.U., Andriano, K.P., et al.: Six bioabsorbable polymers: in vitro acute toxicity of accumulated degradation products. J. Appl. Biomater. **5**(2), 151–157 (1994)
110. McDevitt, T.C., Woodhouse, K.A., Hauschka, S.D., et al.: Spatially organized layers of cardiomyocytes on biodegradable polyurethane films for myocardial repair. J. Biomed. Mater. Res. A **66**(3), 586–595 (2003)
111. Siepe, M., Giraud, M.N., Liljensten, E., et al.: Construction of skeletal myoblast-based polyurethane scaffolds for myocardial repair. Artif. Organs **31**(6), 425–433 (2007)
112. Siepe, M., Giraud, M.N., Pavlovic, M., et al.: Myoblast-seeded biodegradable scaffolds to prevent post-myocardial infarction evolution toward heart failure. J. Thorac. Cardiovasc. Surg. **132**(1), 124–131 (2006)

113. Shin, M., Ishii, O., Sueda, T., et al.: Contractile cardiac grafts using a novel nanofibrous mesh. Biomaterials **25**(17), 3717–3723 (2004)
114. Ishii, O., Shin, M., Sueda, T., et al.: In vitro tissue engineering of a cardiac graft using a degradable scaffold with an extracellular matrix-like topography. J. Thorac. Cardiovasc. Surg. **130**(5), 1358–1363 (2005)
115. Ashton, R.S., Banerjee, A., Punyani, S., et al.: Scaffolds based on degradable alginate hydrogels and poly(lactide-co-glycolide) microspheres for stem cell culture. Biomaterials **28**(36), 5518–5525 (2007)
116. Dai, N.T., Williamson, M.R., Khammo, N., et al.: Composite cell support membranes based on collagen and polycaprolactone for tissue engineering of skin. Biomaterials **25**(18), 4263–4271 (2004)
117. Mei, N., Chen, G., Zhou, P., et al.: Biocompatibility of poly(epsilon-caprolactone) scaffold modified by chitosan–the fibroblasts proliferation in vitro. J. Biomater. Appl. **19**(4), 323–339 (2005)
118. Pankajakshan, D., Krishnan, V.K., Krishnan, L.K.: Vascular tissue generation in response to signaling molecules integrated with a novel poly(epsilon-caprolactone)-fibrin hybrid scaffold. J. Tissue Eng. Regen. Med. **1**(5), 389–397 (2007)
119. Ma, Z., He, W., Yong, T., et al.: Grafting of gelatin on electrospun poly(caprolactone) nanofibers to improve endothelial cell spreading and proliferation and to control cell orientation. Tissue Eng. **11**(7–8), 1149–1158 (2005)
120. Chen, Q.Z., Ishii, H., Thouas, G.A., et al.: An elastomeric patch derived from poly(glycerol sebacate) for delivery of embryonic stem cells to the heart. Biomaterials **31**(14), 3885–3893 (2010)

Inherently Bio-Active Scaffolds: Intelligent Constructs to Model the Stem Cell Niche

Paolo Di Nardo, Marilena Minieri, Annalisa Tirella, Giancarlo Forte and Arti Ahluwalia

Abstract The oft-abused phrase "genes load the gun, environment pulls the trigger" can be applied to stem cells and stem cell niches as well as to cell–material interfaces. Much is known about cell–material interaction in general, perhaps a little less about how these interactions condition cell phenotype. With the increasing interest in stem cells and, in particular, their applications in tissue regeneration, the regulation of the stem cell microenvironment through modulation of intuitive or smart materials and structures, or what we term IBAS (Inherently Bio-Active Scaffolds) is poised to become a major field of research. Here, we discuss how cardiac regeneration strategies have undergone a gradual shift from the emphasis on biochemical signals and basic biology to one in which the material or scaffold plays a major role in establishing an equilibrium state. From being a constant battle or tug-of-war between the cells and synthetic environments, we conceive IBAS as intuitively responding to the cell's requirements to instate a sort of equilibrium in the system.

P. D. Nardo (✉) · M. Minieri · A. Tirella · G. Forte
Laboratorio di Cardiologia Molecolare e Cellulare, Dipartimento di Medicina Interna, Università di Roma Tor Vergata, Via Montpellier, 1, 00133 Rome, Italy
e-mail: dinardo@uniroma2.it

P. D. Nardo · M. Minieri · G. Forte
Istituto Nazionale per le Ricerche Cardiovascolari (INRC), Bologna, Italy

P. D. Nardo · M. Minieri · G. Forte
Japanese-Italian Tissue Engineering Laboratory (JITEL), Tokyo Women's Medical University-Waseda University Joint Institution for Advanced Biomedical Sciences (TWIns), Tokyo, Japan

A. Tirella · A. Ahluwalia
Centro Interdipartimentale di Ricerca "E. Piaggio", Università di Pisa, Pisa, Italy

1 Introduction

The heart, with its romantic associations as well as its fundamental role as the machine of human life, is one of the most well-known and studied organs in the body. But its failure or damage still causes the highest number of deaths per year in developed countries. Indeed, the conventional clinical approach and treatment (early diagnosis + drugs) has failed so far in curing heart failure, a complex pathophysiological state in which the organ contractile activity is no longer able to deliver an adequate quantity of blood and nutrients to match tissue requirements. Furthermore, heart transplants, given the lack of donors and the low cost-effectiveness, is an option open to a tiny percentage of patients. For these reasons, cardiomyocyte regeneration through the use of stem cells has received unparalleled attention both in the press and within the professional academic and medical fields. However, there is a desperate lack of fundamental knowledge on cell regeneration in general, and cardiac tissue regeneration in particular. Our understanding of cardiomyocyte origin and potency is limited; there is a huge amount of controversy and mis-information regarding cardiac stem cells, their isolation, their identification and their expansion. Moreover, despite decades of experimental research on biomaterials and scaffolds, neither the ideal material nor the architecture for cardiac cell proliferation and colonization has been identified.

In 1994, the possibility of regenerating the heart was thoroughly discussed, for the first time, in a Congress in Viterbo (Italy) that gathered the most prominent international scientists of that time [14], but the technology was not adequate to start experimentation and envisage clinical applications. Nevertheless, after a few years, advancements in cell biology and material sciences provided the impetus to start preliminary experiments, in which mostly mature cardiomyocytes were: (i) injected into damaged experimental animal and human heart walls or (ii) used to fabricate very rudimentary parts of myocardium. This approach, as a whole, was unsuccessful. Scientists thus turned their attention to stem cells but, once again, the injection of different stem cell populations into damaged myocardium was inefficient. Many researchers proposed fabricating ex vivo portions of myocardium to be engrafted into injured hearts using a biomaterial support on which stem cells can be grown. Attention was thus drawn to biocompatible materials already in use in the clinical setting, and to the identification of reliable and well-characterized sources of cardiogenic stem cells.

Numerous experiments and options have since been evaluated, in order to identify the best materials and the optimum technologies to design and process scaffolds. However, a consistent design strategy, and well-founded engineering principles and models have been sorely missing from the research scenario. Properly designed scaffolds should deliver signals perceived as "suitable" by stem cells and capable of mimicking the natural 3D environment of the heart. The design of smart biomaterials for tissue regeneration is critically dependent on the fundamental understanding of how cells coordinate their functions in the in vivo environment and in engineered matrices. Therefore, matrix/cell and scaffold/cell interactions must first be properly investigated through both numerical modeling and experimental assessment, and only then can they be validated in vitro and ultimately in vivo.

There is no doubt that current materials and methods cannot deliver functional heart tissue as required for cardiac regeneration, and a change in perspective and strategy is required. Therefore, it is necessary to stimulate the generation of novel biomimicking strategies for inducing cardiac self-repair by proposing and promoting new approaches both in material design and modeling strategies.

2 Towards a Paradigm Shift

Not surprisingly stem cell-based regeneration of the heart has elicited much scientific and public attention, but has also generated commercial interests. Currently, post-infarction myocardial revascularization protocols include the administration of mesenchymal stem cells, either by intravascular or intramyocardial injection. Although a number of controlled clinical trials have been performed, the results have so far been controversial [12, 23, 29, 77]. The small improvements reported have been mostly ascribed to stem cell paracrine effects or to their pro-angiogenic activity rather than to a direct contribution to cardiac muscle repair. Besides the fact that the types of cells suitable for heart regeneration in man remain to be defined [49, 53], it is clear that generating new cardiomyocytes may be not sufficient to efficiently repair the texture of myocardial tissue, which also includes fibroblasts, smooth muscle cells, endothelial cells and adipocytes, among others. In healthy human hearts, only 10–20% of the cell complex constituting the whole organ are cardiomyocytes [56, 68] and at the age of 25 years no more than 1% of them are annually substituted by progenitor cells, with the percentage falling to less than 0.5% at the age of 75. In total, less than 50% of all cardiomyocytes are renewed during a normal human life span [7]. In addition, it is likely that there is a natural limitation to progenitor cell capability to repair injured myocardium, considering that infarction as well as widespread myocardial diseases progress to fibrosis and not to newly generated contractile tissue. Nevertheless, features so far documented in reliably isolated and manipulated progenitor cells resident within the cardiac tissue (cardiac progenitor cells, CPC), although fragmentary and often controversial, do suggest that innovative treatments to repair damaged myocardium could have promising clinical applications.

3 Who and Where are the Stem Cells?

Cardiac progenitor cells and bone-marrow-derived stem cells share antigenic markers. Thus, it has been suggested that bone marrow could be the source of all stem cell populations having mesenchymal-like features resident within different tissues [6]. However, CPC appear within the myocardium during ontogeny and remain quiescent within their niches for short or long periods of time [39]. However, in spite of this unifying view, various CPC classifications have been so far proposed and an unambiguous interpretation is still lacking. As shown in Table 1, a number of cell markers have been credited to identify CPC, but they are not exclusive and their expression pattern can diverge among species. In addition, the existence of progenitor cells self-organizing into cardiospheres has been challenged by [3], who have demonstrated that the cardiosphere cells are not cardiomyogenic and their apparently spontaneous beating is produced by contaminations from cardiac tissue. Therefore, at the present level of knowledge, the identification of the progenitor cells is only based on their capability to self-renew and to produce differentiated progeny [18].

In the past, most studies agreed that progenitor cells (single or clustered) are settled in specific anatomic and functional locations (niches), where cells are protected by depletion and, in turn, the host tissue by their over-proliferation. Niches mediate signals maintaining progenitor cell self renewal and multipotency or inducing their commitment in response to tissue specific needs [64]. To date, though, cardiac progenitor cell niches have never been exhaustively identified or described, very likely because they do not have anatomical dimensions in all organs [25, 82].

4 Unstable Niches

The stem cell niche is generally considered as an anatomic and functional unit, in which stem cells are confined in a semi-quiescent state. In the heart we hypothesize that niches have only functional and temporal dimensions and are characterized by a dynamically symmetric array of signals delivered by the neighboring microenvironment, as depicted in Fig. 1. In our opinion, the signal symmetry maintains progenitor cells in a *metastable* quiescent state, even for long periods of time, preventing aging processes and preserving their multipotency [4]. According to this interpretation, stem cells are inherently unstable (i.e., prone to chaotically adopt multiple phenotypes) and only a complex array of competing symmetric signals confines them in a metastable equilibrium (Fig. 2). A very small modification in local or long-range extracellular signaling can attract progenitor cells towards one of many possible stable states. All this occurs within a highly stable global context, the myocardium. Once the symmetry of signals maintaining the metastable equilibrium has been broken by internal and/or external factors, progenitor cells engage in differentiating pathways and tissue assembly processes in an ever-changing environment characterized by an array of qualitatively and quantitatively diverse factors (biochemical, physico-chemical and mechano-structural) arranged along the time axis to generate a dynamic multicomponent template of microenvironmental symmetry. The progenitor cell metastable state might biologically correspond to fluctuating levels of transcription factors with

Table 1 Cardiac progenitor cell phenotypes identified within the myocardium

	Lin-/c-kit+	Lin-/Sca-1+	Islet-1	Side population	Cardiospheres
Markers	c-kit+, MEF-2c^{low}, Sca-1^{low}, CD34−, GATA-4^{low}, CD45−	Sca-1+, c-kit−, CD34−, CD45−, GATA-4^{low}, Nkx-2.5−, MEF-2c+	Islet-1+, Nkx-2.5+, Sca-1−, GATA-4+, c-kit−, CD31−	ABCG2+, c-kit^{low}, Nkx-2.5−, CD45^{low}, CD34^{low}, GATA-4−	c-kit+, CD31^{low}, Sca-1+, cTnI^{low}, CD34+, MHC+, FLk-1+, CD105+
Source	Human, rat, dog	Rat, mouse	Mouse, rat, human	Mouse	Human, mouse, pig, guinea pig
Clonal activity	Yes	Not determined	Yes	Not determined	Yes
Potency	Cardiomyocytes, smooth muscle, endothelial cells	Cardiomyocytes, osteoblasts, adipocytes	Cardiomyocytes, smooth muscle, endothelial cells	Cardiomyocytes	Cardiomyocytes, smooth muscle, endothelial cells
Frequency	1/10^4 cardiomyocytes	1/3.3 × 10^4 cardiomyocytes	198 ± 35 OT, 137 ± 23 RA, 67 ± 15 RV, 25 ± 7 LV	2% of total cadiac cells	10% of total cardiac cells
Protocol	Spontaneous	FGF-2 physical stimuli	Co-culture with neonatal cardiomyocytes	Co-culture with cardiac main population oxytocin	Spontaneous, co-culture with neonatal cardiomyocytes, electromaganetic field
Ref.	Beltrami et al. [6]	[62, 24]	[36]	[45, 48]	[50, 26]

Fig. 1 The symmetric array of signals which contribute to defining the cardiac stem cell niche

Fig. 2 The unstable progenitor cells are in metastable equilibrium while confined in a niche featured by an appropriate signal symmetry. The evolution of the niche signal symmetry drives progenitor cell differentiation

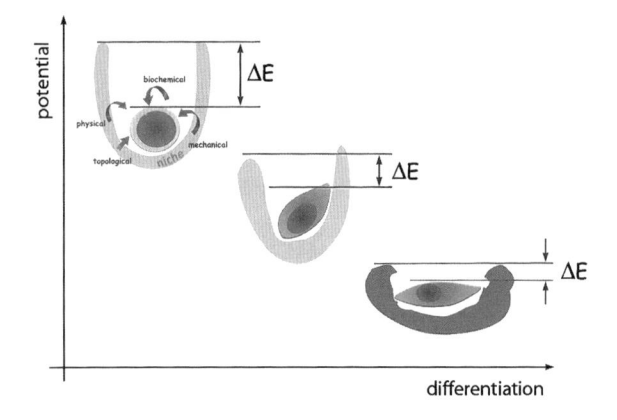

threshold-dependent commitment, similar to that described in the haemopoietic system [16]. This implies that niches are *protempore* activated loci, where the natural progenitor cell instability is confined (*native niche*) or is governed in favor of the acceptance of the required phenotype (*differentiating niche*) and tissue architecture (*architectural niche*). These different niches do not identify specific regions within the myocardium, but correspond to a *continuum* of different critical levels of interplay adopted by a wide array of multiparametric signals in response to local tissue requirements. We therefore prefer to think of them as states, or energy levels, as schematized in Fig. 2. Differentiating progenitor cells undergo a labile transition state, while migrating to niches characterized by a more stable, although more segregating, level of symmetry. This process occurs through environment-controlled differential segregation of progenitor cell "sensors" (membrane and intracellular proteins), so that only those specifically correlated with a definite step are active in differentiating progenitor cells.

The elusive goal we are all aiming to is to replicate this continuum of niches in vitro. Consistently, we need to shift from the emphasis on a trial and error based biochemical tuning of cardiac phenotype, towards more rigorous and well defined engineering and modeling of the stem cell microenvironment.

5 Engineering Factors Which Determine Phenotype

Cell function and fate are determined by three clusters of factors in the microenvironment: (i) the biochemical microenvironment, including ligands, signaling molecules and other cells, (ii) the physico-chemical environment, which comprises gradient-dependent factors, such as surface properties, oxygen tension, pH and temperature, and, finally, (iii) the mechano-structural environment. The mechano-structural environment is the architecture in 2 and 3 dimensions as well as mechanical forces, such as stress and strain, all of which act in a non-linear, but fairly constant manner. Together the three elements represent what we define as the tripartite axes of cues, as depicted in Fig. 3. Cell fate and function also depend strongly on time; even stem cells are equipped with a biological clock which ticks constantly.

Much of the information on cardiac stem cell biology to date has been obtained with respect to their responsiveness to a few soluble factors (growth factors/ cytokines, epigenetic factors, etc.) in vitro. In most cases, the signalling pathways involved in stem cell growth and differentiation are still not clearly identified and methods used to differentiate cells are determined empirically on the basis of trial and error. Till recently, the concept that stem cell fate is mostly governed by

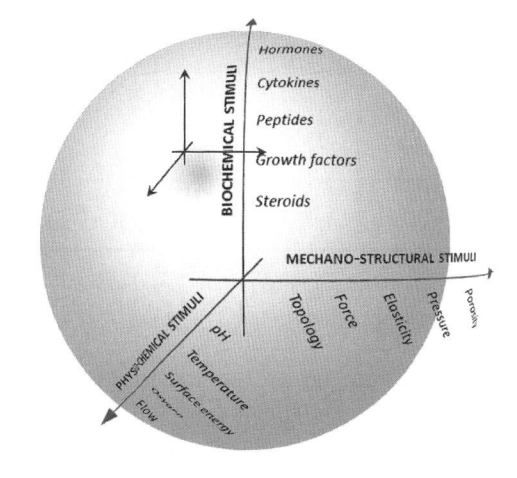

Fig. 3 The three main groups of stimuli which condition the cell microenvironment. The sphere represents time, a constant and irreversible parameter which stretches out in all three dimensions

soluble factors has been passively accepted by investigators. As a consequence, all attempts at replicating in vitro the niche conditions and at activating both in vitro and in vivo stem cell lineage specification have been based on the assumption that a few soluble factors, sometimes injected in either the bloodstream or the damaged tissue, can govern the whole process leading stem cells to full differentiation, i.e. stem cell specification and determination, sarcomere assembly and cardiomyocyte integration into the myocardial architecture [17, 80]. This simplistic vision has been recently challenged by the demonstration that a pivotal role to accurately control cardiac stem cell fate in vitro is played by mechano-structural and physico-chemical cues that, among others, are much easier to engineer [24, 61].

As represented in Fig. 3, the mechano-structural axis comprises those cues which condition the static nature of the cell habitat. Here, we include 3D architecture, surface physical features, such as roughness, as well as bulk properties, such as elastic modulus as well as stress, strain and force. For example, the most striking and repeatable differences are found between 2 and 3 dimensional (3D) environments. In fact, there is a dramatic change in habitat when cells are removed from an in vivo context to a petri dish, and in particular stem cells find themselves in a highly non-symmetrical context. Therefore, several reports have shown that encapsulated cells or spheroids are capable of maintaining stemness, and this could be because the microenvironment does not condition the cells to express differentiation proteins for coupling to an extracellular environment. Although there are few systematic studies on the comparative effects of 2 and 3D structures, there are clear indications that cells respond quite differently to the 3D environment which allows greater cell–cell interaction, and maintenance of spherical morphologies. In this context, hydrogels, such as those derived from alginate, collagen and hyaluronic acid, have been shown to be quite promising—they provide a homogeneous, structure-less soft 3D environment, which is probably ideal for stem cell proliferation and maintenance, as well as for differentiation into softer tissues, such as neural or hepatic [5, 10].

Three dimensional environments can also provide more controlled spatial information to cells, if they are architectured using techniques such as microfabrication based on rapid prototyping. In this case, the 3D system is usually nominated as a "scaffold", and provides a rigid and porous framework for cells to adhere on and spread in 3 dimensions. One of the first reports on architectured scaffolds employed for stem cell engineering describes a random pore (250–500 µm) salt leached synthetic polymer scaffold (a blend of polylactide (PL) and polylactide-co-glycolide (PLGA) seeded with embryonic stem cells (ESC) [38]. The cells generated complex capillary-like organized features which cannot be formed in 2D. Moreover the porous scaffold permitted the ESC to organize and orient, whereas a homogeneous isotropic soft gel did not, suggesting that both stiffness and a structured topology are important features to which the cells respond.

Further studies by Liu and Roy [41, 42] also confirmed that a porous rigid scaffold (tantalum in this case) promotes differentiation of ESC into hematopoietic cells with respect to classical 2D cultures. Here, the cells were observed to interact

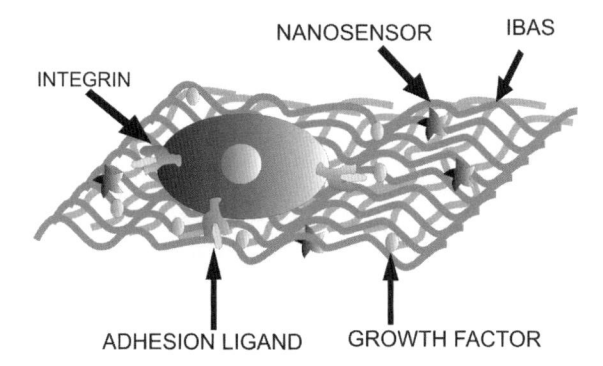

Fig. 4 IBAS (inherently bio-active scaffold) is a three dimensional architecture fabricated from intuitive composite materials and able to deliver stimuli perceived as biologically suitable by progenitor cells independently of the presence of soluble biologically active factors, such as growth factors, ligands, etc. IBAS is not a mere mechanical support for cells and bio-active proteins, but concurs with soluble factors to sustain the signals symmetry driving the progenitor cell fate

with the scaffolds forming smaller aggregates, and so increasing cell-substrate rather than cell–cell contact, probably leading to increased expression of ECM and adhesion proteins. An important aspect of the mechano-structural environment is the chemical nature of the substrate, that is the nature of the biomaterial used as an interface for cell adhesion and a structural support for tissues. Recent work has demonstrated that biomaterials, i.e. matrices, scaffolds, and culture substrates, can present key regulatory signals to create artificial surrogate microenvironments that control stem cell fate. Although little is known about the influence of specific biomaterial features, factors, such as ligand density, and material mechanical properties have also been show to play a role in determining phenotype [22, 24]. We would argue that the biomaterial cue belongs to the mechano-structural axis because it is intimately bound with the elastic modulus, although, if the biomaterial is degradable, it can be dramatically modified by the cells themselves, as discussed in the following section. A number of reviews listing the different biomaterials used in stem cell engineering in general are available [11, 30, 63]. However, very little comparative information can be found regarding the performance and effect of materials on stem cell differentiation, since most approaches use a single material and then test various inducing media to assess the differentiating stimulus provided by the material.

In the context of biomaterials for cardiac tissue engineering, probably the best studied mechano-structural feature is the elastic modulus or stiffness of a substrate or scaffold and its effect on the lineage specification. Engler et al. [22] have pioneered studies on matrix elasticity and its effect on mesenchymal stem cells (MSC). Through the use of a well defined elastically tunable polyacrylamide gel (nominally 2D since cells did not penetrate the substrate but adhered on the surface), they observed lineage specification independent of inductive biochemical factors. Softer gels (0.1–1 kPa) were neurogenic, the hardest (24–40 kPa) were osteogenic, while gels with intermediate elastic moduli (8–17 kPa) were myogenic. In all three cases the elastic modulus matches that of the corresponding native tissue [21]. Studies conducted on substrates with different stiffness show that in cardiac myocytes the development of aligned sarcomers and stress fibers is higher on substrates with elastic moduli of approximately 10 kPa [32]. In other reports, the range of elastic moduli suitable for sustaining cardiac morphology and function of neonatal rat cardiomyocytes was found to be around 22–50 kPa [8]. Therefore, gel-like materials with elastic moduli between 5 and 50 kPa are most suitable for cardiac tissue engineering, although most studies home into the native elastic modulus of normal tissue which is around 10–20 kPa.

The potency of scaffold stiffness and topology in driving cardiac stem cell differentiation in a three-dimensional culture context was confirmed by Forte et al. [24]. Cardiac stem cells adopted the cardiomyocytic phenotype only when cultured in strictly controlled conditions characterized by a critical combination of chemical, biochemical and physical factors, and emulating the inner myocardial environment. In these studies, the emulation of myocardial environment was achieved by fine-tuning the array of growth factors solved in the culture medium and, above all, the chemistry, topology and stiffness of three-dimensional supports on which stem cells were seeded. The absence of one or more appropriate growth factors or the presence of a polymeric scaffold with stiffness higher than that passively expressed by myocardium did not trigger the differentiating cascade leading to the cardiomyocytic phenotype. Here, scaffold stiffness was modulated by changing the topology of the structure, using a rapid prototyping technique known as PAM (Pressure Assisted Microsyringe, [71]. The optimal mechanical properties to induce cardiomyocyte differentiation was a scaffold stiffness <30 kPa on scaffolds with square pores of about 100–150 μm [24].

Shear stress induced by flow is also an important physical stimulus, particularly for driving cells towards vascular phenotypes. Flow at physiological wall shear stresses typical of blood vessels (0.1–1.5 Pa) induces differentiation towards endothelial-like characteristics [51, 72, 78]. Adamo et al. [1] demonstrated that shear stresses of the order of 0.5 Pa increase haematopoietic colony-forming potential and expression of haematopoietic markers in mouse embryos. Therefore, fluid shear stresses also influences the differentiation pathways of cells indirectly associated with the vascular system. Like scaffold stiffness, wall shear stress is a parameter which can be easily controlled and modulated using engineering design tools and technology.

6 Current Material-Based Cardiac Regeneration Strategies

In the past 20 years or so, a number of methods and materials have been employed in an attempt to generate neo-cardiac tissue in vitro as well as in vivo. Traditionally, tissue engineering (TE) strategies to create systems able to replace or repair damaged tissues employ a "top-down" or a "bottom-up" approach as

Table 2 The top-down and bottom up approaches to engineering the cell microenvironment

Top-down		Bottom-up	
Methods	Micro stimulus	Methods	Macro construct
Surface chemistry	Laminin [28], fibronectin [27], RGD [44]	Micro-molding	Microwell templates [33] microchannels [35]
Mechanical properties	Elasticity [21], anisotropy [40]	Self-assembling	Self-assemled aggregation [54]
Mechanical environment	Cyclic strain [9], electrical stimulation [43]	Digital printing	Direct printing [52]
Soluble factor	Morphogenetic proteins [34], VEGF [66], morphogenetic proteins	Cell sheet	Sheet based modules [46], cell laden hydrogel [59]
Architecture design	Topography and shape [24]		
Communication	Cell–cell contact [74]		

Table 3 Materials used in cardiac regeneration

Materials used in cardiac tissue regeneration		
Synthetic		Natural
Solid	Hydrogel	Injectable hydrogel
Poly(glycolic acid) (PGA) [24]	Poly (ethylene glycol)(PEG) [37]	Fibrin [2]
Poly (L-lactic acid) (PLLA) [24]	Poly (N-isopropylacrylamide) (PNIPAAm)	Collagen [20]
Poly (lactic-co-glycolic acid) (PLGA) [24]		Matrigel
Polycaprolactone (PCL) [24]		Self-assembly protein [59] alignate [19]
Poly (urethane) [69]		Hyaluronic acid (HA) [67]

shown in Table 2. These methods aim to recreate the complexity of the biological tissues from two different points of view. The "top-down" strategy attempts to control cell fate modeling the interface between cells and polymeric scaffold to mimic the natural extracellular matrix (ECM); while "bottom-up" approach aims to reproduce biomimetic structures using micromodular blocks built together to recreate larger tissues. Many materials have been studied and characterized using the first strategy starting with synthetic polymers, such as poly(glycolic acid) (PGA), poly(L-lactic acid) (PLLA), poly(lactic-co-glycolic acid) (PLGA), poly(ethylene glycol) (PEG) and polycaprolactone (PCL), or natural derived polymers, such as alginate, hyaluronic acid, collagen, fibrin, etc. Some of the most widely reported materials used for investigating cardiac regeneration are reported in Table 3. Their mechanical properties have been widely investigated and they have been processed with different techniques to reproduce ECM elasticity and deformability. In addition, surface modification techniques have been introduced to immobilize specific factors onto the material surfaces to control cell behavior

Table 4 Commonly used microfabrication techniques for cardiac engineering

Microfabrication techniques
Electrospinning [69]
PAM [71]
Salt leaching [60]
Photo polymerisation [76]
Micro molding templates [33, 35]
Cell encapsulation
Tissue printing [52]
Cell sheet
Micro patterning (soft lithography)
Microfluidics devices [58]

and fate. A key role in designing and realizing a scaffold able to mimic the complexity of the ECM is played by the fabrication technique, some of which are reported in Table 4. In general, the fabrication technique should preserve the properties of the processed material and be capable of producing scaffolds with controlled dimensions (at the nano- and micro-scale) and topology (architecture and shape), as well as high porosity. In the "bottom-up" approach different micromodular structures are designed and built to reproduce tissue engineered constructs with physiological microarchitectural features, providing guidance for tissue morphogenesis at the cellular level. Basically the modules can be created with different microfabrication methods (Table 4) and then assembled randomly or stacked layer-by-layer.

So far, few attempts to reproduce the natural environment for the maintenance and regulation of cardiac stem cells have been carried out. Tissue fabrication techniques have improved a great deal in the past 2 decades and scaffolds can be used to finely tune the chemistry and micro-architecture at the interface between biopolymers and CPC. In particular laminin, fibronectin or RGD sequences can be linked to polymer chains to reproduce ECM signaling allowing precise control of CPC proliferation, survival, differentiation, adhesion and organization. As discussed in Sect. 5, a parameter which can be easily tuned is the elasticity and deformability of scaffolds. These properties are related to the material used, the fabrication process, the porosity and the final scaffold topology. In the case of synthetic polyesters, the porosity, dimensions and topology of scaffold are the key parameters which define mechanical properties, while in hydrogels the cross-linking agent and density as well as polymer characteristics are responsible for material elasticity and stiffness. In fact cell response to substrate stiffness can be elegantly decoupled from biochemical factors using tunable materials such as polyacrylamide or PEG gels thus making it possible to assess which portion of cellular behavior is attributable to cell mechanics [55]. But, while the static mechanical properties of the scaffolds can be easily tuned, the dynamical remodeling of these properties depends on CPC responses, and calls for more responsive materials [65].

There is clearly however a synergy between mechanical and chemical cues as demonstrated by Pagliari et al. [57] and Jacot et al. [31] and reviewed by Tenney and Discher [70]. Pagliari and co-workers demonstrated that biological and mechanical signals must cooperate to drive cardiac stem cell differentiation. CPCs only expressed α-actinin and displayed sarcomeric bands in the presence of PLA scaffolds with an elastic modulus of < 30 kPa as well as neonatal cardiomyoctes, but not when only one of the two signals were absent. Jacot et al. investigated the correlation between substrate stiffness and chemical pathways. Cells on substrates with the same elastic modulus as healthy cardiac tissue (10 kPa) develop aligned sarcomeres whereas stiffer substrates typified by a fibrotic scar induce differentiation towards a more myoblastic phenotype with long, large stress fibers and low contractile forces. The researchers demonstrated that blocking the RhoA/ROCK pathway, which controls calcium transients during contraction, cell traction on all substrates was significantly increased. The effect of substrate stiffness was therefore overruled by inhibiting the formation of actin stress fibers. It is evident however that substrate stiffness and tension driven matrix assembly are intricately linked as protein function is conditioned by form, which may change during deformation. A clear example is the unfolding of the latent TGF-β1 complex, which when activated through stress, releases TFG-β1 and induces the initiation of cell contractile machinery which can result in the formation of fibrous stiff tissue.

In the context of cardiac tissue engineering, the "bottom-up" approach can be used to control stem cell response and cell-to-cell electrophysiological interactions using microscale technologies, such as surface topography and micropatterning, or other methods listed in Table 4. Despite the developments in materials science and processing, however, we are still a long way away from identifying a set of materials and rules for processing them for cardiac stem cell maintenance and differentiation. Here too, a paradigm shift, which focuses on biomimetics, complexity and smart, intuitive materials is necessary. In the following section we discuss how to address complexity by fabricating inherently bio-active scaffolds (IBAS).

7 Smart Materials Strategies: The IBAS Concept

Electronic and information technologies have made leaps and bounds as far as down-sizing, multiplexing and processing rates are concerned. In stark contrast, technologies which involve interfacing with the biological world are far behind. Materials interfacing with the human body in any context—be it chemical sensors for biological analytes, systems for drug testing or biomaterials—have made remarkably little progress in comparison with say silicon technology. The list of materials approved for in vivo use has remained practically unchanged for decades. Biodegradable polyesters continue to be proposed for tissue engineering applications despite reports on undesired inflammatory response and mechanical failure simply because no viable alternatives are available.

Currently, biomaterials are prepared for cell adhesion by the addition of adhesion ligands as proteins or peptides. Let us imagine what happens when cells interact with a biomaterial in vitro; the cells find themselves in an environment which besides the presence of adhesion ligands, possesses zero responsiveness. There is a constant tug-of-war between cells and scaffold, usually with the cells losing out. They either give up and die, or secrete a large quantity of adhesive proteins which act as a barrier between the cells and the material, and the cells go it alone. The material on the other hand has remained passive and unresponsive. How would a responsive material behave? How can we make a material respond to its environment by endowing it with intuition?

In our opinion, to answer the questions, one of the key factors that must be addressed is the complexity of the human body and the interactions between different organs, systems and tissues, as well as the profound role the environment and external chemical or physical factors have on the body's response. No material in the human body is monocomponent and all senses—be they chemical or physical—are integrated, adaptive and fuzzy. Moreover, the materials in our bodies are intuitive, because they react to cues from their environment, and they do so because sensing is an integral part of their structure. Taking a cue from biology, or using the much abused term "biomimetics", the complexity must be challenged and met with complex materials and sensing systems which exploit the plethora of knowledge acquired in chemistry and physics to create new hybrid composite materials and alternative transduction modalities. These materials then need to be shaped into three dimensional scaffolds to produce what we call inherently bioactive scaffolds, or IBAS. IBAS are scaffolds which are able to mimic the tissue microenvironment and address stem cell adhesion, growth and differentiation in the absence of biological factors or any biological functionalization. To emulate tissue complexity, IBAS should be composed of interconnected modules having distinct features, which could be recognized by CPC at the nano- and microscale level (Fig. 4).

Until now, the cell has been considered as an entity doing its job independently from the others, and from the scaffold or substrate, which is unrealistic in vivo. In vivo, not only is the cell in equilibrium with its neighbors, but also with the extracellular matrix. Moreover, the matrix itself is in equilibrium with the cells as it too responds to and is modified by the cells. Therefore, each cell's strategy, as well as the matrix's can be independent, but the overall result will lead to an equilibrium state. Using this hypothesis, an IBAS could be engineered to coerce the system towards an equilibrium consisting of myocardial tissue.

Some attempts have been made to create complex material systems, although these are often based on mixtures of polymers [13]. Carbon nanotubes are also becoming popular additives to classic biomaterials [47], but their main function is to increase the elastic modulus rather than confer sensing or other properties. However, as Whulanza et al. [75] show, CNT loaded scaffolds can sense cellular behavior through changes in the electrical properties of the material. Other, more ambitious, solutions could be explored such as that reported by [73] who developed a water and nanoparticle based hydrogel which has the strength to remain self supporting, as well as self-repairing. Other methods to render scaffolds intuitive involve integration of nanosenors into the materials, either during processing [15] or synthesis [81]. On the other hand, as suggested and demonstrated by Schneider and Strongin [65], polymers can even be designed so that a particular chemical substance or supramolecular binding site reacts selectively, reversibly and non-covalently to change the properties of the polymer, even at a local scale. In the majority of cases, the changes are conformational, and based on the exclusion or inclusion of water molecules. This is exactly how sensors in biology work.

8 Conclusion

In conclusion, materials need to be rethought, reshaped and reassembled using a bottom-up reverse engineering approach, in which the end goal is a purpose or a function irrespective of composition. New approaches should consider the mechanical, chemical and topological signals and design new complex multi-shell [79] or nanocomposite materials able to encapsulate cells, which not only allow induction or control of cell fate, but also enable dynamic remodeling of the cell microenvironment. Using intuitive and adaptive complex materials, or IBAS, different microsystems can be assembled into a modular engineered tissue. The IBAS can be thought of as an intuitive scaffold which naturally leads CPC through a set of equilibrium states, ending with functional cardiac tissue.

References

1. Adamo, L., Naveiras, O., Wenzel, P.L., et al.: Biomechanical forces promote embryonic haematopoiesis. Nature **459**, 1131–1135 (2009)
2. Ahmed, T.A., Dare, E.V., Hincke, M.: Fibrin: a versatile scaffold for tissue engineering applications. Tissue Eng. Part B **14**, 199–215 (2008)
3. Andersen, D.C., Andersen, P., Schneider, M., et al.: Murine "cardiospheres" are not a source of stem cells with cardiomyogenic potential. Stem Cells **27**, 1571–1581 (2009)
4. Anversa, P., Kajstura, J., Leri, A., et al.: Life and death of cardiac stem cells: a paradigm shift in cardiac biology. Circulation **113**, 1451–1463 (2006)
5. Baharvand, H., Hashemi, S.M., Ashtiani, S.M., et al.: Differentiation of human embryonic stem cells into hepatocytes in 2D and 3D culture systems in vitro. Int. J. Dev. Biol. **50**, 645–652 (2006)
6. Beltrami, A.P., Cesselli, D., Bergamin, N., et al.: Multipotent cells can be generated in vitro from several adult human organs (heart, liver, and bone marrow). Blood **110**, 3438–3446 (2007)
7. Bergmann, O., Bhardwaj, R.D., Bernard, S., et al.: Evidence for cardiomyocyte renewal in humans. Science **324**, 98–102 (2009)
8. Bhana, B., Iyer, R.K., Chen, W.L.K., et al.: Influence of substrate stiffness on the phenotype of heart cells. Biotechnol. Bioeng. **105**(6), 1148–1160 (2010)
9. Boublik, J., Park, H., Radisic, M., et al.: Mechanical properties and remodelling of hybrid cardiac constructs made from heart cells, fibrin, and biodegradable, elastomeric knitted fabric. Tissue Eng. **11**, 1122–1132 (2005)

10. Brännvall, K., Bergman, K., Wallenquist, U., et al.: Enhanced neuronal differentiation in a three-dimensional collagen–hyaluronan matrix. J. Neurosci. Res. **85**, 2138–2146 (2007)
11. Burdick, J.A., Vunjak-Novakovic, G.: Engineered microenvironments for controlled stem cell differentiation. Tissue Eng. Part A **15**, 205–219 (2009)
12. Chen, S.L., Fang, W.W., Ye, F., et al.: Effect on left ventricular function of intracoronary transplantation of autologous bone marrow mesenchymal stem cell in patients with acute myocardial infarction. Am. J. Cardiol. **94**, 92–95 (2004)
13. Chiono, V., Pulieri, E., Vozzi, G., et al.: Genipin-crosslinked chitosan/gelatin blends for biomedical applications. J. Mater. Sci. Mater. Med. **19**, 889–898 (2008)
14. Claycomb, W.C., Di Nardo, P. (eds.): Cardiac Growth and Regeneration, vol. 752. New York Academy of Sciences, New York (1995)
15. Coupland, P.G., Briddon, S.J., Aylott, J.W.: Using fluorescent pH-sensitive nanosensors to report their intracellular location after Tat-mediated delivery. Integr. Biol. **1**, 318–323 (2009)
16. Cross, M.A., Enver, T.: The lineage commitment of haemopoietic progenitor cells. Curr. Opin. Genet. Dev. **7**, 609–613 (1997)
17. Dawn, B., Stein, A.B., Urbanek, K., et al.: Cardiac stem cells delivered intravascularly traverse the vessel barrier, regenerate infarcted myocardium, and improve cardiac function. Proc. Natl. Acad. Sci. USA **102**, 3766–3771 (2005)
18. Di Nardo, P., Forte, G., Ahluwalia, A., et al.: Cardiac progenitor cells: potency and control. J. Cell. Physiol. **224**, 590–600 (2010)
19. Drury, J.L., Dennis, R.G., Mooney, D.J.: The tensile properties of alginate hydrogels. Biomaterials **25**, 3187–3199 (2004)
20. Eghbali, M., Weber, K.T.: Collagen and the myocardium: fibrillar structure, biosynthesis and degradation in relation to hypertrophy and its regression. Mol. Cell. Biochem. **96**, 1–14 (1990)
21. Engler, A.J., Carag-Krieger, C., Johnson, C.P., et al.: Embryonic cardiomyocytes beat best on a matrix with heart-like elasticity: scar-like rigidity inhibits beating. J. Cell Sci. **121**, 3794–3802 (2008)
22. Engler, A.J., Sen, S., Sweeney, H.L., et al.: Matrix elasticity directs stem cell lineage specification. Cell **126**, 677–689 (2006)
23. Fischer-Rasokat, U., Assmus, B., Seeger, F.H., et al.: A pilot trial to assess potential effects of selective intracoronary bone marrow-derived progenitor cell infusion in patients with nonischemic dilated cardiomyopathy: final 1-year results of the transplantation of progenitor cells and functional regeneration enhancement pilot trial in patients with nonischemic dilated cardiomyopathy. Circ. Heart Fail. **2**, 417–423 (2009)
24. Forte, G., Carotenuto, F., Pagliari, F., et al.: Criticality of the biological and physical stimuli array inducing resident cardiac stem cell determination. Stem Cells **26**, 2093–2103 (2008)
25. Fuchs, E., Tumbar, T., Guasch, G.: Socializing with the neighbors: stem cells and their niche. Cell **116**, 769–778 (2004)
26. Gaetani, R., Ledda, M., Barile, L., et al.: Differentiation of human adult cardiac stem cells exposed to extremelylow-frequencies electromagnetic fields. Cardiovasc. Res. **82**, 411–420 (2009)
27. Gelain, F., Bottai, D., Vescovi, A., et al.: Designer self-assembling peptide nanofiber scaffolds for adult mouse neural stem cell 3-dimensional cultures. PLoS ONE **1**, e119 (2006)
28. Genove, E., Shen, C., Zhang, S., et al.: The effect of functionalised self-assembling peptide scaffolds on human aortic endothelial cell function. Biomaterials **26**, 3341–3351 (2005)
29. Hare, J.M., Traverse, J.H., Henry, T.D., et al.: A randomized, double-blind, placebo-controlled, dose-escalation study of intravenous adult human mesenchymal stem cells (prochymal) after acute myocardial infarction. J. Am. Coll. Cardiol. **54**, 2277–2286 (2009)
30. Hwang, N.S., Varghese, S., Elisseeff, J.: Controlled differentiation of stem cells. Adv. Drug Deliv. Rev. **60**, 199–214 (2008)
31. Jacot, J.G., Kita-Matsuo, H., Wei, K.A., et al.: Cardiac myocyte force development during differentiation and maturation. Ann. NY Acad. Sci. **1188**, 121–127 (2010)

32. Jacot, J.G., McCulloch, A.D., Omens, J.H.: Substrate stiffness affects the functional maturation of neonatal rat ventricular myocytes. Biophys. J. **95**, 3479–3487 (2008)
33. Karp, J.M., Yeh, J., Eng, G., et al.: Controlling size, shape and homogeneity of embryoid bodies using poly(ethylene glycol) microwells. Lab Chip **7**, 786–794 (2007)
34. Keegan, B.R., Feldman, J.L., Begemann, G., et al.: Retinoic acid signalling restricts the cardiac progenitor pool. Science **307**, 247–249 (2005)
35. Khademhosseini, A., Eng, G., Yeh, J., et al.: Microfluidic patterning for fabrication of contractile cardiac organoids. Biomed. Microdevices **9**, 149–157 (2007)
36. Laugwitz, K.L., Moretti, A., Lam, J., et al.: Postnatal isl1$^+$ cardioblasts enter fully differentiated cardiomyocyte lineages. Nature **433**, 647–653 (2005)
37. Leslie-Barbick, J.E., Moon, J.J., West, J.L.: Covalently-immobilized vascular endothelial growth factor promotes endothelial cell tubulogenesis in poly(ethylene glycol) diacrylate hydrogels. J. Biomater. Sci. **20**, 1763–1779 (2009)
38. Levenberg, S., Huang, N.F., Lavik, E., et al.: Differentiation of human embryonic stem cells on three-dimensional polymer scaffolds. Proc. Natl. Acad. Sci. **100**, 12741–12746 (2003)
39. Li, L., Xie, T.: Stem cell niche: structure and function. Annu. Rev. Cell Dev. Biol. **21**, 605–631 (2005)
40. Linask, K.K., Han, M., Cai, D.H., et al.: Cardiac morphogenesis: matrix metalloproteinase coordination of cellular mechanism underlying heart tube formation and directionality of looping. Dev. Dyn. **233**, 739–753 (2005)
41. Liu, H., Lin, J., Roy, K.H.: Effect of 3D scaffold and dynamic culture condition on the global gene expression profile of mouse embryonic stem cells. Biomaterials **27**, 5978–5989 (2006)
42. Liu, H., Roy, K.: Biomimetic three-dimensional cultures significantly increase hematopoietic differentiation efficacy of embryonic stem cells. Tissue Eng. **11**, 319–330 (2005)
43. Ma, W., Liu, Q.Y., Jung, D., et al.: Central neuronal synapse formation on micropatterned surfaces. Brain Res. Dev. Brain Res. **111**, 231–243 (1998)
44. Mann, B.K., Gobin, A.S., Tsai, A.T., et al.: Smooth muscle cell growth in photopolymerised hydrogels with cell adhesive and proteolytically degradable domains: synthetic ECM analogs for tissue engineering. Biomaterials **22**, 3045–3051 (2001)
45. Martin, C.M., Meeson, A.P., Robertson, S.M., et al.: Persistent expression of the ATP-binding cassette transporter, Abcg2, identifies cardiac SP cells in the developing and adult heart. Dev. Biol. **265**, 262–275 (2004)
46. Masuda, S., Shimizu, T., Yamato, M., et al.: Cell sheet engineering for heart tissue repair. Adv. Drug Del. Rev. **60**, 277–285 (2008)
47. Mattioli-Belmonte, M., Vozzi, G., Seggiani, M., et al.: Tuning polycaprolactone-carbon nanotube composites for bone tissue engineering scaffolds. Mater. Sci. Eng. Biomimetic Mater. (2010, in press)
48. Matsuura, K., Nagai, T., Nishigaki, N, et al.: Adult cardiac Sca-1-positive cells differentiate into beating cardiomyocytes. J. Biol. Chem. **279**, 11384–11391 (2004)
49. Menasché, P.: Cell therapy: results in cardiology. Bull. Acad. Natl. Med. **193**, 559–568 (2009)
50. Messina, E., De Angelis, L., Frati, G., et al.: Isolation and expansion of adult cardiac stem cells from human and murine heart. Circ. Res. **95**, 911–921 (2004)
51. Metallo, C.M., Vodyanik, M.A., de Pablo, J.J., et al.: The response of human embryonic stem cell-derived endothelial cells to shear stress. Biotechnol. Bioeng. **100**, 830–837 (2008)
52. Mironov, V., Boland, T., Trusk, T., et al.: Organ printing: computer-aided jet-based 3D tissue engineering. TRENDS Biotechnol. **21**, 157–161 (2003)
53. Murry, C.E., Soonpaa, M.H., Reinecke, H., et al.: Haematopoietic stem cells do not transdifferentiate into cardiac myocytes in myocardial infarcts. Nature **428**, 664–668 (2004)
54. Napolitano, A.P., Chai, P., Dean, D.M., et al.: Dynamics of the self-assembly of complex cellular aggregates on micromolded nonadhesive hydrogels. Tissue Eng. **13**, 2087–2094 (2007)
55. Nemir, S., West, J.L.: Synthetic materials in the study of cell response to substrate rigidity. Ann. Biomed. Eng. **38**, 2–20 (2010)

56. Oh, H., Bradfute, S.B., Gallardo, T.D., et al.: Cardiac progenitor cells from adult myocardium: homing, differentiation, and fusion after infarction. Proc. Natl. Acad. Sci. USA **100**, 12313–12318 (2003)
57. Pagliari, S., Vilela-Silva, A.C., Forte, G.C., Pagliari, F., Mandoli, C., Vozzi, G., Pietronave, S., Prat, M., Licoccia, S., Ahluwalia, A., Traversa, E., Minieri, M., Di Nardo, P., Cooperation of biological and mechanical signals in cardiac progenitor cell differentiation. Adv. Mater. (2010) 12 Nov 2010. doi:10.1002/adma.201003479.
58. Paguirigan, A.L., Beebe, D.J.: Protocol for the fabrication of enzymatically crosslinked gelatin microchannels for microfluidic cell culture. Nat Protocols **2**, 1782–1788 (2007)
59. Palmer, L.C., Stupp, S.I.: Molecular self-assembly into one-dimensional nanostructures. Acc. Chem. Res. **41**, 1674–1684 (2008)
60. Pego, A.P., Siebum, B., Van Luyn, M.J., et al.: Preparation of degradable porous structures based on 1, 3-trimethylene carbonate and D,L-lactide (co)polymers for heart tissue engineering. Tissue Eng. **9**, 981–994 (2003)
61. Reilly, G.C., Engler, A.J.: Intrinsic extracellular matrix properties regulate stem cell differentiation. J. Biomech. **43**, 55–62 (2010)
62. Rosenblatt-Velin, N., Lepore, M.G., et al.: FGF-2 controls the differentiation of resident cardiac precursors into functional cardiomyocytes. J. Clin. Invest. **115**, 1724–1733 (2005)
63. Saha, K., Pollock, J.F., Schaffer, D.V., et al.: Designing synthetic materials to control stem cell phenotype. Curr. Opin. Chem. Biol. **11**, 381–387 (2007)
64. Scadden, D.T.: The stem-cell niche as an entity of action. Nature **441**, 1075–1079 (2006)
65. Schneider, H.J., Strongin, R.M.: Supramolecular interactions in chemomechanical polymers. Acc. Chem. Res. **42**, 1489–1500 (2010)
66. Seliktar, D., Zisch, A.H., Lutolf, M.P., et al.: MMP-2 sensitive, VEGF-bearing bioactive hydrogels for promotion of vascular healing. J. Biomed. Mater. Res. A **68**, 704–716 (2004)
67. Serban, M.A., Prestwich, G.D.: Modular extracellular matrices: solutions for the puzzle. Methods **45**, 93–98 (2008)
68. Smith, R.R., Barile, L., Cho, H.C., et al.: Regenerative potential of cardiosphere-derived cells expanded from percutaneous endomyocardial biopsy specimens. Circulation **115**, 896–908 (2007)
69. Stankus, J.J., Guan, J., Fujimoto, K., et al.: Microintegrating smooth muscle cells into a biodegradable, elastomeric fiber matrix. Biomaterials **27**, 735–744 (2006)
70. Tenney, R.M., Discher, D.E.: Stem cells, microenvironment mechanics, and growth factor activation. Curr. Opin. Cell Biol. **21**, 630–635 (2009)
71. Vozzi, G., Previti, A., De Rossi, D., et al.: Microsyringe-based deposition of two-dimensional and three-dimensional polymer scaffolds with a well-defined geometry for application to tissue engineering. Tissue Eng. **8**, 1089–1098 (2002)
72. Wang, H., Riha, G.M., Yan, S., et al.: Shear stress induces endothelial differentiation from a murine embryonic mesenchymal progenitor cell line. Arterioscler. Thromb. Vasc. Biol. **25**, 1817–1823 (2005)
73. Wang, Q., Mynar, J.L., Yoshida, M., et al.: High-water-content mouldable hydrogels by mixing clay and a dendritic molecular binder. Nature **463**, 339–343 (2010)
74. White, S.M., Claycomb, W.C.: Embryonic stem cells form an organised, functional cardiac conduction system in vitro. Am. J. Physiol. Heart Circ. Physiol. **288**, H670–H679 (2005)
75. Whulanza, Y., Ucciferri, N., Vozzi, G., Domenici, C.: Sensing scaffolds to monitor cellular activities using impedance measurements. Biosensors Bioelectronics (in press)
76. Williams, C.G., Malik, A.N., Kim, T.K., et al.: Variable cytocompatibility of six cell lines with photoinitiators used for polymerising hydrogels and cells encapsulation. Biomaterials **26**, 1211–1218 (2005)
77. Wollert, K.C., Meyer, G.P., Lotz, J., et al.: Intracoronary autologous bone-marrow cell transfer after myocardial infarction: the BOOST randomised controlled clinical trial. Lancet **364**, 141–148 (2004)

78. Wu, C.C., Chao, Y.C., Chen, C.N., et al.: Synergism of biochemical and mechanical stimuli in the differentiation of human placenta-derived multipotent cells into endothelial cells. J. Biomech. **41**, 813–821 (2008)
79. Yao, H., Dao, M., Imholt, T., et al.: Protection mechanisms of the iron-plated armor of a deep-sea hydrothermal vent gastropod. Proc. Natl. Acad. Sci. USA **107**, 987–992 (2010)
80. Zampetaki, A., Kirton, J.P., Xu, Q.: Vascular repair by endothelial progenitor cells. Cardiovasc. Res. **78**, 413–421 (2008)
81. Zhang, G., Palmer, G.M., Dewhirst, M.W., et al.: A dual-emissive-materials design concept enables tumour hypoxia imaging. Nat. Mater. **8**, 747–751 (2009)
82. Zipori, D.: The stem state: plasticity is essential, whereas self-renewal and hierarchy are optional. Stem Cells **23**, 719–726 (2005)

Strategies for Myocardial Tissue Engineering: The Beat Goes On

Payam Akhyari, Mareike Barth, Hug Aubin and Artur Lichtenberg

Abstract The striving for the ability to build a piece of tissue that resembles the functional features of the heart muscle has been part of the scientific work for almost a century. More than any other organ, the constantly beating pump that drives the blood circulation, has been the focus of spectacular experimental breakthrough reports and subject to significant public interest. Beside the culture-related and religious reasons for the especial public attention that lies upon most medical and scientific work related to the heart, there are numerous matter-of-fact motivations to pay attention to the causes and the cure of heart disease. This work reviews the current strategies for the in vitro or in vivo reconstitution of myocardial tissue, including current concepts or the generation and implementation of extracellular matrix components, discussion of promising sources and culture techniques for the generation of cardiomyocytes or other cell populations necessary for the assembly of a functional myocardium, and finally limitations and perspectives of established myocardial tissue engineering models.

P. Akhyari (✉) · A. Lichtenberg
Department of Cardiovascular Surgery, Duesseldorf University Hospital,
Duesseldorf, Germany
e-mail: payam.akhyari@med.uni-duesseldorf.de

P. Akhyari · H. Aubin · A. Lichtenberg
Experimental Surgical Research Group, Duesseldorf University Hospital,
Duesseldorf, Germany

M. Barth
Helmholtz Group for Cell Biology, German Cancer Research Center,
Heidelberg, Germany

1 Background and Clinical Demand

It has been a longstanding dream of humankind to cure heart disease by the creation and application of an artificial cardiac muscle. However, although numerous reports periodically focus the attention of scientific and popular media on ongoing attempts to re-build the 'motor of life', a true breakthrough is yet out of reach. There are a number of reasons for the high level of interest in cardiac regeneration. Beside the scientific challenges associated with cardiac tissue engineering, the permanently urgent clinical demand for heart donations and the limitations of alternative therapy options (such as the implantation of a mechanical cardiac support) underline the globally important role of this research field. Although ethical issues involved in this topic are subject to a regionally and ethnically varying perception, research for a cure of heart disease receives a high level of public approval and attention across most countries, irrespective of the predominant religious and cultural background of the respective society.

1.1 Adult Cardiac Disease

In the adult population, cardiovascular disorders account for more lethal conditions than any other disease [1]. In the industrialized world, this has been the case for many years. But also among the population of developing countries, the number of deaths due to cardiovascular disease (CVD) has been rising in the past years, turning cardiovascular disease to a global health burden [2]. More interestingly, among all cardiovascular diseases, coronary heart disease (CHD) is the most relevant cause of mortality and morbidity, both, in the industrialized and the emerging countries [3].

Since the introduction of saphenous vein grafts for coronary bypass surgery by René G. Favaloro in the late 1960s [4], coronary artery bypass graft surgery (CABG) and interventional procedures to treat CHD, including percutaneous transluminal coronary angioplasty (PTCA) [5], later also performed with coronary stent implantation, have tremendously improved the outcome of patients suffering from CHD. However, CABG and PTCA both aim at preventing the progression of CHD in the acute or chronic condition by ensuring sufficient blood flow to viable but undersupplied myocardial muscle. In contrast, a pre-existing myocardial damage, e.g. due to a significant loss of cardiomyocytes during previous myocardial infarction(s), remains unaffected by these therapy options.

Despite significant improvement of treatment options and even in the presence of the best medical, interventional and surgical therapy, in a significant proportion of patients with CVD disease progression to a condition that is called end stage heart failure cannot be hindered. But also other cardiac diseases, e.g. valvular heart disease or idiopathic cardiomyopathy may lead to end stage heart failure. For patients with end-stage heart failure, the remaining options for a definitive cure lie either in organ transplantation, or in implantation of a mechanical assist device, supporting their diminished cardiac function [6]. Both therapy options harbour significant limitations. Donor organ shortage has limited organ transplantation to only a small fraction of all patients suffering from heart failure, and even after successful transplantation, lifelong medication, particularly immunosuppression inducing medication, and related co-morbidities may diminish life time expectancy and quality of life [7, 8]. In the case of the emerging options of a mechanical assist device therapy, main drawbacks lie in thromboembolic or bleeding complications and infection of the alloplastic implants [9]. Hence, alternatives are warranted to overcome the limitations of the currently best available options for the cure of a patient with diminished cardiac function.

More recently, cell transplantation has been introduced to treat patients with significant reduction of myocardial function after large myocardial infarctions. On the structural level, these patients suffer from a substantial and irreversible loss of myocardial tissue that results in functional limitations. In the early era of experimental and clinical cell transplantation, the theory of a functional integration and active force development by the cells that were injected into scarred myocardial regions dominated the scientific forums [10–12]. Accordingly, myogenic cells, mainly neonatal cardiomyocytes in the experimental setups and skeletal myoblasts in the clinical setting were applied [13]. However, reports on comparable functional benefits after myocardial transplantation of cells lacking working contractile elements shed doubts on the initial hypothesis. Soon the currently still valid theory of paracrine effects that may be triggered by transplanted cells was established [14]. In the recent years, further studies with novel findings on different administration routes and time points with respect to a preceding myocardial damage were presented. Most of these studies have added new aspects to the old story of cell transplantation which once started as a relatively simple and straightforward intervention. Now cell transplantation appears more like a complex of dynamic and cross-talking sequences of biological events that may be triggered by more than one cell type. The observation that after cell transplantation, even the sole physical persistence of the applied cell mass is not necessary for a lasting effect has widened the spectrum of possible explanations for the underlying mechanisms that account for the overall clinical effect of this therapy. The loss of the transplanted cells has very recently urged scientists to search for a method to enhance cell engraftment and survival at the target area [15, 16]. The latter trend brings the scientific focus closer to what may be generally called *tissue engineering*.

1.2 Congenital Cardiac Disease

In the younger population, several congenital cardiac malformations necessitate an operative correction. Congenital malformations affect about 1% of all neonates. The severity of the malformation may vary significantly, ranging from atrial or ventricular septal defects (ASD and VSD, respectively) to more complex pathologic anatomy, including the rather frequent tetralogy of Fallot or the severe malformations of univentricular or hypoplastic left ventricle hearts. The necessity

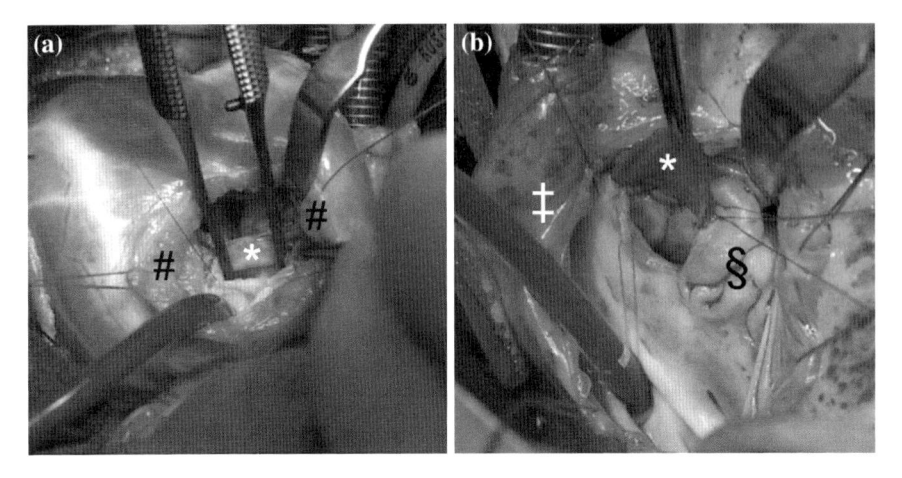

Fig. 1 Clinical use of pericardial tissue in surgical correction of congenital cardiac defects. Intraoperative images demonstrating the use of pericardial tissue as patch material (*asterisk*) to repair a ventricular septal defect (VSD) via incision of the right ventricular free wall (**a**; *hash* marks the incision plane of the ventricular wall) or via incision of the tricuspid valve (paragraph) through the right atrium (*double cross* in **b**; *asterisk* marks the patch material)

and the urgency of an operative correction is largely dictated by the involved pathology. Often, the reconstruction of missing or diseased cardiac structures with prosthetic implants is necessary. Based on currently applied techniques and the involved materials subsequent re-operations at later time points are frequently inevitable due to age-related growth of structures. The structures of particular concern may involve heart valves or parts of the myocardial tissue. In the past decades, several attempts to find a definite cure for the most relevant congenital cardiac malformations have been undertaken. However, the need for repetitive operative intervention and the progression to heart failure moderate the overall outcome for many of these patients. Biological solutions, be there autologous materials like the pericardial tissue, or allogenic donor tissue, e.g. valve-bearing vascular allografts (homografts) are generally preferred (Fig. 1). But no current implant fulfils all criteria of an ideal graft. Novel implants with superior functional outcome and durability for the time period of no less than a patient life span are needed to improve the outcome for the concerned population. This indication has actually been the initial driving motivation for scientifically active clinicians to strive for novel organ or tissue substitutes, forming a scientific field that was termed *tissue engineering* two decades ago [17].

1.3 In Vitro Drug Testing

Finally, with growing public conscious for animal rights and the awareness of the large number of laboratory animals subjected to experimental studies, a public call for limiting animal testing arose. Moreover, motivated by cost effectiveness calculations in the era of growing economic pressure, pharmaceutical companies are searching for reproducible alternatives to expensive animal test series, yet relying on physiologically relevant models. Hence, organ and tissue culture have become attractive models for in vitro drug testing. This way, methods that may allow the in vitro generation of the tissues and organs, summarized as *tissue engineering*, have emerged to a novel field of investigation for health care providers, their associated basic science institutions, and also for pharmaceutical companies.

2 Current Strategies in Myocardial Tissue Engineering

Numerous experimental protocols have been introduced to substitute or assist myocardial tissue for in vitro or in vivo applications. Depending on the general intention and the type of further use, different approaches for myocardial tissue engineering may be promising. To date, these strategies may be categorized into four principal subgroups.

2.1 Myocardial Tissue Engineering According to the Traditional Paradigm

Following the traditional definition, this approach comprises the creation of a tissue or organ fragment in vitro by culturing cellular components in the vicinity of a scaffold under defined conditions. The scaffold component may be of a solid macrostructure for a secondary cell repopulation by surface seeding or other cell transfer techniques, e.g. injection into the central vicinity (Fig. 1, *Traditional Paradigm*, left column). Alternatively, it may also be applied as a liquid precursor mixture, where the conjunction of the liquid scaffold components with a suspension of the cellular components leads to a gelling process resulting in a solid three-dimensional construct of scaffold embedded cells (Fig. 2, *Traditional Paradigm*, right column).

The scaffold provides structural stability and protection for the seeded cells during the in vitro culture period and even more so after a possible in vivo application. In this setting the role of the scaffold as a carrier for the loaded cell population is of predominant importance. Between cell seeding onto the scaffold and the in vivo application of the final cell-scaffold-construct a variable time span may be used to perform a preconditioning of the construct in vitro. This culture period allows the optimization of the construct. Typically main culture conditions and a variety of culture stimuli are governed by an automated bioreactor system. Controlled physical and biological stimulation may be used for an enhancement of numerous biological events, including cell proliferation and differentiation, establishment of cell–cell and cell-scaffold interaction, the initiation of production

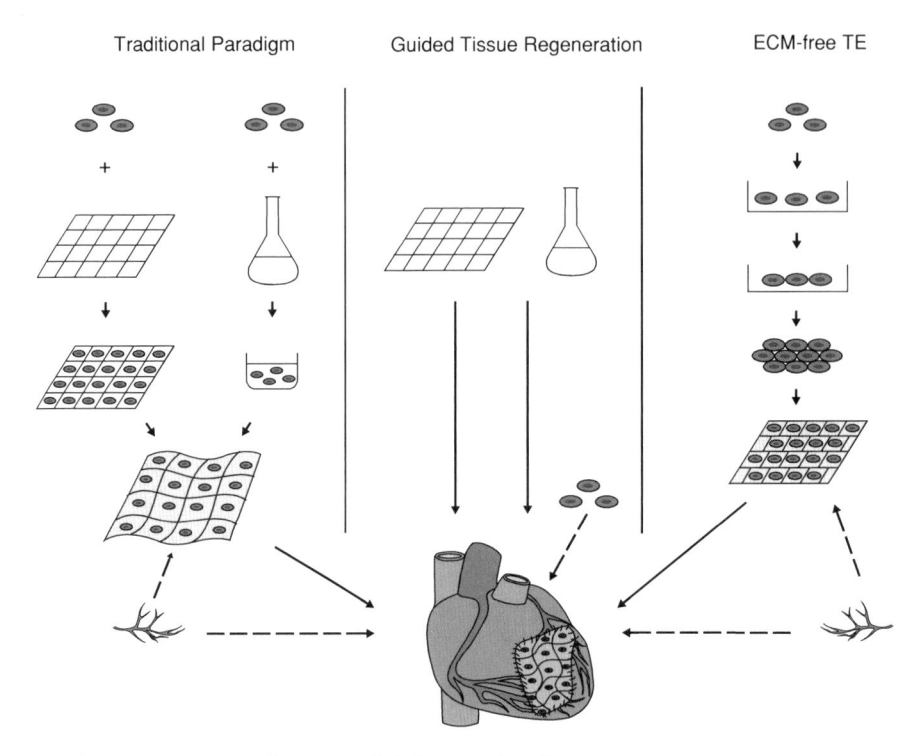

Fig. 2 Current strategies for myocardial tissue engineering

and secretion of bioactive messengers and extracellular matrix precursors, and many more. Not surprisingly, tissue formation appears highly dependent on the scale of biological interactions that are supported by the scaffold material. Hence, this specific quality, which may be summarized as biocompatibility, is a crucial characteristic of the scaffold. Furthermore, the establishment of a convection system that offers a balanced supply of nutrition and oxygen as well as a sufficient drainage of metabolic products is a prerequisite of sustained construct survival and development. The latter topic, namely the induction of neo-*vascularization and -angiogenesis*, represents a complex research area of its own (see Sect. 3.2). In most of the current tissue engineering models (which follow the traditional paradigm) a vascular supply to the construct is closely linked to or integrated part of the scaffold. On the cellular side, the general potency to fulfil the specific functional requirements of the implant tissue is warranted for an optimal functional performance of the construct (see Sect. 3.1). Main issues for myocardial constructs include the development of contractile elements for mechanical performance and junction proteins for electrophysiological integration into the recipient's myocardium (see Sect. 3.3).

In the following, only a brief summary of different aspects of tissue engineering according to the traditional paradigm are recollected, for a detailed review, refer to Chap. XXX, and the recently published review literature [18–22].

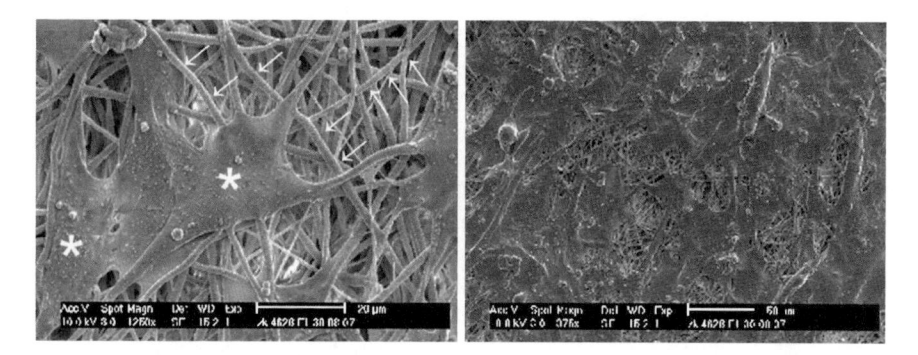

Fig. 3 Scanning electron microscopic images demonstrating the pattern of adhesion and spreading of mouse skeletal myoblasts (C2C12; *asterisk*) grown on electrospun, non-woven, microsized (approximately 2 μm) fibers of polycaprolactone (PCL; *arrows*). Scale bar = 20 μm (*left*) and 50 μm (*right*). Images provided by Marie-Noëlle Giraud and Hendrik T. Tevaearai, Department of Cardiovascular Surgery, Inselspital, Berne, Switzerland

2.1.1 Tissue Engineering Based on Solid Scaffolds

Typically, constructs that are engineered on solid scaffolds according to the traditional paradigm, display initial biomechanical characteristics and physical dimensions of the involved scaffold component. The ideal tissue engineered myocardial graft should be capable of active force generation for a significant contribution to the global pump function after cardiac implantation. In this context, active force development is regarded as a function to be fulfilled by the living, e.g. cellular, components. The passive biomechanical characteristics however, are more predominantly defined by the scaffold. The scaffold materials that have been applied until now, may be categorized according to a variety of characteristics. One of the general categorizations distinguishes chemically synthesized scaffold materials (Fig. 3) from biologically derived scaffolds or composite materials combining elements of the aforementioned.

Synthetic Scaffold Materials

Synthetic scaffolds have gained increasing interest among the tissue engineering community because of a number of advantages in comparison to their biological counterparts. The definite chemical composition avoids the possibility of contaminating donor material. Thereby the risk of pathogen transmission, which has been anticipated for scaffolds derived from animal tissue products, is abolished. Also, the technical possibility of designing and manufacturing synthetic scaffolds with a defined profile regarding the macro- and micro-architecture as well as biomechanical characteristics has contributed to the popularity of synthetic scaffolds.

Poly-L-lactic acid (PLLA) and Polyglycolic acid (PGA) and their co-polymers were among the first scaffold materials to be investigated for tissue engineering purposes [23, 24], other more modern polymer scaffolds, such as Polyurethane (PU) [25, 26] or Polycaprolactone (PCL) [27] and their co-polymers [28] have been utilized to engineer extracellular matrix (ECM) materials with improved biocompatibility and higher predictive values for scaffold performance in vivo. While some groups have utilized scaffolds with random fibre orientation which are produced by electrospinning, others work on scaffolds with knitted or woven fibres or with scaffolds that contain interconnected pores of variable size and spatial distribution [29–31]. Based on the current dynamic state of scientific work on synthesized biomaterials, a further improvement of existing concepts in the near future is expected, diminishing some of the current limitations and increasing the potency of synthetic scaffolds for tissue engineering purposes. Improvement is warranted for a higher diversification of scaffold based elements for a biological interaction with cells and host tissue, as well as for an increased control over the maturation and remodelling after implantation.

Biological Scaffold Materials

A convincing body of evidence supports the general suitability of biological scaffolds for in vitro studies as well as for therapeutic in vivo applications.

Novel biologically derived scaffolds are synthesized like the synthetic materials and integrate the high biocompatibility benefits of multiple single native ECM proteins (collagens, elastins, proteoglycans, etc.) and polysaccharides with a controlled micro and nanostructure, finally allowing for tailored biodegradability and predictable biomechanical performance.

Decellularized Biological Scaffolds

The use of decellularized biological scaffolds is an exciting field of research that has gained an increasing amount of interest in the past few years, although pioneering experiments date more than two decades back from today. This group of native ECM materials comprises scaffolds that are obtained from a donor tissue followed by the extraction of the donor cellular components. Because of the preservation of the general macro-architecture and the absence of extensive industrial biochemical production steps, decellularized ECM materials (dECM) may be categorized as a distinct entity.

The rationale for the recovery of dECM is based on the conception that a native ECM represents the best niche for culture and growth of cell populations. Furthermore, it has been hypothesized that clinically observed adverse immunological reactions that occur upon transplantation of allogenic tissue fragments or organs are mainly related to the donor cellular antigens. Based on this hypothesis, the ECM components of different organisms may be well tolerated with regards to clinically relevant immunological rejection upon allogenic implantation.

Decellularized ECM has the potential of providing a preserved functional vascular network with native configuration and optimal spatial distribution within the ECM. Furthermore, dECM harbours all benefits of biological scaffolds, including the generally high biocompatibility. Moreover, for the first time, organ specific ECM models may be established, reflecting a molecular biological ECM composition that closely mimics the native in vivo situation.

2.1.2 Tissue Engineering with Liquid- and Gel-Based Scaffolds

Collagen and collagen-based gel constructs have achieved tremendous attention by proving a fruitful base for 3D culture of cardiac cells and in vivo tissue engineering studies. Indeed, some of the most sophisticated in vitro models of myocardial tissue engineering in terms of structural maturation and functional performance of the embedded cells are based on these materials [32–34], and also, animal studies involving myocardial infarction models have well documented the efficacy of constructs that are generated according to this approach for cell transfer and heart failure therapy [35, 36].

To address the problem of physical strength and to further enhance the exiting models based on liquid ECM, other ECM materials, such as fibrin and its precursor fibrinogen [37, 38], which are recognized by putative recipient cells and allowing for pronounced cell–matrix interactions are investigated [39–43].

Beside the above mentioned models, other liquid scaffold materials, such as alginate, a polysaccharide derived from seaweed [44], are under current investigation [45–47].

A continuously growing number of synthetic hydrogels are currently reported as scaffolds for myocardial tissue engineering, e.g. polyethyleneglycol (PEG). A comprehensive review of these materials goes beyond the focus of this article, however, the interested reader is referred to other chapters, where different aspects of synthetic hydrogels are discussed in detail.

2.2 Guided Tissue Regeneration

Guided tissue regeneration circumscribes a concept that relies on the naturally in vivo occurring adaptive tissue responses upon introduction of ECM components or other bioactive materials. The hypothesis of guided tissue regeneration is founded on the observation that a certain regenerative capacity exists in most tissues and organs throughout life. The scientific debate on the extent and generating source for the observed regenerative capacity is still ongoing, while a growing number of progenitor or stem cell types are discovered throughout the adult human body, even in the heart. Some of these cell entities display characteristics barely credible according to the traditional understanding of terminally differentiated cells. Today, the question is rather concerning the extent of the observed stem and progenitor cell capacity than their bare existence. Interestingly, a certain regenerative capacity persists in vivo, even in the incidence of local pathological degeneration, when the balance of regenerative capacity to adverse pro-degenerative events is reduced. Based on these theoretical considerations, a significant therapeutic intervention may become feasible by providing a change of the local environment to support the aforementioned internal mechanisms of regeneration. As one of few options, this intervention may be done by the delivery of a healthy extracellular scaffold and inherent bioactive stimuli, enabling a physiological tissue turnover inside the novel scaffold. Following delivery, the scaffold will be prone to consecutive spontaneous repopulation in vivo, eventually to result in a new tissue fragment, a concept that is termed guided tissue regeneration. In the cardiovascular field, heart valve tissue engineering has experienced significant success by adopting the principles of guided tissue regeneration. In the recent years, several reports by different groups confirmed the great potential of an adult organism to repopulate a suitable heart valve scaffold under the working conditions of heart valves that are present in vivo [48–52]. These implants prove successful with merely an endothelial layer or even without any cellular components. An in vivo repopulation process resuming after implantation will progressively restore the cellular components according to the principles of guided tissue regeneration. Today, a growing body of evidence demonstrates the success of replacing degenerated heart valves by decellularized counterparts. Unfortunately, in the case of myocardial guided tissue regeneration, comparable results have not been accessible, so far. However, a few promising studies encourage further exploitation.

Very recently, a report on the chronic transmural reconstruction of the right ventricular free wall by an autologous, small bowel-derived implant was published [53]. The employed biomaterial offers the advantage of an autologous and perfectly vascularised platform with apparently sufficient biomechanical characteristics for right ventricular full thickness repair. At the time of implantation, no contractile elements beyond the smooth muscle fraction of the jejunal medial layer were present within the graft tissue. However, after one month in vivo, cells displaying several markers of cardiomyocytes were found in a scattered pattern within the explants. The predominantly perivascular occurrence of these cells with cardiomyogenic features suggests a potential role for the vascular supply, as decellularized pericardial control implants contained no cells with morphological features of cardiomyocytes. Beyond the practical success of this model to substitute ventricular myocardium without rupture or dilation for up to one month, the results of this study may shed a new light on previous findings concerning adult cardiac stem cells and their residual regenerative potential [54, 55].

More frequent models of guided tissue regeneration employ liquid and gelling scaffold materials injected directly into the target area of the myocardium. Among the exploited materials, collagen-based scaffolds and fibrin scaffolds were successful candidates in the setting of an ischemic myocardial injury [56, 57].

In several studies involving a myocardial infarction model, the injection of the latter biological scaffold materials has resulted in an inhibition of left ventricular structural and functional deterioration and ultimately in improved overall outcome. Examination of explants indicates an enhanced neo-vascularisation and limited expansion of the adverse post infarction remodelling as possible underlying events. A progressive in vivo repopulation of the biocompatible scaffolds is another observation that serves as a scientific ground for a cell centred hypothesis regarding the mechanistic events in guided tissue regeneration.

The involved fibrin scaffolds offer the advantage of the possibility of ease incorporation and presentation of natural, ECM-bound factors for an enhanced control of cell growth and tissue development. Due to its natural role as a provisional matrix after injury, fibrin is prone to a rather rapid degradation and replacement by a new, collagen dominated ECM [58, 59]. This might be either useful or rather contra-productive, depending on the nature of the targeted application. For instance, the mere role of a transport vehicle and initial anchor material for cell transfer procedures may be well fulfilled by fibrin-based scaffolds. However, the general possibility of mixing a liquid solution of the fibrin precursor fibrinogen together with other components, e.g. cell suspensions or growth factor solutions, resulting in constructs of desired form and shape, has increased the application possibilities of fibrin [41–43]. Moreover, a wide range of variation of ECM characteristics is accessible through the modification of fibrinogen concentration and the concentration of thrombin, the biological activator of fibrin polymerization [57]. Interestingly, although under naturally occurring circumstances a purely fibrin-based ECM is unlikely as an environment for cell differentiation, under in vitro conditions human mesenchymal cells isolated from bone marrow mononuclear cells displayed viability rates above 95% when encapsulated in a fibrin matrix of 2% w/v fibrinogen and 20 U/ml thrombin. However, the elucidation of cell differentiation characteristics of progenitor or stem cells under these artificial ECM conditions will demand further detailed studies. On the other hand, a wide range of possibilities to design a certain biomechanical behaviour makes the use of fibrin an attractive model for studies on cell application [57]. With the introduction of systems for the ease of separation of fibrin from whole blood donations, the clinical perspective of an autologous scaffold has emerged [60–62]. Today, the isolation of sufficient amounts of a liquid fibrin ECM (generally in the range of a few millilitres) out of 100–200 ml of peripheral blood is feasible. The former volume easily covers the clinically desired or experimentally applied volume of ECM for cell delivery purposes. This way, one of the main drawbacks of biological scaffolds, the possibility of pathogen transmission, has been removed for fibrin based scaffolds. In a recent study, 3D cultures of contractile cardiac cells in shape of a pouch were obtained by using an initially liquid ECM containing fibrin. Portraying comparable functional performance as previously reported patch or ring shaped counterparts, these beating 3D constructs yet virtually demonstrate the possibility of constructing a whole beating chamber [34]. Based on these results, fibrin-based 3D cultures of cardiac cells (initially introduced by Eschenhagen and co-workers as a mainly collagen-based bioartificial construct termed Engineered Heart Tissue [32]) were transplanted in a small animal model [63].

Following the speed of chemical engineering progress, novel biomaterials, presented as versatile hydrogels are introduced as competing tools for myocardial regeneration. Similar to solid synthetic materials, a wide range of physical and functional characteristics may be designed or at least well predicted. Myocardial regeneration is more and more influenced by developments in the field of synthetic biomaterials. In a very recent study conducted in rodents undergoing experimental myocardial infarction, the intramyocardial injection of a temperature-sensitive and injectable hydrogel was able to prevent left ventricular deterioration with respect to geometry and function at 6 weeks after hydrogel injection, although at this time the implanted material was already degraded [64]. The employed aliphatic polyester hydrogel was synthesized from ethyleneglycol and d-valerolactone [(Poly (d-valerolactone)-block-poly (ethylene glycol)- block-poly (d-valerolactone) (PVL-b-PEG-b-PVL)], resulting in thermosensitive characteristics with gelation at 37°C and dissolution in water at room temperature. More interestingly, the addition of cytokines was achieved and in vivo, the application of vascular endothelial growth factor (VEGF, at a dose of 80–100 ng/kg recipient body weight) further improved the outcome of rats with a one-week-old myocardial infarction as compared to saline injection or injection of the hydrogel alone. However, the best results were achieved with a hydrogel preparation including covalently bound VEGF, which resulted in the best left ventricular functional outcomes among all groups, the best myocardial morphometric analysis and highest microvascular density scores. This study shows another promising path of cardiac regeneration that relies on a temporal alteration of the local tissue remodelling by in vivo application of growth factor conjugated biomaterials.

Although in several trials an overall functional benefit has been appreciated for the application of biocompatible ECM components into infarcted myocardial regions, the generation of functional myocardial tissue has not been demonstrated so far. A further improvement may be achieved by integrating the concept of ECM-based guided tissue regeneration with the principle of cell transplantation. In an iridescent manner, the hereby in vivo achieved mass consisting of cells entrapped in a gelling ECM environment has been termed *injectable myocardium*. The practical crossing point between the path of traditional tissue engineering science and the regenerative movement applying cell transplantation techniques is of particular interest for investigators with either background. It has been well known for a long time that cell survival and development is crucially dependent on the extracellular environment. In this context, the ischemic myocardium most probably resembles a hostile environment for isolated and freshly transplanted cells. Cell engraftment and persistence at the target zone of injection are further shortcomings of previous cell transplantation studies. At the same time, the quest for improved methods of in vitro cell incorporation into ECM scaffolds, along with a continuous search for methods allowing superior and seamless integration of cell embedded scaffolds into host myocardium have paved the previous path of tissue engineering. Finally, guided tissue regeneration with scaffolds and cells applied in vivo simultaneously or in a sequential time manner might offer a better solution to the problems of tissue engineering and cell transplantation [65]. A simultaneous application of cells and scaffold material narrows the range of suitable ECM materials. The biocompatibility demands concerning the product material and the involved precursors are comparable to those that apply to in vitro tissue engineering with liquid scaffold materials. However, after injection of the liquid scaffold components, the scaffold will have to solidify within the vicinity of a working myocardium. Specific measures of precaution will have to guide the design and selection process of an appropriate ECM, which will have to solidify in a narrow time window to allow for a thorough mixture of the liquid ECM and cellular components in vivo, but also will prevent wash out or leakage phenomena by timely solidification and entrapment within the native tissue.

Specific application routes, e.g. via transvascular catheters, might impose further limitations on the scaffold solidification kinetics which in turn might have an impact on the clinical practice. The identification of transcatheter application routes may become more important in the future, as recent studies with an intracoronary application of biomaterials have proven efficacy in the treatment of early post myocardial infarction treatment. By intracoronary injection of 2 ml alginate solution (2 or 0.6% weight/volume sodium alginate) 4 days after an acute myocardial infarction in a pig model, a substantial reversal of adverse myocardial remodelling has been observed. Remarkably, no coronary occlusion or peripheral embolism in remote organs was detected. Instead, histological examination of explanted hearts suggests a transvascular transportation of the biomaterial into the infarcted area with deposition of the alginate in the ventricular wall. The resulting biomechanical support and the subsequent spontaneous cellular infiltration of the biomaterial with myofibroblasts result in an increased scar thickness and favourable outcome at 60 days, although no de novo generation of genuine myocardial tissue was demonstrated [66]. Further optimization of this concept may involve tailoring of the degradation kinetics of the applied biomaterial, but also an increased control over the target area. Also, the implementation of bioactive components, such as growth factors may lead to a further improvement of the results. The latter approach has been already integrated with a direct EM injection model, as referred to earlier [64]. Moreover, a general comparison of different biomaterial classes, e.g. native ECM derived biomaterials vs. synthetic polymeric biomaterials, vs. composite biomaterials, may be helpful to identify the most promising biomaterial platform for a growth factor-supplemented or cell-enhanced regenerative therapy. Accumulating data suggest a differential response of the recipient upon injection or implantation of a specific biomaterial. Ideally, the implanted material undergoes repopulation and thereby becomes an autologous vital tissue fragment. In this context, it is likely that the cell population that initially repopulates the introduced biomaterial will crucially affect the further tissue remodelling and thereby influence the overall organ function in the long term. It will be interesting to observe detailed aspects of the in vivo repopulation process, specifically the composition and arrangement of the cell types that migrate into an implanted material. Here, a recent study on myocardial implantation of decellularized native ECM from xenogenic small intestine segments as a treatment of myocardial infarction shows a functional benefit for treated animals at up to 6 weeks. Moreover, higher indices of angiogenesis and neo-vascularisation were observed. For the first, an enhanced recruitment of a cell population that reveals positive for a putative stem cell marker (c-kit) is demonstrated, along with increased tissue levels of stem cell factor, a key player in the regulation of differentiation of stem and progenitor cells and a potential inductor of the cardiac lineage [67]. C-kit$^+$-cells have been demonstrated to be activated by an ischemic trigger, migrate and transiently appear in the vicinity of ischemic myocardium [68]. And growing evidence suggests that a number of resident or remote progenitor cells, such as c-kit$^+$-cells may have the potential to contribute to myocardial regeneration after an ischemic infarction [69–71]. These findings together with the new insight in stem cell recruitment and retention via ECM injection spark greater interest in the concept of guided tissue engineering, particularly using decellularized native ECM materials [72].

Future studies will have to prove the long term benefits of currently proven short term benefits of guided tissue engineering using natural or synthetic ECM components and functional cardiac recovery will most probably remain the prime target. In the second line, evidence of true regenerative capacity will determine the success of the currently promising concepts of guided tissue engineering.

2.3 Starter Matrix-Free Myocardial Tissue Engineering

By utilizing thermo-responsive membranes, cell sheet engineering has opened the door to a whole new field of engineering multicellular aggregates. The chemical base of this technology is founded on a polymer that changes its molecular configuration upon modification of the surrounding temperature, i.e. poly(N-isopropylacrylamide) (PIPAAm) [73–75]. By reducing the temperature from 37 to values below 32°C, the hydrophobic polymer surface changes to hydrophilic. Thereby, the cell adhesion complexes that bind the cell monolayer to their culture ground are detached from the surface of the polymer [76]. In contrast to the conventional enzymatic process of cell passaging in case of thermo thickness responsive polymers the monolayer detachment takes place without a general destruction of the cell–matrix and cell–cell adhesion components. As a consequence, instead of single cells or oligocellular units complete monolayer units may be detached from the tissue culture plate, which contain intact intercellular junctions and will attach onto a new surface within a significantly shorter time period than enzymatically detached cells. The conservation of cell–cell junctions is one of the major advantages of this technique. More interestingly, beyond the cell surface components, also the delicate ECM structures are preserved and may serve as a substrate for the construction of a three dimensional, multilayered culture [77]. This approach is independent of the nature of the cultured cells. Today, the efficacy of cell sheet engineering to promote organ specific tissue engineering has been proven for multiple organ systems and tissues, including the cornea [78], the gastrointestinal tract [79], the respiratory airway system [80], and periodontal tissue [81]. Not surprising, cell sheet engineering has been a frequent tool to produce multicellular constructs that may serve as a cardiac patch.

Using this technology, multiple cardiomyocyte layers have been stacked on each other, resulting in a multilayered, densely organized cardiomyocyte aggregate, developing sarcomeres, desmosomes and gap junctions [82]. This technology requires no initial scaffold material, because single cell layers are vertically added to each other without an intermediate space-holder. Instead, most of the cells that are used for myocardial tissue engineering are likely to express and secrete components of the ECM when cultured according to this technique, e.g. cardiomyocytes, cardiac fibroblasts or endothelial cells [77]. This may result in a de novo synthesized, cell-derived ECM and resembles another advantage of cell sheet engineering. The layer-by-layer application of single monolayer units allows the exact spatial configuration of different cell types, e.g. one endothelial cell layer on top of every two cardiomyocyte layers and so forth. Moreover, the interaction of multiple cell types and their respective role for the in vitro generation of organized multicellular constructs may be studied by this technique in a controlled experimental frame, e.g. the role of non-cardiomyogenic cells for the electrophysiological performance of 3D cardiomyocyte cultures. Cell sheet engineering has now been applied to a variety of cell types aiming at tissue engineering of various organ systems. However, one principle limitation that is shared by most of the aforementioned concepts particularly applies to cell sheet engineering: Without an effective system of oxygen and nutrition supply the maximum tissue thickness, or in other measure, the maximum number of layered cell sheets is limited to the merits of medium perfusion. Compared to constructs that are engineered according to other concepts, e.g. based on solid porous scaffolds, cell sheet engineering results in constructs with an extremely high cell density. As a consequence, only very few layers of the metabolically active cardiomyocytes may be stacked without relevant perfusion deficits in the core regions of the developing monolayer stack [83]. Needless to say that for tissue engineering purposes, cell sheets will have to develop significant passive and active biomechanical strength, which in turn is correlated to the maximum number of the involved cell layers. In one of the early studies to overcome these limitations, a multistep application of respective few layers thick components has been suggested to allow for a gradual in vivo ingrowth of microvessels. To achieve a native tissue derived neo-vascularization, multiple implantation procedures are necessary, a process that has been termed *polysurgery* by the authors of the study [84]. Another, very recent solution to the question of perfusion is delivered by a combination of cell sheet engineering with a biological, vascularized ECM, BioVaM (see Sect. 2.1.1) [83]. In this concept, the advantages of cell sheet engineering regarding the cellular characteristics are paired with the highly biocompatible ECM characteristics and the vascular network of a decellularized native scaffold. This composite approach is a unique solution to many aforementioned concepts and may gain increasing attention as it combines the benefits of different isolated models and surpasses some of their limitations (Fig. 4).

Recently, by integrating the vast data on myocardial recovery by myoblast transplantation and the advantages of cell sheet engineering, novel options for an autologous therapy for diseased myocardial tissue have been suggested.

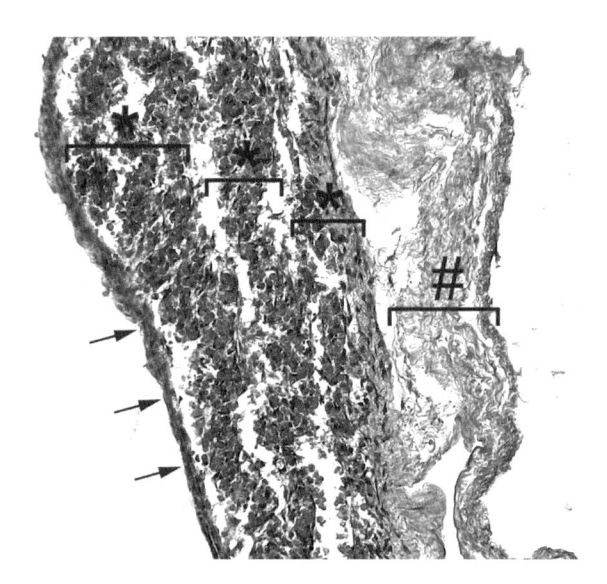

Fig. 4 Combination of cardiomyocyte sheets with a biological, vascularized ECM (BioVaM, decellularized porcine intestinal submucosa) to integrate the functional advantages of cell sheet engineering with the vascular network of a decellularized native scaffold. Cross-section of three cardiomyocyte sheets (*brackets* and *asterisk*) cultured on BioVAM (*brackets* and *hash*). A compact outer cell sheet covers the culture after in vitro culture. H&E staining, magnification: 50x. Image provided by Dr. Hilfiker, Department of Transplantation, Thoracic and Cardiovascular Surgery, Hannover Medical School, Germany; related studies are published by Hata et al. [83]

This approach is particularly interesting because of the relatively long-standing experience with cell transplantation using myoblasts, which dates back to the mid 1990s of the past century [85]. Although the results of two clinical trials (MAGIC I and II) could not entirely fulfill the initial hopes for a new potent cure of ischemic cardiomyopathy, and despite the disclosure of the arrhythmogenic risk of myoblast transplantation, cardiac regenerative therapy via myoblast adiministration remained somewhat en vogue [13, 86, 87]. Similarly, tissue engineering concepts according to the traditional paradigm have utilized skeletal myoblasts for successful in vitro engineering of myogenic tissue (Fig. 5). There are some good reasons for the continued interest in myoblast-based cardiac therapy: The feasibility of isolation and in vitro expansion of myoblasts to huge numbers from a small muscle biopsy of 10–15 g (to numbers ranging between 10^8 and 10^9 cells) clearly speaks for clinical suitability of this procedure. Moreover, the implantation of an implantable cardioverter defibrillator (ICD), which has become obligatory due to the observed arrhythmogenic potential of transplanted myoblasts, appears to be of less significance, as patients suffering from ischemic cardiomyopathy and largely reduced left ventricular function are at greater risk for ventricular arrhythmias in any way, with or without a cell therapy. The above-mentioned findings have built the scientific ground to plan for a multi-center clinical trial

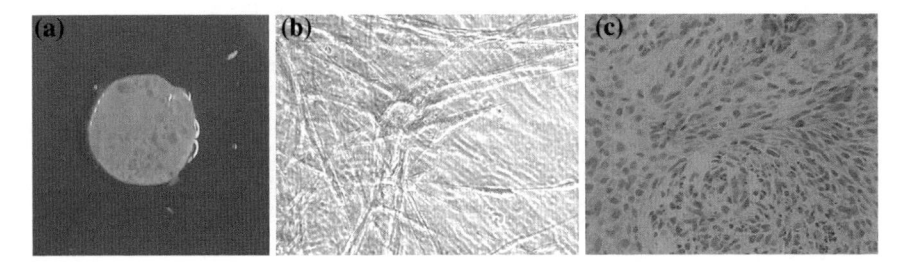

Fig. 5 Skeletal myoblasts as a potential cell source for cardiac tissue engineering and regenerative therapy. In vitro cultures of myoblasts using collagen and matrigel as matrix components result in solid macroscopic constructs (diameter 0.8 cm, thickness ranging from 1 to 4 mm) (**a**) containing a dense network of branched, multi-nucleated myotubes (**b**, contrast phase) in a densely populated three-dimensional culture (**c**, H&E). Magnification = ×200 in **b** and ×400 in **c**. Images of Engineered skeletal muscle graft (ESMG) provided by Marie-Noëlle Giraud and Hendrik T. Tevaearai, Department of Cardiovascular Surgery, Inselspital, Berne, Switzerland

investigating the effect of myoblast cell sheet transplantation, which is expected to start in the near future. Among the most interesting questions that future studies have to address is the identification of the patient collective that will most effectively benefit from a cell sheet transplantation, e.g. patients with ischemic cardiomyopathy, or those with a dilatative cardiomyopathy.

2.4 Alternative Concepts

Contradictory to what is aimed for by the concept of guided tissue engineering, the injection of larger amounts of collagen has been performed in the setting of an ischemic injury. The intramural ECM injection into infarcted myocardial regions achieves a large acellular mass, and surprisingly provokes a similar functional outcome as other concepts relying on mixed scaffold and cellular components. Several clinically relevant parameters, such as enddiastolic and endsystolic LV volumes and myocardial wall motion improve within six weeks, when the treatment is applied early after a large myocardial infarction (e.g. 7 days). But in contrast to results from guided tissue engineering, on the microscopic level no neo-angiogenesis or interstitial cell repopulation inside the bulgy ECM mass are observed up to six weeks after injection in rats after an experimental myocardial infarction and subsequent intramyocardial ECM injection [88]. Furthermore, also functional in vivo measurements indicate that the involved benefit appears to be independent from neo-vascularization events or overall blood supply to the injured myocardial region. Similarly, the injection of a mixture of a-cyclodextrin and a triblock polymer (poly(ethylene glycol)-b-polycaprolactone-(dodecanedioic acid)-polycaprolactone-poly(ethylene glycol)

[MPEG-PCL-MPEG]) one week after experimental myocardial infarction resulted in attenuated deterioration of left ventricular geometry and function at 28 days [89]. Although the myocardial injection of the polymer mixture did not induce a significant inflammatory response, and also could not induce a significantly increased microvessel density, the infarct size was still reduced as compared to the control group receiving saline injection. Most striking, this study showed an increase in left ventricular ejection fraction as a surrogate of active mechanical work performed by the myocardial mass of left ventricle. This appears surprising, because of the short time window of action (28 days) and due to the absence of any evidence for a myocardial regeneration within the injected hydrogel mass. Hence, it may be concluded that the primarily active mode of action is the reduction of the infarct size and maintaining the left ventricular geometry. These basic principles may readily be applied when different ECM materials of biological or synthetic origin are utilized.

Based on these findings, a novel approach to treat post MI adverse remodelling has emerged. By injection of a suitable biomaterial into the injured myocardial region, an internal mechanical support may be installed. This is rather a mechanical intervention, and yet it seems effective enough to prevent a later progressive wall thinning, ventricular dilation and development of dyskinetic wall regions that all together collectively lead to the vicious circle of end stage heart failure [88]. The internal support by injection of large amounts of ECM represents a completely novel approach that will have to be validated in larger series of long term studies. If this concept ultimately should prove as effective as it appears in the current early stage of development, the previous rule of the thumb, after which neo-vascularization and introduction of an appropriate cellular mass are top priority goals of the treatment, has to be re-evaluated. ECM injection then becomes a competing strategy to cell transplantation. Indeed, in more than one in vivo study, improved functional outcome with this approach has been demonstrated without any histological and functional cue for increased vascular density or cell repopulation [88, 89].

Further theoretical consideration leads to conclusions that are even more contradictory to the common understanding of the principles of cardiac regeneration: one might speculate that the observed benefits of acellular biomaterial injection would even be diminished by a significant cellular repopulation, because cell driven ECM degradation might reverse the mechanical effects of the injected scaffold. Concerns about the long term durability of this concept remain to be dissolved by further studies. From the clinical perspective, the idea of mechanically changing the LV geometry to improve cardiac function is not new. Other previous attempts with initially promising results have garnered only transient attention. One of most prominent approaches involves a reorganization of the LV configuration by surgical means to reduce LV wall stress and prevent LV dilatation for maintaining global pump function [90, 91], a procedure that has lost much of its popularity only a few years after the initial introduction [92].

Another tissue engineering concept that is designed for the support of myocardial function aims at generation of heterotopic contractile myocardial tissue. By injection of a significant number of cardiomyocytes into the peri-aortic region, a contractile mass is achieved that encompasses the aorta in a circular configuration. By simultaneous contraction upon electrical pacing an external compression of the aorta may be achieved, which may lead to a slight, but yet measurable increase in the intra-aortic pressure profile. This approach has certain unique advantages. The circumvention of an intrapericardial intervention may be particularly suited for patients with previous heart surgical procedures, e.g. previous CABG. Also, due to an area of intervention that is remote from the heart, the employed cellular components are protected from a range of adverse conditions that are present in the milieu of a failing or ischemic myocardium. Moreover, for achieving a therapeutic effect, mainly an active contractile performance is needed and passive mechanical strength might remain of secondary importance. Furthermore, the native aortic wall resembles the interface between the implant and the circulating blood. Heterotopic myocardial tissue engineering circumscribes the principle that is applied here. In another study in vitro fabricated beating cell sheets were implanted in the peri-aortic region, resulting in a beating construct in a tube configuration around the native aorta. Consequently, measurable effects on the haemodynamic parameters in vivo were observed, confirming the proof of this principle by means of cell sheet engineering [93].

Another recently described approach combines the most up-to-date advantage of polymer science with previous attempts of dynamic cardiomyoplasty. In a rat model of myocardial infarction, an electromechanically active nanopolymer consisting of poly-vinyl-alcohol (PVA) combined with chitosan was externally wrapped around the heart and electrically stimulated. Invasive hemodynamic measurements showed an increase of stroke volume and dp/dt in treated animals as compared to control animals after myocardial infarction [94]. Further studies for the proof of long term benefits and durability of such an external cardiac support concept are warranted. However, the introduced model gives way to further concepts and directions of future developments in the field of cardiovascular regenerative therapy. When aforementioned electromechanically active nanopolymers are applied heterotopically, e.g. as an external wrapping of the aorta, an extravascular support system may become available, with similar theoretic effects as exerted by the clinically established cardiac support through intraaortic balloon counterpulsation (IABP) [95, 96]. The latter therapy option uses a inflatable balloon that is inserted via the groin vessels to be positioned in the proximal descending thoracic aorta. By using ECG signals and the arterial blood pressure line, a software-based algorithm governs the cyclic inflation and deflation of the balloon in such manner that inflation occurs during the diastole. The rapid diastolic inflation of the intraaortic balloon leads to a diastolic increase of the blood pressure, while the deflation shortly before the begin of systole leads to a slight decrease of the peak systolic pressure, thereby decreasing both, myocardial afterload and effective myocardial workload. In the clinical routine, this therapeutic option is applied in patients with critical cardiac function, particularly when coronary perfusion is suspected to be marginal. Applying electromechanically active nanopolymers as active external support systems may one day abolish a number of problems of current circulatory support systems, such IABP or cardiac assist devices, which result from the activation of the coagulation system at the blood-implant interface leading to thrombembolic events.

3 Current Problems

3.1 The Seemingly Endless Search for the Ideal Cell Source

Above the numerous challenges that tissue engineering across all disciplines are confronted with, the lack of an ideal cell source is currently the most significant limitation to myocardial tissue engineering. A true progress towards a clinically relevant concepts for a wide spread use is depending on an appropriate cell source. Hence, the identification of a reliable and safe source for a reproducible isolation of suitable adult cardiomyocytes or their progenitors is subject to intense scientific investigation. Unfortunately, up to the current time no definitive solution to this question has been found. As a consequence, from the beginning era of three dimensional cell cultures for cardiac tissue engineering even up to now, successful concepts of myocardial tissue engineering have been dependent on the utilization of robust, but clinically non-relevant cellular components, e.g. neonatal rat cardiomyocytes [84, 97–101] or fetal cardiomyocytes [99].

In the late 1990s an increasing interest in alternative sources for the generation of differentiated somatic cells led to the development of protocols for the culture embryonic stem cells (ESC) of murine and also human origin [102–104]. Successive scientific achievements included the optimization of targeted differentiation of ESC towards cardiomyocytes, the avoidance of potentially pathogenic agents or components during the culture process, and the enhancement of the quantitative yield in terms of maximum total number of cells that may be generated within a reasonable time. However, concerns about the tumorigenic potential of ESC-derived cell populations were generated by transplantation studies in rodent models of myocardial injury. Furthermore, the initially anticipated hypothesis of an immunologically privileged status of allogenic ESC upon transplantation was virtually overshadowed by the observation that the initiation of differentiation also triggers immunological response patterns in allogenic recipients. More importantly, a gain in lineage commitment appears to be correlated with an increase in the expected immunological reaction in vivo [105–107]. These technical hurdles joined the ever since persisting and yet unanswered public ethical concerns and urged the scientific efforts to search for alternative strategies.

The principle presence of cardiomyocyte renewal in the adult heart has been indicated by different findings [108–110]. In rodent models of myocardial infarction, different groups have demonstrated that certain intrinsic capacity for regeneration after injury is maintained even in the adult myocardium [55, 111]. Even more striking, isolation and in vitro culture as well as clonal expansion of cardiac cells was demonstrated, which display typical features of stem and progenitor cells and possess the potential of giving raise to functional cardiomyocytes.

The term *adult cardiac stem cell* (aCSC) was born and gave a dramatic boost to the investigative efforts in myocardial regeneration [54]. At this point, concepts of myocardial tissue engineering were gifted with the first perspective on a practical solution for question of the appropriate cell source—without any major ethical concerns. However, quantitatively, the beneficial effect of aCSC are not sufficient to prevent cardiomyocyte loss related cardiac failure in vivo, e.g. after an extensive myocardial infarction. And more importantly from the clinical point of view, the number of CSC seems to even decrease in chronic conditions of myocardial injury [112]. Likewise, despite the application of the most advanced cell culture techniques, the adult human heart is problematic as a source for the isolation of cardiomyocytes for myocardial tissue engineering and until now a targeted application of aCSC that may result in their successful utilization for therapeutic purposes is at an early stage. Small safety trials for adult heart cells began in 2010, with cells taken from heart biopsies and grown in the laboratory to provide larger numbers, then re-injected (CArdiosphere-Derived aUtologous Stem CElls to Reverse ventricUlar dySfunction [CADUCEUS]). Similarly, pluripotent stem cells of mesenchymal origin have been studied for their regenerative capacity under in vivo conditions and a better overall outcome was described in the initial reports. But more detailed analysis of the fate of these cells could not prove their long term survival, and a true de novo development of cardiac muscle, cardiomyogenesis, was absent. More disappointing results were obtained by studies that were performed exclusively in vitro: No concept has been able to induce the in vitro differentiation of adult mesenchymal stem cells to cardiomyocytes within a qualitatively and quantitatively convincing scale.

Hence, the initial enthusiasm about autologous adult cell sources has sobered. In a broader sense, some of the old problems that have brought other previous cell sources to a failure also form a firm barrier for a clinical introduction and breakthrough of autologous adult stem cell sources: (1) generally, only a low frequency of autologous stem cells is observed in the adult patient, (2) ironically, with disease progression and increased need for cell based therapy, a further diminishing of the frequency and quality of the cell populations of interest occurs, and (3) immunologic incompatibilities with cells other than those derived from the recipient himself are persistent limitations for allogenic cell sources of various developmental stages.

Therefore, other, more immature cell populations are in the focus of ongoing investigation for a potential source for the generation of autologous cardiomyocytes. Among these cell entities, stem cells derived from cord blood or induced pluripotent stem cells currently hold the biggest promise. Cord blood is a clinically well known source for the isolation of undifferentiated cell colonies that are used in established therapeutic protocols to restore haematopoietic cell populations after an allogenic transplantation [113]. Their therapeutic success has led to the development and establishment of central cord blood banks that are specialized in the allocation of recipient matching stem cells according to relevant antigen profile specifications [114]. Recently, a small fraction of the mononuclear cells in cord blood donations has been identified and termed unrestricted somatic stem cells (USSC) [115]. Although of perinatal origin, these cells retain high stem cell capacities, including self renewal and potential of giving raise to diverse tissue specific cell lines, including neuronal, haematopoietic and endothelial precursors [116–118]. A cardiomyogenic potential has been anticipated for USSC, supported by preliminary in vivo results [119, 120]. More detailed studies are warranted to determine the value of USSC for in vivo regenerative concepts. More importantly, the in vitro feasibility of USSC differentiation to cardiomyocytes remains a subject to further investigation. The use of elaborate three dimensional cell culture techniques with a highly instructive extracellular environment and the utilization of co-culture models may lead to novel instruments of targeted cell differentiation for an increased cardiomyocyte yield based on old and new undifferentiated cell sources.

Finally, induced pluripotent stem cells (iPS) have been developed by the addition of defined factors to adult somatic cells [121, 122]. The young history of iPS as novel magic players on the field of regenerative medicine resembles a similar evolution as experienced for ESC. Their introduction was celebrated as a breakthrough and their potency held no limits [123]. With growing number of applications, including in vivo transplantation of iPS derived differentiated cell populations, several serious concerns that were typically related to the application of ESC derived cells, also proved as relevant for iPS based concepts. These concerns include the risk of tumor development, ethical issues and the practical feasibility of a targeted differentiation in a large scale [124].

3.2 At the Crossroad: The Vital Importance of Establishing a Functional Vascular Network

Vascularization offering nutrition and oxygen supply is a true limitation to most of the commonly known tissue engineering concepts. Guided tissue regeneration may rely on slow, but relevant angiogenesis in vivo, and a further promotion of this process may be achieved by the employment of a wide range of molecular cues, including growth factors, bioactive peptides, or local gene expression modification of the applied cells. In contrast, tissue engineering concepts following the classic paradigm without a pre-existing vascular network will most likely be prone to an ischemic injury after implantation. This may even occur in vitro when a clinically relevant size and cell density is attempted. Hence, different strategies have been developed to meet the requirements of the growing three dimensional cultures of metabolically highly active cardiac cells.

Taking advantage of naturally grown structures, a principle that is implemented into decellularization concepts, may be a reasonable way of achieving functional perfusion networks integrated with a biofunctional ECM [83, 100, 125, 126] (Figs. 4, 6). Further improvement is needed for the prevention of thrombotic degeneration of preserved vascular structures within decellularized scaffolds after their in vivo implantation and connection to the recipient circulation. An in vitro step for a thorough re-endothelialization appears indispensable at the current time.

Micropatterning technology has demonstrated the feasibility of constructing in vivo pre-vascularization of scaffolds for a subsequent implantation of the cellular

Strategies for Myocardial Tissue Engineering 71

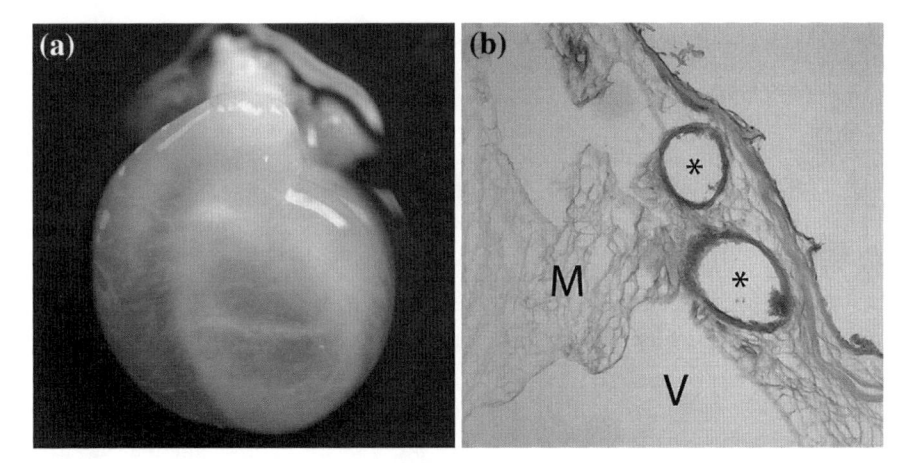

Fig. 6 Native myocardial extracellular matrix (ECM) as a pre-vascularized biological matrix. Decellularization leads to a macroscopic (**a**) and microscopic (**b**) removal of the majority of cellular components, resulting in a porous matrix with preserved vascular structures (*asterisks* indicating coronary vessels). **a** Macroscopic image of a rat heart after detergent based decellularization; **b** H&E staining of a histological cross-section. *V* ventricular cavity; *M* = ECM representing the decellularized myocardial tissue

components may be one approach. Alternatively, cell seeded constructs may be transiently preconditioned in an in vivo environment that is expected to be less hostile than the ultimate target region, which is prone to ischemic or inflammatory remodelling [127]. Modification of the scaffold components by introducing pro-angiogenetic molecular cues, e.g. RGD sequences that preferentially attract endothelial cells and their progenitors, might resemble another path to prevent ischemia within engineered cardiac constructs. The latter concept has been evaluated by modification of alginate scaffolds for an in vivo attenuation of adverse post MI remodelling [128, 129]. Further concepts rely on technical advances in microconfiguration of biomaterials. By designing microscopic structures, such as channels or interconnected culture chambers, the construction of cavernous structures of desired configuration, e.g. a vascular network, becomes feasible [53, 130–133]. The translation of these technological tools in comprehensive systems to produce multicellular constructs of a relevant scale remains as a target for future experimental work.

Other concepts that rely on the utilization of native autologous vasculature may suffer from significant concerns around the issue of donor site morbidity. Experimental in vivo has demonstrated promising outcome of these concepts. A staged preconditioning of scaffold embedded cells either around surgically created arterio-venous shunts [134] or around native peripheral vessels [135] results in optimized remodelling of the tissue, presumably due to more appropriate oxygen and nutrition support in close vicinity of functional vasculature. However, in these models a completion of the procedure involves the transfer of the scaffold together with the enclosed vascular segments to the heart. A significant morbidity and risk of the development of associated complications has to be considered.

In another approach, a small bowel segment has been utilized as a cardiac patch, with preserved perfusion via native vasculature that may be connected via surgical anastomosis at the implantation site, e.g. right ventricular free wall [136]. Although these concepts have been proven as effective and may be technically feasible, a broad clinical introduction most probably will be restricted to highly selected cases where due to a desperate clinical situation a salvage intervention is attempted.

3.3 Out of Our Hands: How to Control the Fate of a Tissue Engineered Graft After Implantation?

The long term benefit of an engineered implant tissue may be defined by the persistence and maturation of the tissue or by a sequence of events that are triggered by the implantation process. The latter explanation is most likely involved many functional effects of cell transplantation. However, in case of a transmural myocardial reconstruction a persisting tissue that maintains physical strength and supports versatile biological interaction possibilities, e.g. electrophysiological properties, is of great importance. Some of the aforementioned topics, such as the choice of the embedded cell population, or the utilization of pre-existing vascular network etc., are intrinsic determinants of the long term development of an engineered myocardial construct. But above the initial components, in vitro optimization strategies and also in vivo measures of biological implant protection deserve particular consideration.

The in vitro conditioning of an engineered cardiac construct may be a potent tool to promote almost all aspects of the developing culture that are relevant to a later tissue performance in vitro or in vivo. These aspects include passive [137] and active [138] biomechanical performance, de novo ECM synthesis [137, 139], intracellular organization [33, 140, 141], cell–cell and cell–ECM interaction [29, 142] and electrophysiological features [143, 144]. Most of these characteristics are best controlled in a set-up that monitors and regulates basic culture conditions, e.g. by using a bioreactor system. The development of a system that mimics the ideal in vivo environment and thereby provides for a valuable culture niche is the main gaol of bioreactor technology [143, 145–148]. Meanwhile many different bioreactor systems for the culture of engineered constructs, particularly for cardiac constructs, have been introduced. The most advanced systems allow for software based monitoring and control of multiple culture conditions. The utilization of a potent bioreactor system enhances the quality of the resulting constructs with respect to many functional parameters. The later in vivo performance of such a construct is likely to be improved by a prior bioreactor-based conditioning.

After implantation, a change in the local environment may result in a decreased survival rate and functional deterioration of the construct. To address this problem, additional strategies that protect the engineered construct in vivo have to be implemented. Local administration of growth factors by scaffold modification or by genetic cell manipulation may be an option. Further therapeutic instruments may be provided by systemically active agents that are used to interrupt an ongoing vicious cycle, e.g. to interrupt a post-MI inflammatory process [149]. Long term remodelling of engineered constructs will also have to consider degenerative events such as fibrosis and calcification along with cellular repopulation and other events, all of which may occur weeks or months after the initial application [150].

In summary, myocardial tissue engineering has become an emerging field of basic and translational investigation. The involved concepts cover a wide spectrum of biochemical, biotechnological, molecular biological and surgical techniques, which is depicted by the various promising applications reported until today. The number of principle strategies is likely to increase, although human application may need considerable investigative efforts and time for a routine establishment.

References

1. Lloyd-Jones, D.M., et al.: Defining and setting national goals for cardiovascular health promotion and disease reduction: the American Heart Association's strategic impact goal through 2020 and beyond. Circulation **121**, 586–613 (2010)
2. Liu, L.: Cardiovascular diseases in China. Biochem. Cell Biol. **85**, 157–163 (2007)
3. Gaziano, T.A., Bitton, A., Anand, S., Abrahams-Gessel, S., Murphy, A.: Growing epidemic of coronary heart disease in low- and middle-income countries. Curr. Probl. Cardiol. **35**, 72–115 (2010)
4. Favaloro, R.G.: Saphenous vein autograft replacement of severe segmental coronary artery occlusion: operative technique. Ann. Thorac. Surg. **5**, 334–339 (1968)
5. Gruntzig, A.R., Senning, A., Siegenthaler, W.E.: Nonoperative dilatation of coronary-artery stenosis: percutaneous transluminal coronary angioplasty. N. Engl. J. Med. **301**, 61–68 (1979)
6. Hunt, S.A., et al.: Focused update incorporated into the ACC/AHA 2005 Guidelines for the Diagnosis and Management of Heart Failure in Adults: a report of the American College of Cardiology Foundation/American Heart Association Task Force on Practice Guidelines: developed in collaboration with the International Society for Heart and Lung Transplantation. Circulation **119**, e391–479 (2009)
7. Kofler, S., et al.: Long-term outcomes after 1000 heart transplantations in six different eras of innovation in a single center. Transpl. Int. **22**, 1140–1150 (2009)
8. Kobashigawa, J.A., Patel, J.K.: Immunosuppression for heart transplantation: where are we now? Nat. Clin. Pract. Cardiovasc. Med. **3**, 203–212 (2006)
9. Wilson, S.R., Givertz, M.M., Stewart, G.C., Mudge Jr., G.H.: Ventricular assist devices the challenges of outpatient management. J. Am. Coll. Cardiol. **54**, 1647–1659 (2009)
10. Marelli, D., Desrosiers, C., el-Alfy, M., Kao, R.L., Chiu, R.C.: Cell transplantation for myocardial repair: an experimental approach. Cell Transplant. **1**, 383–390 (1992)
11. Scorsin, M., et al.: Can grafted cardiomyocytes colonize peri-infarct myocardial areas? Circulation **94**, II337–II340 (1996)
12. Taylor, D.A., et al.: Regenerating functional myocardium: improved performance after skeletal myoblast transplantation. Nat. Med. **4**, 929–933 (1998)
13. Menasche, P., et al.: Autologous skeletal myoblast transplantation for severe postinfarction left ventricular dysfunction. J. Am. Coll. Cardiol. **41**, 1078–1083 (2003)
14. Menasche, P.: Current status and future prospects for cell transplantation to prevent congestive heart failure. Semin. Thorac. Cardiovasc. Surg. **20**, 131–137 (2008)
15. Giraud, M.N., et al.: Hydrogel-based engineered skeletal muscle grafts normalize heart function early after myocardial infarction. Artif. Organs **32**, 692–700 (2008)

16. Hamdi, H., et al.: Cell delivery: intramyocardial injections or epicardial deposition? A head-to-head comparison. Ann. Thorac. Surg. **87**, 1196–1203 (2009)
17. Cima, L.G., et al.: Tissue engineering by cell transplantation using degradable polymer substrates. J. Biomech. Eng. **113**, 143–151 (1991)
18. Bursac, N.: Cardiac tissue engineering using stem cells. IEEE Eng. Med. Biol. Mag. **28**, 80–82, 84–86, 88–89 (2009)
19. Gaetani, R., et al.: Cardiospheres and tissue engineering for myocardial regeneration: potential for clinical application. J. Cell Mol. Med. **14**, 1071–1077 (2010)
20. Jawad, H., Lyon, A.R., Harding, S.E., Ali, N.N., Boccaccini, A.R.: Myocardial tissue engineering. Br. Med. Bull. **87**, 31–47 (2008)
21. Miyagawa, S., Roth, M., Saito, A., Sawa, Y., Kostin, S.: Tissue-engineered cardiac constructs for cardiac repair. Ann. Thorac. Surg. **91**, 320–329 (2011)
22. Vunjak-Novakovic, G., et al.: Challenges in cardiac tissue engineering. Tissue Eng. Part B Rev. **16**, 169–187 (2010)
23. Bursac, N., et al.: Cardiac muscle tissue engineering: toward an in vitro model for electrophysiological studies. Am. J. Physiol. **277**, H433–H444 (1999)
24. Ke, Q., et al.: Embryonic stem cells cultured in biodegradable scaffold repair infarcted myocardium in mice. Sheng Li Xue Bao **57**, 673–681 (2005)
25. Siepe, M., et al.: Myoblast-seeded biodegradable scaffolds to prevent post-myocardial infarction evolution toward heart failure. J. Thorac. Cardiovasc. Surg. **132**, 124–131 (2006)
26. Fujimoto, K.L., et al.: An elastic, biodegradable cardiac patch induces contractile smooth muscle and improves cardiac remodeling and function in subacute myocardial infarction. J. Am. Coll. Cardiol. **49**, 2292–2300 (2007)
27. Shin, M., Ishii, O., Sueda, T., Vacanti, J.P.: Contractile cardiac grafts using a novel nanofibrous mesh. Biomaterials **25**, 3717–3723 (2004)
28. Jin, J., et al.: Transplantation of mesenchymal stem cells within a poly(lactide-co-epsilon-caprolactone) scaffold improves cardiac function in a rat myocardial infarction model. Eur. J. Heart Fail. **11**, 147–153 (2009)
29. Boublik, J., et al.: Mechanical properties and remodeling of hybrid cardiac constructs made from heart cells, fibrin, and biodegradable, elastomeric knitted fabric. Tissue Eng. **11**, 1122–1132 (2005)
30. Torikai, K., et al.: A self-renewing, tissue-engineered vascular graft for arterial reconstruction. J. Thorac. Cardiovasc. Surg. **136**, 37–45 (2008). 45 e31
31. Ozawa, T., et al.: Histologic changes of nonbiodegradable and biodegradable biomaterials used to repair right ventricular heart defects in rats. J. Thorac. Cardiovasc. Surg. **124**, 1157–1164 (2002)
32. Eschenhagen, T., et al.: Three-dimensional reconstitution of embryonic cardiomyocytes in a collagen matrix: a new heart muscle model system. FASEB J. **11**, 683–694 (1997)
33. Zimmermann, W.H., et al.: Tissue engineering of a differentiated cardiac muscle construct. Circ. Res. **90**, 223–230 (2002)
34. Yildirim, Y., et al.: Development of a biological ventricular assist device: preliminary data from a small animal model. Circulation **116**, I16–I23 (2007)
35. Zimmermann, W.H., et al.: Cardiac grafting of engineered heart tissue in syngenic rats. Circulation **106**, I151–I157 (2002)
36. Zimmermann, W.H., et al.: Engineered heart tissue grafts improve systolic and diastolic function in infarcted rat hearts. Nat. Med. **12**, 452–458 (2006)
37. Haverich, A., Walterbusch, G., Borst, H.G.: The use of fibrin glue for sealing vascular prostheses of high porosity. Thorac. Cardiovasc. Surg. **29**, 252–254 (1981)
38. Goerler, H., Oppelt, P., Abel, U., Haverich, A.: Safety of the use of Tissucol Duo S in cardiovascular surgery: retrospective analysis of 2149 patients after coronary artery bypass grafting. Eur. J. Cardiothorac. Surg. **32**, 560–566 (2007)
39. Lichtenberg, A., et al.: A multifunctional bioreactor for three-dimensional cell (co)-culture. Biomaterials **26**, 555–562 (2005)

40. Kofidis, T., et al.: Pulsatile perfusion and cardiomyocyte viability in a solid three-dimensional matrix. Biomaterials **24**, 5009–5014 (2003)
41. Johnson, P.J., Parker, S.R., Sakiyama-Elbert, S.E.: Controlled release of neurotrophin-3 from fibrin-based tissue engineering scaffolds enhances neural fiber sprouting following subacute spinal cord injury. Biotechnol. Bioeng. **104**, 1207–1214 (2009)
42. Johnson, P.J., Tatara, A., Shiu, A., Sakiyama-Elbert, S.E.: Controlled release of neurotrophin-3 and platelet-derived growth factor from fibrin scaffolds containing neural progenitor cells enhances survival and differentiation into neurons in a subacute model of SCI. Cell Transplant. **19**, 89–101 (2010)
43. Lee, Y.B., et al.: Bio-printing of collagen and VEGF-releasing fibrin gel scaffolds for neural stem cell culture. Exp. Neurol. **223**, 645–652 (2010)
44. Leor, J., et al.: Bioengineered cardiac grafts: a new approach to repair the infarcted myocardium? Circulation **102**, III56–III61 (2000)
45. Mukherjee, R., et al.: Targeted myocardial microinjections of a biocomposite material reduces infarct expansion in pigs. Ann. Thorac. Surg. **86**, 1268–1276 (2008)
46. Veriter, S., et al.: In vivo selection of biocompatible alginates for islet encapsulation and subcutaneous transplantation. Tissue Eng. Part A **16**, 1503–1513 (2010)
47. Wang, T., et al.: An encapsulation system for the immunoisolation of pancreatic islets. Nat. Biotechnol. **15**, 358–362 (1997)
48. Baraki, H., et al.: Orthotopic replacement of the aortic valve with decellularized allograft in a sheep model. Biomaterials **30**, 6240–6246 (2009)
49. Cebotari, S., et al.: Clinical application of tissue engineered human heart valves using autologous progenitor cells. Circulation **114**, I132–I137 (2006)
50. Lichtenberg, A., et al.: Preclinical testing of tissue-engineered heart valves re-endothelialized under simulated physiological conditions. Circulation **114**, I559–I565 (2006)
51. Vincentelli, A., et al.: In vivo autologous recellularization of a tissue-engineered heart valve: are bone marrow mesenchymal stem cells the best candidates? J. Thorac. Cardiovasc. Surg. **134**, 424–432 (2007)
52. Akhyari, P., et al.: In vivo functional performance and structural maturation of decellularised allogenic aortic valves in the subcoronary position. Eur. J. Cardiothorac. Surg. **38**, 539–546 (2010)
53. Maidhof, R., Marsano, A., Lee, E.J., Vunjak-Novakovic, G.: Perfusion seeding of channeled elastomeric scaffolds with myocytes and endothelial cells for cardiac tissue engineering. Biotechnol. Prog. **26**, 565–572 (2010)
54. Messina, E., et al.: Isolation and expansion of adult cardiac stem cells from human and murine heart. Circ. Res. **95**, 911–921 (2004)
55. Beltrami, A.P., et al.: Adult cardiac stem cells are multipotent and support myocardial regeneration. Cell **114**, 763–776 (2003)
56. Kofidis, T., et al.: Novel injectable bioartificial tissue facilitates targeted, less invasive, large-scale tissue restoration on the beating heart after myocardial injury. Circulation **112**, I173–I177 (2005)
57. Martens, T.P., et al.: Percutaneous cell delivery into the heart using hydrogels polymerizing in situ. Cell Transplant. **18**, 297–304 (2009)
58. Aper, T., Teebken, O.E., Steinhoff, G., Haverich, A.: Use of a fibrin preparation in the engineering of a vascular graft model. Eur. J. Vasc. Endovasc. Surg. **28**, 296–302 (2004)
59. Flanagan, T.C., et al.: In vivo remodeling and structural characterization of fibrin-based tissue-engineered heart valves in the adult sheep model. Tissue Eng. Part A **15**, 2965–2976 (2009)
60. Velada, J.L., Hollingsbee, D.A.: Physical characteristics of Vivostat patient-derived sealant. Implications for clinical use. Eur. Surg. Res. **33**, 399–404 (2001)
61. Buchta, C., et al.: Fibrin sealant produced by the CryoSeal FS system: product chemistry, material properties and possible preparation in the autologous preoperative setting. Vox Sang. **86**, 257–262 (2004)

62. Buchta, C., Hedrich, H.C., Macher, M., Hocker, P., Redl, H.: Biochemical characterization of autologous fibrin sealants produced by CryoSeal and Vivostat in comparison to the homologous fibrin sealant product Tissucol/Tisseel. Biomaterials **26**, 6233–6241 (2005)
63. Conradi, L., et al.: Fibrin-based engineered heart tissue for cardiac repair: Initial results after transplantation. in: 40th Annual meeting of the German Society for Thoracic and Cardiovascular Surgery. Stuttgart, Germany (2011)
64. Wu, J., et al.: Infarct stabilization and cardiac repair with a VEGF-conjugated, injectable hydrogel. Biomaterials **32**, 579–586 (2011)
65. Wang, F., et al.: Injectable, rapid gelling and highly flexible hydrogel composites as growth factor and cell carriers. Acta Biomater. **6**, 1978–1991 (2010).
66. Leor, J., et al.: Intracoronary injection of in situ forming alginate hydrogel reverses left ventricular remodeling after myocardial infarction in Swine. J. Am. Coll. Cardiol. **54**, 1014–1023 (2009)
67. Kamrul Hasan, M., et al.: Myogenic differentiation in atrium-derived adult cardiac pluripotent cells and the transcriptional regulation of GATA4 and myogenin on ANP promoter. Genes Cells **15**, 439–454 (2010)
68. Limana, F., et al.: Myocardial infarction induces embryonic reprogramming of epicardial c-kit(+) cells: role of the pericardial fluid. J. Mol. Cell. Cardiol. **48**, 609–618 (2010)
69. Miyamoto, S., et al.: Characterization of long-term cultured c-kitÂ +Â cardiac stem cells derived from adult rat hearts. Stem Cells Dev. **19**, 105–116 (2010)
70. Degeorge Jr., B.R., et al.: BMP-2 and FGF-2 synergistically facilitate adoption of a cardiac phenotype in somatic bone marrow c-kit+/Sca-1+ stem cells. Clin. Transl. Sci. **1**, 116–125 (2008)
71. Xaymardan, M., et al.: c-Kit function is necessary for in vitro myogenic differentiation of bone marrow hematopoietic cells. Stem Cells **27**, 1911–1920 (2009)
72. Zhao, Z.Q., et al.: Improvement in cardiac function with small intestine extracellular matrix is associated with recruitment of C-kit cells, myofibroblasts, and macrophages after myocardial infarction. J. Am. Coll. Cardiol. **55**, 1250–1261 (2010)
73. Okano, T., Yamada, N., Sakai, H., Sakurai, Y.: A novel recovery system for cultured cells using plasma-treated polystyrene dishes grafted with poly(N-isopropylacrylamide). J. Biomed. Mater. Res. **27**, 1243–1251 (1993)
74. Okano, T., Yamada, N., Okuhara, M., Sakai, H., Sakurai, Y.: Mechanism of cell detachment from temperature-modulated, hydrophilic-hydrophobic polymer surfaces. Biomaterials **16**, 297–303 (1995)
75. Yamato, M., et al.: Thermo-responsive culture dishes allow the intact harvest of multilayered keratinocyte sheets without dispase by reducing temperature. Tissue Eng. **7**, 473–480 (2001)
76. Kikuchi, A., Okuhara, M., Karikusa, F., Sakurai, Y., Okano, T.: Two-dimensional manipulation of confluently cultured vascular endothelial cells using temperature-responsive poly(N-isopropylacrylamide)-grafted surfaces. J. Biomater. Sci. Polym. Ed. **9**, 1331–1348 (1998)
77. Kushida, A., et al.: Decrease in culture temperature releases monolayer endothelial cell sheets together with deposited fibronectin matrix from temperature-responsive culture surfaces. J. Biomed. Mater. Res. **45**, 355–362 (1999)
78. Nishida, K., et al.: Corneal reconstruction with tissue-engineered cell sheets composed of autologous oral mucosal epithelium. N. Engl. J. Med. **351**, 1187–1196 (2004)
79. Ohki, T., et al.: Treatment of oesophageal ulcerations using endoscopic transplantation of tissue-engineered autologous oral mucosal epithelial cell sheets in a canine model. Gut **55**, 1704–1710 (2006)
80. Kanzaki, M., et al.: Dynamic sealing of lung air leaks by the transplantation of tissue engineered cell sheets. Biomaterials **28**, 4294–4302 (2007)
81. Hasegawa, M., Yamato, M., Kikuchi, A., Okano, T., Ishikawa, I.: Human periodontal ligament cell sheets can regenerate periodontal ligament tissue in an athymic rat model. Tissue Eng. **11**, 469–478 (2005)

82. Shimizu, T., et al.: Fabrication of pulsatile cardiac tissue grafts using a novel 3-dimensional cell sheet manipulation technique and temperature-responsive cell culture surfaces. Circ. Res. **90**, e40 (2002)
83. Hata, H., et al.: Engineering a novel three-dimensional contractile myocardial patch with cell sheets and decellularised matrix. Eur. J. Cardiothorac. Surg. **38**(4), 450–455 (2010)
84. Shimizu, T., et al.: Polysurgery of cell sheet grafts overcomes diffusion limits to produce thick, vascularized myocardial tissues. FASEB J. **20**, 708–710 (2006)
85. Murry, C.E., Wiseman, R.W., Schwartz, S.M., Hauschka, S.D.: Skeletal myoblast transplantation for repair of myocardial necrosis. J. Clin. Invest. **98**, 2512–2523 (1996)
86. Menasche, P., et al.: The Myoblast Autologous Grafting in Ischemic Cardiomyopathy (MAGIC) trial: first randomized placebo-controlled study of myoblast transplantation. Circulation **117**, 1189–1200 (2008)
87. Haider, H., Lei, Y., Ashraf, M.: MyoCell, a cell-based, autologous skeletal myoblast therapy for the treatment of cardiovascular diseases. Curr. Opin. Mol. Ther. **10**, 611–621 (2008)
88. Dai, W., Wold, L.E., Dow, J.S., Kloner, R.A.: Thickening of the infarcted wall by collagen injection improves left ventricular function in rats: a novel approach to preserve cardiac function after myocardial infarction. J. Am. Coll. Cardiol. **46**, 714–719 (2005)
89. Jiang, X.J., et al.: Injection of a novel synthetic hydrogel preserves left ventricle function after myocardial infarction. J. Biomed. Mater. Res. A **90**, 472–477 (2009)
90. Gordon, D.: In rural Brazil a surgeon uses a revolutionary and controversial method of repairing too big a heart. Time **150**, 34–37 (1997)
91. Batista, R.J., et al.: Partial left ventriculectomy to improve left ventricular function in end-stage heart disease. J. Card. Surg. **11**, 96–97 (1996); discussion 98
92. Kawaguchi, A.T., et al.: Left ventricular volume reduction surgery: the 4th International Registry Report 2004. J. Card. Surg. **20**, S5–S11 (2005)
93. Sekine, H., Shimizu, T., Yang, J., Kobayashi, E., Okano, T.: Pulsatile myocardial tubes fabricated with cell sheet engineering. Circulation **114**, I87–I93 (2006)
94. Ruhparwar, A., Ghodsizad, A., Piontek, P., Karck, M.: Electrically contractile tissue assembled from polymer nanofibers for myocardial tissue engineering enables systolic augmentation in an ischemic rat heart model. In: 39th Annual Meeting of the German Society for Cardiovascular and Thoracic Surgery, vol. 58. The Thoracic and Cardiovsacular Surgeon, Stuttgart (2009)
95. Miceli, A., et al.: A clinical score to predict the need for intraaortic balloon pump in patients undergoing coronary artery bypass grafting. Ann. Thorac. Surg. **90**, 522–526 (2010)
96. Baskett, R.J., Ghali, W.A., Maitland, A., Hirsch, G.M.: The intraaortic balloon pump in cardiac surgery. Ann. Thorac. Surg. **74**, 1276–1287 (2002)
97. Kofidis, T., et al.: A novel bioartificial myocardial tissue and its prospective use in cardiac surgery. Eur. J. Cardiothorac. Surg. **22**, 238–243 (2002)
98. Kofidis, T., et al.: Bioartificial grafts for transmural myocardial restoration: a new cardiovascular tissue culture concept. Eur. J. Cardiothorac. Surg. **24**, 906–911 (2003)
99. Sakai, T., et al.: The fate of a tissue-engineered cardiac graft in the right ventricular outflow tract of the rat. J. Thorac. Cardiovasc. Surg. **121**, 932–942 (2001)
100. Ott, H.C., et al.: Perfusion-decellularized matrix: using nature's platform to engineer a bioartificial heart. Nat. Med. **14**, 213–221 (2008)
101. Shimizu, T., et al.: Long-term survival and growth of pulsatile myocardial tissue grafts engineered by the layering of cardiomyocyte sheets. Tissue Eng. **12**, 499–507 (2006)
102. Evans, M.J., Kaufman, M.H.: Establishment in culture of pluripotential cells from mouse embryos. Nature **292**, 154–156 (1981)
103. Martin, G.R.: Isolation of a pluripotent cell line from early mouse embryos cultured in medium conditioned by teratocarcinoma stem cells. Proc. Natl. Acad. Sci. U. S. A. **78**, 7634–7638 (1981)
104. Thomson, J.A., et al.: Embryonic stem cell lines derived from human blastocysts. Science **282**, 1145–1147 (1998)

105. Drukker, M., et al.: Human embryonic stem cells and their differentiated derivatives are less susceptible to immune rejection than adult cells. Stem Cells **24**, 221–229 (2006)
106. Li, L., et al.: Human embryonic stem cells possess immune-privileged properties. Stem Cells **22**, 448–456 (2004)
107. Swijnenburg, R.J., et al.: Immunosuppressive therapy mitigates immunological rejection of human embryonic stem cell xenografts. Proc. Natl. Acad. Sci. U. S. A. **105**, 12991–12996 (2008)
108. Hierlihy, A.M., Seale, P., Lobe, C.G., Rudnicki, M.A., Megeney, L.A.: The post-natal heart contains a myocardial stem cell population. FEBS Lett. **530**, 239–243 (2002)
109. Martin, C.M., et al.: Persistent expression of the ATP-binding cassette transporter, Abcg2, identifies cardiac SP cells in the developing and adult heart. Dev. Biol. **265**, 262–275 (2004)
110. Matsuura, K., et al.: Adult cardiac Sca-1-positive cells differentiate into beating cardiomyocytes. J. Biol. Chem. **279**, 11384–11391 (2004)
111. Oh, H., et al.: Cardiac progenitor cells from adult myocardium: homing, differentiation, and fusion after infarction. Proc. Natl. Acad. Sci. U. S. A. **100**, 12313–12318 (2003)
112. Urbanek, K., et al.: Cardiac stem cells possess growth factor-receptor systems that after activation regenerate the infarcted myocardium, improving ventricular function and long-term survival. Circ. Res. **97**, 663–673 (2005)
113. Kogler, G., et al.: Simultaneous cord blood transplantation of ex vivo expanded together with non-expanded cells for high risk leukemia. Bone Marrow Transplant. **24**, 397–403 (1999)
114. Brand, A., et al.: Cord blood banking. Vox Sang. **95**, 335–348 (2008)
115. Kogler, G., et al.: A new human somatic stem cell from placental cord blood with intrinsic pluripotent differentiation potential. J. Exp. Med. **200**, 123–135 (2004)
116. Greschat, S., et al.: Unrestricted somatic stem cells from human umbilical cord blood can be differentiated into neurons with a dopaminergic phenotype. Stem Cells Dev. **17**, 221–232 (2008)
117. Schaap-Oziemlak, A., et al.: MicroRNA hsa-miR-135b regulates mineralization in osteogenic differentiation of human Unrestricted Somatic Stem Cells (USSCs). Stem Cells Dev. **19**, 877–885 (2010)
118. Kogler, G., et al.: Cytokine production and hematopoiesis supporting activity of cord blood-derived unrestricted somatic stem cells. Exp. Hematol. **33**, 573–583 (2005)
119. Ghodsizad, A., et al.: Transplanted human cord blood-derived unrestricted somatic stem cells improve left-ventricular function and prevent left-ventricular dilation and scar formation after acute myocardial infarction. Heart **95**, 27–35 (2009)
120. Iwasaki, H., et al.: Therapeutic potential of unrestricted somatic stem cells isolated from placental cord blood for cardiac repair post myocardial infarction. Arterioscler. Thromb. Vasc. Biol. **29**, 1830–1835 (2009)
121. Takahashi, K., et al.: Induction of pluripotent stem cells from adult human fibroblasts by defined factors. Cell **131**, 861–872 (2007)
122. Takahashi, K., Yamanaka, S.: Induction of pluripotent stem cells from mouse embryonic and adult fibroblast cultures by defined factors. Cell **126**, 663–676 (2006)
123. Nishikawa, S., Goldstein, R.A., Nierras, C.R.: The promise of human induced pluripotent stem cells for research and therapy. Nat. Rev. Mol. Cell. Biol. **9**, 725–729 (2008)
124. Blum, B., Benvenisty, N.: The tumorigenicity of diploid and aneuploid human pluripotent stem cells. Cell Cycle **8**, 3822–3830 (2009)
125. Bar, A., et al.: The pro-angiogenic factor CCN1 enhances the re-endothelialization of biological vascularized matrices in vitro. Cardiovasc. Res. **85**, 806–813 (2010)
126. Eitan, Y., Sarig, U., Dahan, N., Machluf, M.: Acellular cardiac extracellular matrix as a scaffold for tissue engineering: In vitro cell support, remodeling and biocompatibility. Tissue Eng. Part C Methods **16**, 671–683 (2009)
127. Dvir, T., et al.: Prevascularization of cardiac patch on the omentum improves its therapeutic outcome. Proc. Natl. Acad. Sci. U. S. A. **106**, 14990–14995 (2009)
128. Yu, J., et al.: Restoration of left ventricular geometry and improvement of left ventricular function in a rodent model of chronic ischemic cardiomyopathy. J. Thorac. Cardiovasc. Surg. **137**, 180–187 (2009)

129. Yu, J., et al.: The effect of injected RGD modified alginate on angiogenesis and left ventricular function in a chronic rat infarct model. Biomaterials **30**, 751–756 (2009)
130. Khademhosseini, A., et al.: Microfluidic patterning for fabrication of contractile cardiac organoids. Biomed. Microdevices **9**, 149–157 (2007)
131. Iyer, R.K., Chiu, L.L., Radisic, M.: Microfabricated poly(ethylene glycol) templates enable rapid screening of triculture conditions for cardiac tissue engineering. J. Biomed. Mater. Res. A **89**, 616–631 (2009)
132. Chang, R., Nam, J., Sun, W.: Direct cell writing of 3D microorgan for in vitro pharmacokinetic model. Tissue Eng. Part C Methods **14**, 157–166 (2008)
133. Khademhosseini, A., et al.: Micromolding of photocrosslinkable hyaluronic acid for cell encapsulation and entrapment. J. Biomed. Mater. Res. A **79**, 522–532 (2006)
134. Morritt, A.N., et al.: Cardiac tissue engineering in an in vivo vascularized chamber. Circulation **115**, 353–360 (2007)
135. Birla, R.K., Dhawan, V., Dow, D.E., Huang, Y.C., Brown, D.L.: Cardiac cells implanted into a cylindrical, vascularized chamber in vivo: pressure generation and morphology. Biotechnol. Lett. **31**, 191–201 (2009)
136. Tudorache, I., et al.: Viable vascularized autologous patch for transmural myocardial reconstruction. Eur. J. Cardiothorac. Surg. **36**, 306–311 (2009); discussion 311
137. Akhyari, P., et al.: Mechanical stretch regimen enhances the formation of bioengineered autologous cardiac muscle grafts. Circulation **106**, I137–I142 (2002)
138. Fink, C., et al.: Chronic stretch of engineered heart tissue induces hypertrophy and functional improvement. FASEB J. **14**, 669–679 (2000)
139. Altman, G.H., et al.: Cell differentiation by mechanical stress. FASEB J. **16**, 270–272 (2002)
140. Radisic, M., et al.: Biomimetic approach to cardiac tissue engineering: oxygen carriers and channeled scaffolds. Tissue Eng. **12**, 2077–2091 (2006)
141. Tandon, N., et al.: Alignment and elongation of human adipose-derived stem cells in response to direct-current electrical stimulation. Conf. Proc. IEEE Eng. Med. Biol. Soc. **2009**, 6517–6521 (2009)
142. Kofidis, T., Balsam, L., de Bruin, J., Robbins, R.C.: Distinct cell-to-fiber junctions are critical for the establishment of cardiotypical phenotype in a 3D bioartificial environment. Med. Eng. Phys. **26**, 157–163 (2004)
143. Bursac, N., et al.: Cultivation in rotating bioreactors promotes maintenance of cardiac myocyte electrophysiology and molecular properties. Tissue Eng. **9**, 1243–1253 (2003)
144. Sekine, H., Shimizu, T., Kosaka, S., Kobayashi, E., Okano, T.: Cardiomyocyte bridging between hearts and bioengineered myocardial tissues with mesenchymal transition of mesothelial cells. J. Heart Lung Transplant. **25**, 324–332 (2006)
145. Altman, G.H., et al.: Advanced bioreactor with controlled application of multi-dimensional strain for tissue engineering. J. Biomech. Eng. **124**, 742–749 (2002)
146. Cimetta, E., Figallo, E., Cannizzaro, C., Elvassore, N., Vunjak-Novakovic, G.: Micro-bioreactor arrays for controlling cellular environments: design principles for human embryonic stem cell applications. Methods **47**, 81–89 (2009)
147. Tandon, N., Marsano, A., Cannizzaro, C., Voldman, J., Vunjak-Novakovic, G.: Design of electrical stimulation bioreactors for cardiac tissue engineering. Conf. Proc. IEEE Eng. Med. Biol. Soc. **2008**, 3594–3597 (2008)
148. Vunjak-Novakovic, G., et al.: Bioreactor cultivation conditions modulate the composition and mechanical properties of tissue-engineered cartilage. J. Orthop. Res. **17**, 130–138 (1999)
149. Martinez, E.C., et al.: Ascorbic acid improves embryonic cardiomyoblast cell survival and promotes vascularization in potential myocardial grafts in vivo. Tissue Eng. Part A **16**, 1349–1361 (2010)
150. De Visscher, G., et al.: The remodeling of cardiovascular bioprostheses under influence of stem cell homing signal pathways. Biomaterials **31**, 20–28 (2010)

Creating Unique Cell Microenvironments for the Engineering of a Functional Cardiac Patch

Tal Dvir, Jonathan Leor and Smadar Cohen

Abstract Tissue engineering is an approach used to create a functional cardiac patch for the purpose of scar support after a myocardial infarct (MI). Cardiac cells, or cells of other sources, are seeded into scaffolds, which provide an artificial biomechanical support until the cells secrete extracellular matrix and regenerate into a functional tissue. In this chapter we describe the creative design of various cell microenvironments, which promote the development of a thick vascularized cardiac patch, ready to face the harsh conditions of the infarcted heart. Among these microenvironments are unique bioreactor systems that increase mass transfer through the developing cardiac tissue at the in vitro engineering stage and the use of various vascularization techniques, including the use of the body as a bioreactor to induce rapid vascularization prior to transplantation on the infarcted heart.

1 Introduction

The heart is essentially a non-regenerating organ. Consequently, the loss of cardiac cells and the formation of scar tissue after extensive myocardial infarction (MI)

T. Dvir and S. Cohen (✉)
Avram and Stella Goldstein-Goren Department of Biotechnology Engineering,
Ben-Gurion University of the Negev, 84105 Beer-Sheva, Israel
e-mail: scohen@bgu.ac.il

T. Dvir
Department of Chemical Engineering, Massachusetts Institute of Technology,
Cambridge, MA 02139, USA

J. Leor
Neufeld Cardiac Research Institute, Sheba Medical Center, Tel Aviv University,
52621 Tel Hashomer, Israel

frequently lead to congestive heart failure [1, 2]. Given the scarcity of cardiac donors, a potential approach to treat the infarcted heart and restore function may be the repopulation of the scar tissue zone by the transplantation of new contracting cells. Various cell types have been transplanted into the area of the infarct, including skeletal myoblasts, embryonic stem cells, cardiac progenitor cells and bone marrow stem cells (BMSCs) [3–8]. Yet, regeneration of the infarcted myocardium by these cellular therapy models has been limited due to the low cell retention and viability in the damaged myocardium.

To address this challenge, a tissue engineering approach has been developed to create a functional cardiac patch for the purpose of scar support and replacement. In this approach, cardiac cells are seeded into scaffolds, which provide artificial biomechanical support until the cells secrete extracellular matrix (ECM) and regenerate into a functional tissue [9]. According to the vision of this strategy, such a cardiac patch is transplanted onto the infarcted area and following its integration with the scar, the ventricle wall is expected to maintain its appropriate thickness, contractility and cardiac function. Our group has been amongst the first to introduce the strategy of the cardiac patch for infarct therapy [10]. The advances made by us and other groups have provided proof-of-concept for the cardiac patch approach, showing that a cardiac patch can be engineered with sizes and contractile properties that can provide considerable support to failing hearts in small animal models [11, 12]. Yet, several hurdles still remain to be overcome before the ex vivo engineered cardiac patches are ready to be implanted in human patients as therapy for congestive heart failure. Among these are (1) when the patch is created in vitro, its thickness is limited to 100–200 µm, due to mass-transfer restrictions; and (2) once transplanted, the vascularization of the patch is slow, leading to poor engraftment of the patch and cell death.

In body tissues, the transport of dissolved oxygen and nutrients is aided by complex vascular networks. These networks serve as transport channels for convective blood flow, thus enabling the diffusive nutrients to transfer more efficiently through the cellular space and ECM. In static cultures of three-dimensional (3D) ex vivo cell constructs, two main engineering challenges exist. First, a boundary layer around the cell constructs decreases the transport of nutrients and dissolved oxygen from the bulk medium to the surface of the constructs. Secondly, the internal mass transfer rate from the surface of the construct into its core is also limited due to lack of inherent vasculature. Therefore, one of the major hurdles in ex vivo tissue engineering is attaining functional tissues that constitute of more than a few layers of viable cells.

The second hurdle arises due to the absence of proper vasculature in the newly developed cardiac patches, leading to cardiomyocyte death soon after implantation due to the absence of perfusing blood. To address this problem by prevascularizing tissues prior to transplantation, researchers in recent years have co-cultured myocytes and endothelial cells derived from non-autologous stem cells [13–15]. In addition to questions raised regarding the clinical relevance of the use of these stem cells, the resultant capillaries were mostly composed of endothelial cells without stabilizing pericytes, and thus were extremely fragile.

In this chapter, we describe the creative design of various cell microenvironments, developed to offer a solution for these setbacks and promote the development of a thick vascularized cardiac patch, ready to face the harsh conditions of the infarcted heart. Such microenvironments implement designed biomaterials and unique bioreactor systems that increase mass transfer through the developing cardiac tissue at the in vitro engineering stage, as well as various vascularization techniques including the use of the body as a bioreactor to induce rapid vascularization prior to transplantation on the infarcted heart.

2 Improving In Vitro Mass Transfer by the Use of Perfusion Bioreactors

Bioreactors are generally defined as devices in which biological and/or biochemical processes develop under closely monitored and tightly controlled environmental and operating conditions (e.g. pH, temperature, pressure, nutrient supply and waste removal). In particular, bioreactors designed for tissue engineering should house multiple cell constructs for monitoring the development of the engineered tissue at different time points.

The use of bioreactors can solve one of the major difficulties facing ex vivo cardiac tissue engineering; how to grow cardiac patches that contain more than a few layers of muscle cells [16]. Due to the a-vascular nature of the cardiac patch, the thickness of the viable tissue is mainly determined by the oxygen diffusion distance, ~ 100 μm. In order to improve mass transport throughout cell constructs, researchers have utilized several bioreactors, exhibiting different patterns of fluid dynamics and vessel geometry. Among these bioreactors are the spinner flask [17] and the rotary cell culture system (RCCS) designed by NASA [17, 18]. Although these bioreactors decrease the stagnant layer around cell constructs and provide a homogenous external environment within the vessel, they cannot enhance mass transfer into the developing tissue. Thus, perfusion bioreactors were designed.

A successful perfusion system requires several important characteristics. First and foremost, it must foster the culture medium into the cell construct. Second, the system must allow control over essential parameters such as pH, pO_2, pCO_2, temperature and flow rate. In addition, the system must maintain sterility over prolonged periods. During recent years, several research groups have designed different perfusion systems, which answered all of the criteria above and shared similar basic features [19–23]. In the following section we will describe some of the perfusion bioreactors designed for the engineering of cardiac tissues.

2.1 Perfusion Bioreactors for Cardiac Tissue Engineering

In the attempt to design an efficient bioreactor for the engineering of a thick functional cardiac tissue, one must take into consideration two main issues. First, cardiomyocytes, the muscle cells of the heart are terminally differentiated cells with limited proliferation potential. Therefore, it is crucial to maintain cell viability during cultivation period. Second, being muscle cells, cardiomyocytes have a high oxygen demand, therefore oxygen transport to and within the developing tissue must be satisfactory.

To study the effects of medium perfusion on cardiac cells, Carrier and colleagues utilized perfusion cartridges consisting of 13 mm filter holders in which a stainless steel screen was placed at the cartridge inlet to disperse flow over the construct surface, while a nylon mesh was used to fix the downstream side of the construct. The cartridges were arranged in a series of four and each series was connected to a four-channel peristaltic pump perfusing the medium through the constructs at different flow rates, varying from 0.6 to 3 mL/min. Cultivation under these conditions improved cardiac cell distribution in the construct and enhanced the expression of cardiac-specific markers [24]. In a follow-up study using the same bioreactor, this group reported that control over oxygen concentration provided by the perfusion bioreactor improved the function and structure of the engineered cardiac muscle [25]. Radisic and colleagues were the first to show the positive effects of medium perfusion for 7 days on cell viability within the construct, and the formation of engineered cardiac tissue that contracted synchronously in response to electrical stimulation [26]. Although encouraging results were obtained with the above described perfusion system, the developed engineered cardiac tissue was confined mostly to the periphery of the construct, while in the center the cells were not aligned or elongated.

Several factors affect the homogeneity of the tissue developing within the perfusion bioreactor constructs, as well as between the multiple constructs cultivated in the bioreactor. In most perfusion bioreactors, the cell constructs are held in place by fixing nets, which block in part medium perfusion into the cell constructs. In addition, the character of fluid flow developed in a tube also promotes non-homogeneous flow in perfusion bioreactors. The velocity profile of the developed laminar flow is unequal along the cross-section area of the bioreactor due to frictions of the fluid flow with the walls. Thus, the level of shear stress acting on the cell constructs is different in various locations within the bioreactor. All these factors contribute to the development of a non-homogeneous tissue within the bioreactor.

To address the non-homogeneous milieu within perfusion systems, we designed a bioreactor system with two unique cell construct-fixing nets, with an open area of 95.8% [27]. The net architecture (Fig. 1a) is an assembly of equally spaced, 380-square-based micropyramidal structures, with their heads pointing toward and holding the multiple scaffolds. Four round openings are located at the corners of the square pyramid, enabling medium perfusion from the small holes toward the pyramid head, at an angle of 60°, thus maintaining the fluid flow direction. Using this net architecture, 99.88% of the cell construct volume is perfused by the culture medium (Fig. 1b). In addition, to impose equal shear stress along the bioreactor cross-section, we inserted micromesh designed to disrupt the perfusing media for a split second before entering the cell constructs. The mesh transforms the

Fig. 1 Medium perfusion throughout the cell seeded scaffold enabled by the unique scaffold fixing nets. **a** Architecture of the nets holding the constructs in the bioreactor. **b** The cellular constructs along the bioreactor cross-section are subjected to equal medium velocity. Direction of the velocity vectors is dictated by the net architecture, entering the constructs at 60°. Thus, 99.88% of the cellular constructs are exposed to the medium flow

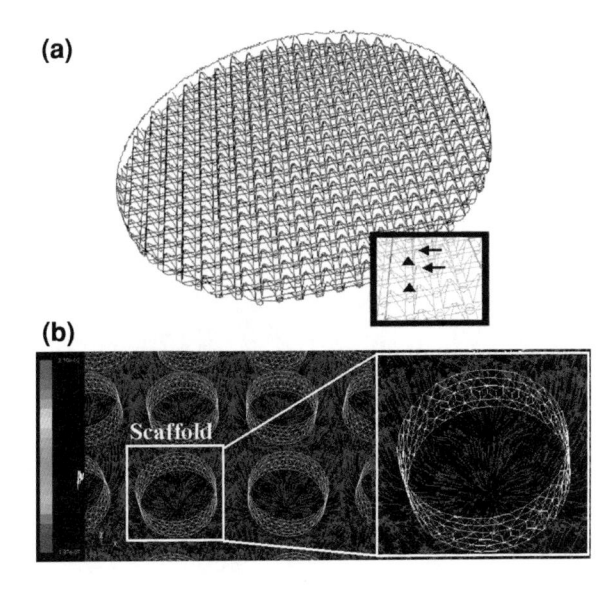

developed laminar fluid flow velocity profile into an undeveloped one, having equal velocity vectors along the bioreactor cross-section and subjecting the cellular constructs to an equal shear stress (Fig. 2).

By employing this advanced perfusion bioreactor, the cardiac cell constructs maintained viability of almost 100% of the seeded cells after a 7-day cultivation period, while less than 60% of the cells in static cultures were viable at the end of this period [27]. Moreover, a homogenous thick (>500 μm) cardiac tissue was generated, composed of elongated and aligned cells with a massive striation. Ultrastructural morphology analyses revealed organized sarcomeres, defined Z-lines and intercalated disks, resembling the native adult heart [28] (Fig. 3).

Our group has further investigated the effects of the unique microenvironment, provided by the perfusion bioreactor described above, on cardiac tissue assembly. For the first time, the various molecular and cellular events taking place in the cell construct, from the cell signaling level up to the ultrastructural phenotype of the regenerated cardiac muscle tissue, have been elucidated [28]. We demonstrated that immediately after operating the bioreactor at a shear stress of 0.6 dynes/cm^2, the extracellular signal-regulated kinase (ERK) 1/2 signal transduction pathway is activated, inducing the synthesis of cardiac contractile proteins (Troponin T and sarcomeric actinin) and cell–cell interaction proteins (Connexin 43 and N cadherin). Furthermore, we showed that prolonged cultivation under this stimulation regime promoted the assembly of a thick cardiac tissue, in many respects resembling the native adult heart. A model describing the proposed mechanism, translating mechanical cues to tissue regeneration, is presented in Fig. 4.

Realizing that electrical stimulation of the developing cardiac cells in the 3D patch is essential to promote their alignment [29], we recently combined an electrical field within the perfusion bioreactor. Cultivation within this bioreactor

Fig. 2 Profile of the fluid flow velocity in the perfusion bioreactor, constructed using the computerized fluid dynamics software Fluent after solving Navier-Stock equations. **a** The velocity profiles in a perfusion bioreactor with no fluid-distributing mesh or **b** when approaching it in a bioreactor with the flow-distributing mesh reveal a well-developed flow with large-velocity vectors (*light gray*) at the center of the bioreactor and smaller ones on the sides (*dark gray*). **c** Once past the mesh, the flow is interrupted and transforms to undeveloped flow, with equal velocity vectors (*dark gray line*) along the bioreactor cross-section. *Arrowhead* indicates the location of the cellular constructs; *thin, long arrows* indicate the location of the flow-distributing mesh 1.5 cm upstream from the construct compartment, and *thick, short arrows* indicate the flow velocity profile. Velocity scale-bar is presented on the left. *Dark blue* represents the smallest vectors

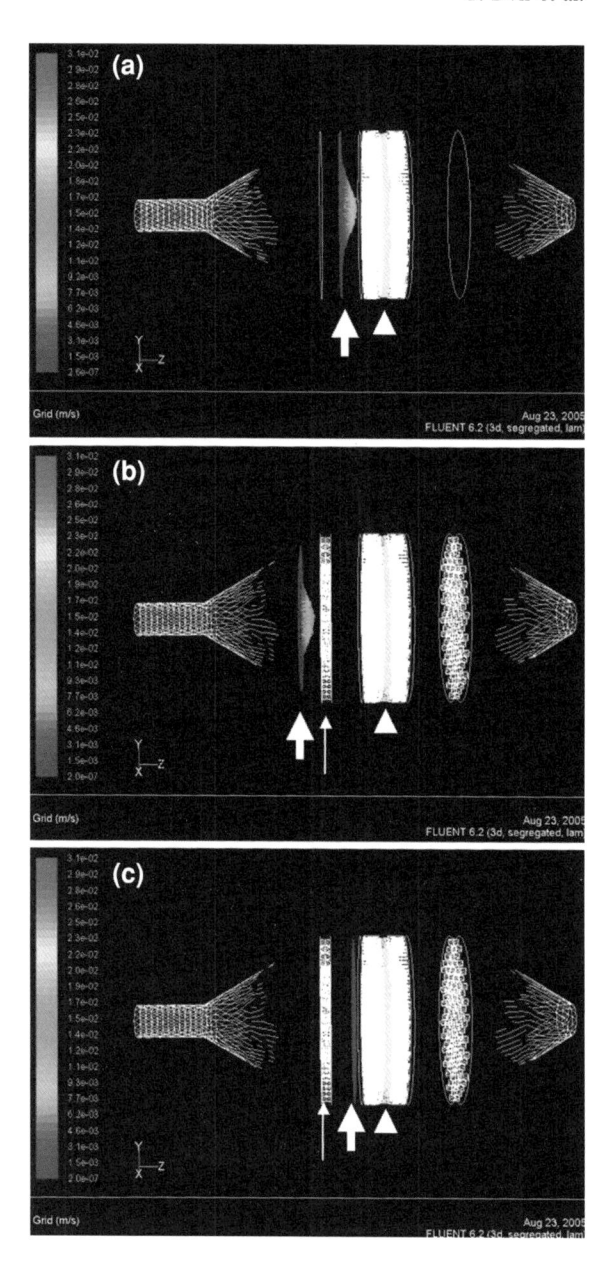

for only 4 days under perfusion and continuous electrical stimulus promoted cell elongation and striation in the cell constructs and enhanced the expression level of connexin-43, the gap junction protein responsible for cell–cell coupling [30].

Once taken out of the bioreactor system and transplanted on the infarcted heart, the cardiac patch loses its nourishing system and with the absence of proper vasculature the cells comprising the patch do not survive. The following section

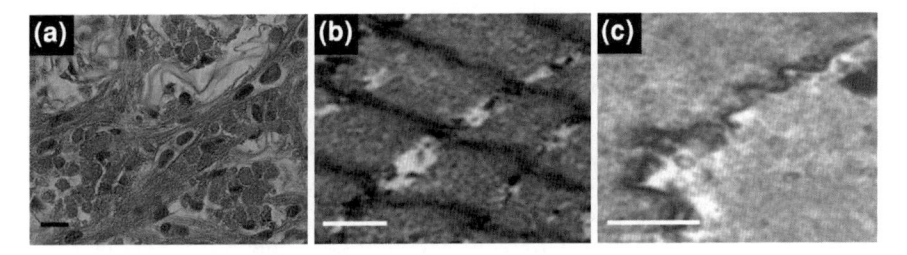

Fig. 3 Histo and ultrastructural morphology of the engineered tissue. **a** H&E staining reveals massive striation and elongation of the cultured cells. **b** TEM images show defined Z-lines and multiple high-ordered sarcomeres and **c** intercalated disks between adjacent cardiomyocytes. Scale bars: **a** 20 μm, **b** 1 μm, **c** 0.5 μm

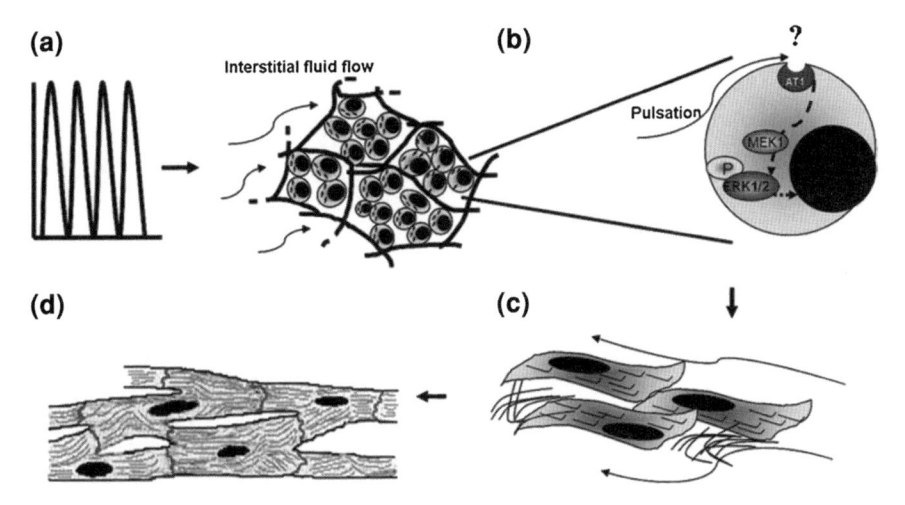

Fig. 4 A proposed model describing the mechanistic effects of pulsatile fluid flow/shear stress on cardiac tissue regeneration. **a** A pulsatile interstitial fluid flow, provided by the perfusion bioreactor subjects the scaffold-seeded cardiac cells to a physiological shear stress of 0.6 dynes/cm^2. **b** The applied shear stress activates the mechanoreceptor, AT1, thus initiating the ERK1/2 cascade. In the presence of the MEK inhibitor, the pulsatile flow cannot activate ERK1/2. **c** Activation of ERK1/2 cascade signals for cardiac cell survival and hypertrophy, as revealed by the enhanced synthesis of contractile cardiac proteins and proteins associated with cell–cell contacts and gap junction responsible for electrical coupling. **d** With the secretion of collagen and its arrangement in fibers around the cardiac myofibers, the cardiac tissue reaches full maturity

will focus on creative solutions to promote vascularization that eventually enhances mass transfer in the transplanted patch and maintains its viability.

3 Prevascularization of Engineered Tissues

Until recently, strategies to promote graft vascularization focused on inducing vasculature formation in the scaffold, in vivo, prior to transplantation of the functional cells. For example, empty scaffolds were transplanted, incorporating angiogenic growth factors to encourage new vessels to sprout from pre-existing vessels into the scaffolds [31, 32]. Once the host blood vessels had infiltrated the scaffold, functional cells were injected into the scaffold, to create a functional, vascularized tissue [33]. This approach had limited success since together with the initial blood vessel ingrowth, other cells also infiltrated, leaving almost no room for the later injected functional cells.

To solve this issue and better control the process, cardiac tissue may be engineered ex vivo and its vascularization performed simultaneously, in vitro or in vivo, prior to its implantation onto the infarcted heart.

3.1 Engineering Microvascular Networks

Most attempts at engineering a microvascular network in vitro have involved seeding empty scaffolds with endothelial cells (ECs) [34, 35]. Once such a construct is implanted, it was thought that the ECs would integrate with the host vasculature. Thus, adequate perfusion through the cellular construct could be achieved in a relatively short period. During recent years, many researchers have employed this method with the aim of constructing microvessel networks [36, 37]. Koike et al. showed that a network of long-lasting blood vessels can be formed in mice by co-implantation of human vascular ECs and mesenchymal precursor cells for capillary stabilization [38]. The authors seeded the co-culture within a collagen type I gel and implanted the construct in mice. The ECs formed long, stable, interconnected tubes with numerous branches that connected to the mouse's circulatory system and became perfused [38]. Mooney and co-workers demonstrated that sustained release of vascular endothelial growth factor (VEGF) from polymer matrices can increase the overall density of blood vessels present in the growing tissue and around the matrix [32]. In addition, these authors suggested that providing a local source of VEGF increased the ability of transplanted ECs to integrate into the host vasculature and form functional blood vessels [32].

3.2 Capillary Formation within Engineered Tissues by Ex Vivo Co-Cultures

The ex vivo approach currently used to develop microvascular network within functional tissues is to seed scaffolds with co-cultures of ECs and functional cells. Thus, concurrently with the assembly of the functional cells into a living tissue, the ECs may organize and form micro-blood vessels between the cells.

Tonello and colleagues prepared endothelialized skin by co-culturing three human cell types, keratinocytes, fibroblasts and endothelial cells obtained from human full thickness skin samples, in scaffolds produced from modified hyaluronic acid [39]. The obtained in vitro engineered skin developed into a well-differentiated upper layer of stratified keratinocytes lining a dermal-like structure, in which fibroblasts, extracellular matrix and a microvascular network were present [39]. Levenberg and co-workers found that the vascular network induced in a tri-culture system consisting of ECs, myoblasts and embryonic fibroblasts, seeded simultaneously, was of a higher density than the network induced in a co-culture of ECs and myoblasts [13]. They further demonstrated that the increase in vessel density is mediated by secretion of VEGF into the scaffolds by peri-vascular cells. Upon implantation at three locations in mice (i.e. subcutaneous, intramuscular and intraabdominal), the in vitro pre-vascularization of the tissue constructs reduced cell apoptosis and increased cell survival. Furthermore, by non-invasive whole-body imaging, these authors demonstrated that pre-vascularization increased blood perfusion [13]. In a different study, Caspi et al. reported the formation of a synchronously contracting engineered cardiac tissue derived from human embryonic stem cells containing endothelial vessel networks. The 3D muscle consisted of cardiomyocytes, endothelial cells (ECs), and embryonic fibroblasts (EmFs). The formed vessels were further stabilized by the presence of mural cells originating from the EmFs [15].

3.3 The Body as a Bioreactor for Vascularization

A different approach to vascularize a tissue is by using the body as a continuous providing source for ECs and pericytes. The body can be used as a bioreactor, serving as a humidified, temperature-controlled incubator. This bioreactor can also promote essential processes during tissue development, such as cell proliferation and differentiation. Previously it was suggested that various locations in the body could supply the various cells, nutrients, growth factors and other necessary components to create new blood vessels [40].

With this in mind, Lee and colleagues quantitatively compared the omentum, mesentery, and subcutaneous space as sites for engraftment of an engineered tissue. According to their results, the omentum is the preferred organ for engraftment and vascularization due to growth factors secreted which affect cell survival, proliferation and migration towards the engineered tissue [40].

Our group has implanted VEGF-releasing alginate scaffolds onto liver lobes for creating hepatocyte-based cell constructs in close proximity to the liver. The results revealed the liver lobe to be a feasible site for scaffold implantation. No serious bleeding or infection was observed at the implantation site. We have also shown that the alginate scaffold integrated well with the adjacent liver tissue; within 3 days of implantation, the graft was covered with a transparent thin membrane that was enriched with newly formed blood vessels [33].

Moreover, in a recent study, Baumert and co-workers took advantage of the rich vasculature of the omentum to reproduce in vivo the functions ensured previously by bioreactors. In a porcine model, the omentum was wrapped around an in vitro engineered bladder, thus permitting in vivo maturation of the seeded scaffolds with the development of dense vascularization, anticipated to prevent fibrosis and [41]. In a similar manner Suzuki and colleagues have shown that the omentum was able to maintain the viability and angiogenesis of transplanted myocardial cell sheets [42].

Cardiac tissue engineers have also utilized the body as an efficient vascularization tool. Shimizu and colleagues noted that after transplantation of cardiac monolayers subcutaneously, vascularization within the constructs began promptly after implantation, creating a well-organized vascular network after a few days in vivo [43].

Our group has demonstrated the use of a peritoneal-generated cardiac patch to replace a full thickness of the ventricle in a rat model of heterotopic heart transplantation [44]. The patch was developed in vitro by seeding of fetal cardiomyocytes onto alginate scaffolds. Implantation of these patches into the rat peritoneum cavity resulted in the continuation of the cell and tissue development and extensive vascularization.

In a different and more recent study, we sought to use the rich blood-vessel environment of the omentum to promote rapid vascularization of a cardiac patch [45] (Fig. 5a). To enhance this process and better protect the cardiac cells within the patch we used a cocktail of pro-survival and angiogenic factors (IGF-1, SDF-1 and VEGF) that were retained and presented by alginate/alginate-sulfate scaffolds, previously shown to retain growth factors (GF) and cytokines via interactions imitating GF binding to heparin [46, 47]. The incorporated SDF-1 and IGF-1 independently initiated activation of the Akt and ERK1/2 cardioprotective signaling pathways in the 3D environment. Collectively, the cocktail was able to maintain 100% cardiac cell viability during the short in vitro cultivation period prior to the patch transplantation onto the omentum. After the transplantation, SDF-1 within the patch was able to recruit BMSCs that differentiated to ECs and formed functional blood vessels (Fig. 5b). Seven days post implantation on the omentum the patches were harvested and after detecting the formation of proper networks of blood vessels the omentum-generated patches were re-transplanted to replace the scar tissue of the infarcted heart [45].

Four weeks post-transplantation onto an infarcted heart, the omentum-generated patches had fully integrated into the scar tissue. The average relative scar thickness and blood vessel density in hearts treated with omentum-generated patches were statistically greater than the control groups, suggesting that these patches, with their full vascular content thickened the scar. Most importantly, by echocardiography and electrophysiology we showed that using the body as a bioreactor for engineering a prevascularized cardiac patch was able to attenuate the deterioration of the heart after infarction [45] (Fig. 5c, d).

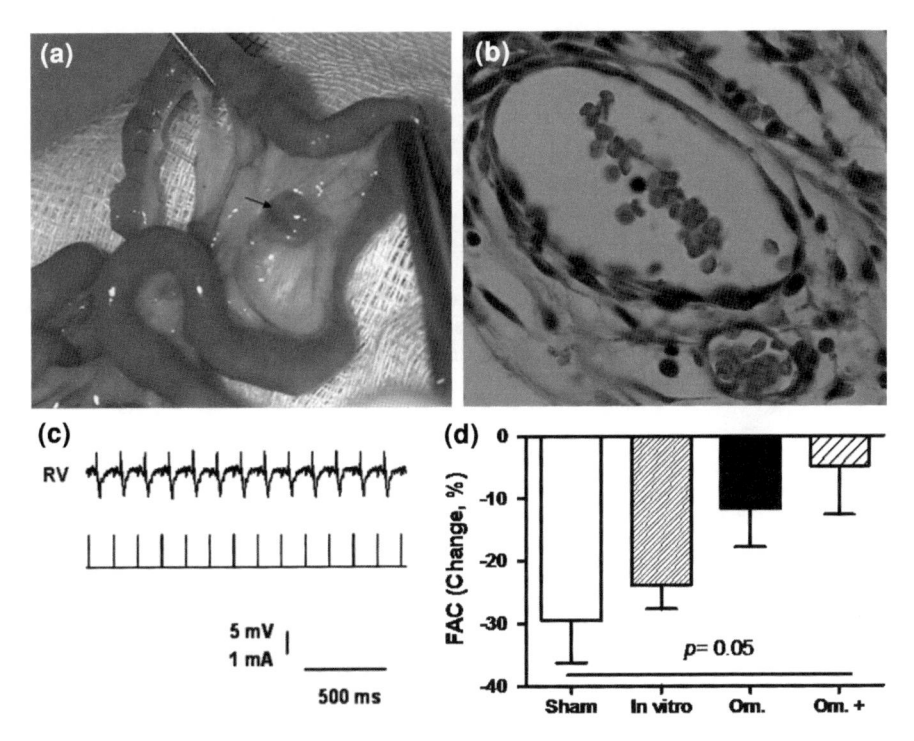

Fig. 5 Prevascularization of a cardiac patch on the omentum improves its therapeutic outcome. **a** The cardiac patch is transplanted on the omentum for 7 days to promote its vasculogenesis. **b** Functional blood vessels within the patch. **c** Twenty-eight days post transplantation of the vascularized patch on the infarcted heart, stimulation of the patch was able to trigger synchronized beating of the healthy right ventricle. **d** Fractional area change (FAC) of infarcted hearts after treatment with a stitch (sham), transplantation of in vitro-grown patch, empty patch grown on the omentum (Om), or omentum-generated cardiac patch (Om+)

4 Concluding Remarks

In this chapter, we describe the design of creative microenvironments, aimed to solve two of the most critical problems in cardiac tissue engineering, the enhancement of mass transfer through a cardiac cell construct to obtain thick tissues, and prevascularization of the tissue prior to its implantation onto the infarcted heart.

In the future, we envision a combined solution for these issues, a microfluidic bioreactor system that provides large surface to volume ratio and therefore high mass transfer rates to maintain cell and tissue viability. The microchannels of such a bioreactor can be perfused with ECs and pericytes getting cues from controlled release systems embedded in the devices, instructing the cells to organize into functional blood vessels.

References

1. Pasumarthi, K.B., Field, L.J.: Cardiomyocyte cell cycle regulation. Circ. Res. **90**, 1044–1054 (2002)
2. Rubart, M., Field, L.J.: Cardiac regeneration: repopulating the heart. Annu. Rev. Physiol. **68**, 29–49 (2006)
3. Steendijk, P., Smits, P.C., Valgimigli, M., van der Giessen, W.J., Onderwater, E.E., Serruys, P.W.: Intramyocardial injection of skeletal myoblasts: long-term follow-up with pressure–volume loops. Nat. Clin. Pract. **3**(Suppl 1), S94–S100 (2006)
4. Caspi, O., Huber, I., Kehat, I., Habib, M., Arbel, G., Gepstein, A., Yankelson, L., Aronson, D., Beyar, R., Gepstein, L.: Transplantation of human embryonic stem cell-derived cardiomyocytes improves myocardial performance in infarcted rat hearts. J. Am. Coll. Cardiol. **50**, 1884–1893 (2007)
5. Leor, J., Gerecht, S., Cohen, S., Miller, L., Holbova, R., Ziskind, A., Shachar, M., Feinberg, M.S., Guetta, E., Itskovitz-Eldor, J.: Human embryonic stem cell transplantation to repair the infarcted myocardium. Heart **93**, 1278–1284 (2007)
6. Orlic, D., Kajstura, J., Chimenti, S., Jakoniuk, I., Anderson, S.M., Li, B., Pickel, J., McKay, R., Nadal-Ginard, B., Bodine, D.M., Leri, A., Anversa, P.: Bone marrow cells regenerate infarcted myocardium. Nature **410**, 701–705 (2001)
7. Rota, M., Kajstura, J., Hosoda, T., Bearzi, C., Vitale, S., Esposito, G., Iaffaldano, G., Padin-Iruegas, M.E., Gonzalez, A., Rizzi, R., Small, N., Muraski, J., Alvarez, R., Chen, X., Urbanek, K., Bolli, R., Houser, S.R., Leri, A., Sussman, M.A., Anversa, P.: Bone marrow cells adopt the cardiomyogenic fate in vivo. Proc Natl Acad Sci USA **104**, 17783–17788 (2007)
8. Urbanek, K., Rota, M., Cascapera, S., Bearzi, C., Nascimbene, A., De Angelis, A., Hosoda, T., Chimenti, S., Baker, M., Limana, F., Nurzynska, D., Torella, D., Rotatori, F., Rastaldo, R., Musso, E., Quaini, F., Leri, A., Kajstura, J., Anversa, P.: Cardiac stem cells possess growth factor–receptor systems that after activation regenerate the infarcted myocardium, improving ventricular function and long-term survival. Circ. Res. **97**, 663–673 (2005)
9. Shachar, M., Cohen, S.: Cardiac tissue engineering, ex vivo: design principles in biomaterials and bioreactors. Heart Fail. Rev. **8**, 271–276 (2003)
10. Leor, J., Aboulafia-Etzion, S., Dar, A., Shapiro, L., Barbash, I.M., Battler, A., Granot, Y., Cohen, S.: Bioengineered cardiac grafts: a new approach to repair the infarcted myocardium? Circulation **102**, III56–III61 (2000)
11. Zimmermann, W.H., Melnychenko, I., Wasmeier, G., Didie, M., Naito, H., Nixdorff, U., Hess, A., Budinsky, L., Brune, K., Michaelis, B., Dhein, S., Schwoerer, A., Ehmke, H., Eschenhagen, T.: Engineered heart tissue grafts improve systolic and diastolic function in infarcted rat hearts. Nat. Med. **12**, 452–458 (2006)
12. Sekine, H., Shimizu, T., Hobo, K., Sekiya, S., Yang, J., Yamato, M., Kurosawa, H., Kobayashi, E., Okano, T.: Endothelial cell coculture within tissue-engineered cardiomyocyte sheets enhances neovascularization and improves cardiac function of ischemic hearts. Circulation **118**, S145–S152 (2008)
13. Levenberg, S., Rouwkema, J., Macdonald, M., Garfein, E.S., Kohane, D.S., Darland, D.C., Marini, R., van Blitterswijk, C.A., Mulligan, R.C., D'Amore, P.A., Langer, R.: Engineering vascularized skeletal muscle tissue. Nat. Biotechnol. **23**, 879–884 (2005)
14. Levenberg, S., Zoldan, J., Basevitch, Y., Langer, R.: Endothelial potential of human embryonic stem cells. Blood **110**, 806–814 (2007)
15. Caspi, O., Lesman, A., Basevitch, Y., Gepstein, A., Arbel, G., Habib, I.H., Gepstein, L., Levenberg, S.: Tissue engineering of vascularized cardiac muscle from human embryonic stem cells. Circ. Res. **100**, 263–272 (2007)
16. Leor, J., Amsalem, Y., Cohen, S.: Cells, scaffolds, and molecules for myocardial tissue engineering. Pharmacol. Ther. **105**, 151–163 (2005)

17. Carrier, R.L., Papadaki, M., Rupnick, M., Schoen, F.J., Bursac, N., Langer, R., Freed, L.E., Vunjak-Novakovic, G.: Cardiac tissue engineering: cell seeding, cultivation parameters, and tissue construct characterization. Biotechnol. Bioeng. **64**, 580–589 (1999)
18. Papadaki, M., Bursac, N., Langer, R., Merok, J., Vunjak-Novakovic, G., Freed, L.E.: Tissue engineering of functional cardiac muscle: molecular, structural, and electrophysiological studies. Am. J. Physiol. Heart Circ. Physiol. **280**, H168–H178 (2001)
19. Bancroft, G.N., Sikavitsas, V.I., Mikos, A.G.: Design of a flow perfusion bioreactor system for bone tissue-engineering applications. Tissue Eng. **9**, 549–554 (2003)
20. Bancroft, G.N., Sikavitsas, V.I., van den Dolder, J., Sheffield, T.L., Ambrose, C.G., Jansen, J.A., Mikos, A.G.: Fluid flow increases mineralized matrix deposition in 3D perfusion culture of marrow stromal osteoblasts in a dose-dependent manner. Proc. Natl Acad. Sci. USA **99**, 12600–12605 (2002)
21. Frohlich, M., Grayson, W.L., Marolt, D., Gimble, J.M., Kregar-Velikonja, N., Vunjak-Novakovic, G.: Bone grafts engineered from human adipose-derived stem cells in perfusion bioreactor culture. Tissue Eng. Part A **16**, 179–189 (2010)
22. Shvartsman, I., Dvir, T., Harel-Adar, T., Cohen, S.: Perfusion cell seeding and cultivation induce the assembly of thick and functional hepatocellular tissue-like construct. Tissue Eng. Part A **15**, 751–760 (2009)
23. Timmins, N.E., Scherberich, A., Fruh, J.A., Heberer, M., Martin, I., Jakob, M.: Three-dimensional cell culture and tissue engineering in a T-CUP (tissue culture under perfusion). Tissue Eng. **13**, 2021–2028 (2007)
24. Carrier, R.L., Rupnick, M., Langer, R., Schoen, F.J., Freed, L.E., Vunjak-Novakovic, G.: Perfusion improves tissue architecture of engineered cardiac muscle. Tissue Eng. **8**, 175–188 (2002)
25. Carrier, R.L., Rupnick, M., Langer, R., Schoen, F.J., Freed, L.E., Vunjak-Novakovic, G.: Effects of oxygen on engineered cardiac muscle. Biotechnol. Bioeng. **78**, 617–625 (2002)
26. Radisic, M., Euloth, M., Yang, L., Langer, R., Freed, L.E., Vunjak-Novakovic, G.: High-density seeding of myocyte cells for cardiac tissue engineering. Biotechnol. Bioeng. **82**, 403–414 (2003)
27. Dvir, T., Benishti, N., Shachar, M., Cohen, S.: A novel perfusion bioreactor providing a homogenous milieu for tissue regeneration. Tissue Eng. **12**, 2843–2852 (2006)
28. Dvir, T., Levy, O., Shachar, M., Granot, Y., Cohen, S.: Activation of the ERK1/2 cascade via pulsatile interstitial fluid flow promotes cardiac tissue assembly. Tissue Eng. **13**, 2185–2193 (2007)
29. Radisic, M., Park, H., Shing, H., Consi, T., Schoen, F.J., Langer, R., Freed, L.E., Vunjak-Novakovic, G.: Functional assembly of engineered myocardium by electrical stimulation of cardiac myocytes cultured on scaffolds. Proc. Natl Acad. Sci. USA **101**, 18129–18134 (2004)
30. Barash, Y., Dvir, T., Tandeitnik, P., Ruvinov, E., Guterman, H., Cohen, S.: Electric field stimulation integrated into perfusion bioreactor for cardiac tissue engineering. Tissue Eng. Part C Methods **16**(6), 1417–1426 (2010)
31. Perets, A., Baruch, Y., Weisbuch, F., Shoshany, G., Neufeld, G., Cohen, S.: Enhancing the vascularization of three-dimensional porous alginate scaffolds by incorporating controlled release basic fibroblast growth factor microspheres. J. Biomed. Mater. Res. A **65**, 489–497 (2003)
32. Peters, M.C., Polverini, P.J., Mooney, D.J.: Engineering vascular networks in porous polymer matrices. J. Biomed. Mater. Res. **60**, 668–678 (2002)
33. Kedem, A., Perets, A., Gamlieli-Bonshtein, I., Dvir-Ginzberg, M., Mizrahi, S., Cohen, S.: Vascular endothelial growth factor-releasing scaffolds enhance vascularization and engraftment of hepatocytes transplanted on liver lobes. Tissue Eng. **11**, 715–722 (2005)
34. Mooney, D.J., Mikos, A.G.: Growing new organs. Sci. Am. **280**, 60–65 (1999)
35. Cohen, S., Leor, J.: Rebuilding broken hearts. Biologists and engineers working together in the fledgling field of tissue engineering are within reach of one of their greatest goals: constructing a living human heart patch. Sci. Am. **291**, 44–51 (2004)

36. McGuigan, A.P., Sefton, M.V.: Vascularized organoid engineered by modular assembly enables blood perfusion. Proc. Natl Acad. Sci. USA **103**, 11461–11466 (2006)
37. Shinoka, T., Shum-Tim, D., Ma, P.X., Tanel, R.E., Isogai, N., Langer, R., Vacanti, J.P., Mayer Jr., J.E.: Creation of viable pulmonary artery autografts through tissue engineering. J. Thorac. Cardiovasc. Surg. **115**, 536–545 (1998). discussion 545–536
38. Koike, N., Fukumura, D., Gralla, O., Au, P., Schechner, J.S., Jain, R.K.: Tissue engineering: creation of long-lasting blood vessels. Nature **428**, 138–139 (2004)
39. Tonello, C., Vindigni, V., Zavan, B., Abatangelo, S., Abatangelo, G., Brun, P., Cortivo, R.: In vitro reconstruction of an endothelialized skin substitute provided with a microcapillary network using biopolymer scaffolds. FASEB J. **19**, 1546–1548 (2005)
40. Lee, H., Cusick, R.A., Utsunomiya, H., Ma, P.X., Langer, R., Vacanti, J.P.: Effect of implantation site on hepatocytes heterotopically transplanted on biodegradable polymer scaffolds. Tissue Eng. **9**, 1227–1232 (2003)
41. Baumert, H., Simon, P., Hekmati, M., Fromont, G., Levy, M., Balaton, A., Molinie, V., Malavaud, B.: Development of a seeded scaffold in the great omentum: feasibility of an in vivo bioreactor for bladder tissue engineering. Eur. Urol. **52**, 884–890 (2007)
42. Suzuki, R., Hattori, F., Itabashi, Y., Yoshioka, M., Yuasa, S., Manabe-Kawaguchi, H., Murata, M., Makino, S., Kokaji, K., Yozu, R., Fukuda, K.: Omentopexy enhances graft function in myocardial cell sheet transplantation. Biochem. Biophys. Res. Commun. **387**, 353–359 (2009)
43. Shimizu, T., Sekine, H., Yang, J., Isoi, Y., Yamato, M., Kikuchi, A., Kobayashi, E., Okano, T.: Polysurgery of cell sheet grafts overcomes diffusion limits to produce thick, vascularized myocardial tissues. FASEB J. **20**, 708–710 (2006)
44. Amir, G., Miller, L., Shachar, M., Feinberg, M.S., Holbova, R., Cohen, S., Leor, J.: Evaluation of a peritoneal-generated cardiac patch in a rat model of heterotopic heart transplantation. Cell Transplant. **18**, 275–282 (2009)
45. Dvir, T., Kedem, A., Ruvinov, E., Levy, O., Freeman, I., Landa, N., Holbova, R., Feinberg, M.S., Dror, S., Etzion, Y., Leor, J., Cohen, S.: Prevascularization of cardiac patch on the omentum improves its therapeutic outcome. Proc. Natl Acad. Sci. USA **106**, 14990–14995 (2009)
46. Freeman, I., Cohen, S.: The influence of the sequential delivery of angiogenic factors from affinity-binding alginate scaffolds on vascularization. Biomaterials **30**, 2122–2131 (2009)
47. Freeman, I., Kedem, A., Cohen, S.: The effect of sulfation of alginate hydrogels on the specific binding and controlled release of heparin-binding proteins. Biomaterials **29**, 3260–3268 (2008)

Intramyocardial Stem Cell Transplantation Without Tissue Engineered Constructs: The Current Clinical Situation

Peter Donndorf and Gustav Steinhoff

Abstract After promising preclinical results, the surgical treatment of chronic ischemic heart disease patients with bone marrow stem cells has been successfully introduced within the last decade in the context of clinical studies and therapy trials. Combining intramyocardial bone marrow stem cell injection with established revascularisation procedures such as bypass surgery seems to offer additional functional benefits compared to standard therapy strategies alone. However, although the safety of intramyocardial stem cell therapy could be demonstrated, the clinical evidence obtained so far is heterogeneous and completion of ongoing Phase III trails is mandatory for conclusive evaluation regarding functional advantages offered by this new therapeutic tool in the cardiac surgeon's hand. Besides ischemic heart disease also other pathologies—e.g. non-ischemic cardiomyopathy—might form indications for cardiac stem cell therapy, yet further clinical as well as preclinical evidence need to be obtained before adequate therapy strategies might be designed. Delivery and retention of adult stem cells remain limiting factors for treatment.

1 Introduction

Chronic ischemic heart disease remains one of the most important causes of morbidity and mortality worldwide. Although revascularisation procedures and conventional drug therapy may delay ventricular remodelling, there is no basic

---

P. Donndorf (✉) and G. Steinhoff
Department of Cardiac Surgery, University of Rostock, Schillingallee 35,
18057 Rostock, Germany
e-mail: peter.donndorf@med.uni-rostock.de

therapeutic regime available to prevent or even reverse this process. Chronic coronary artery disease and heart failure impair quality of life and are associated with subsequent worsening of the cardiac pump function. In the recent past experimental have demonstrated the capacity of bone marrow stem cells in restoring the function of ischemically damaged myocardium [1, 2]. Several clinical trials showed the safety and efficacy of autologous bone marrow stem cell (BMSC) transplantation in the patients with acute myocardial infarction or chronic ischemic heart disease [3, 4]. Today the therapeutic strategy of stem cell administration during cardiac surgery or coronary artery intervention is entering the clinical practice. In the following chapter we will focus on surgical stem cell transplantation in the setting of chronic ischemic heart disease.

2 Background for Intramyocardial Stem Cell Transplantation in Cardiac Surgery

Stem cells have the important properties of self-renewal and differentiation potential [5–8]. Thus, they are ideal candidates for regeneration of damaged myocardial tissue [9], for example, in myocardial infarction or in congestive heart failure. When acute myocardial infarction occurs, heart muscle tissue is regionally destroyed [10, 11]. Myocardial regeneration by direct injection of c-kit positive bone marrow stem cells for treatment of acute heart failure following myocardial infarction in a mouse model was reported the first time by Orlic et al. in 2001[12]. This work initiated a controversial but very inspiring discussion by reporting of cardiac cell type regeneration by more than 50% in the infarcted area. Although a straight reproduction of the reported results failed in some groups [13], plausible confirmation of the principle of cardiac regeneration caused by the transplantation of BMSC in different animal models followed until today [14]. The aim followed with stem cell therapy for ischemic heart disease is to use the body's own stem cells from the bone marrow or from the peripheral blood for cardiac tissue repair. These cells contain, in differing proportions, all three fractions of mononuclear bone marrow cells, haematopoietic, angioblastic and mesenchymal stems cells. By intramyocardial injection these cells are accumulated within the infarct zone and the border zone where they contribute to myocardial regeneration and scar size reduction. The first intramyocardial stem cell injection for treatment of chronic ischemic heart disease due to myocardial infarction applying this rational was performed in 2001 by the Rostock group [15]. Until today several clinical trials investigating safety and efficacy of surgical intramyocardial stem cell injection have followed. In the recent past new evidence suggests that cardiomyocyte renewal takes place in humans, although this process seems to be of low magnitude [16]. Future therapeutic strategies should be aimed to stimulate this process, and adult cardiac stem cell isolation, expansion and transplantation could offer a therapeutic tool to fully exploit this self-renewing capacity of the adult human heart.

3 Cell Types for Intramyocardial Stem Cell Transplantation

3.1 Skeletal Myoblasts

Skeletal myoblasts were among the first cell types considered for cardiac repair. Also called satellite cells, they are found beneath the basal membrane of muscle tissue and start to proliferate when stimulated by muscle injury or disease [17]. Skeletal myoblasts are of special interest for cardiac repair as they can differentiate into nonmuscle cell types [18] and are resistant to ischaemia [19], which is an obvious obstacle to the function of other stem cells in injured myocardium. Animal studies in cardiac disease models have been performed with encouraging results. However, skeletal myoblasts do not fully differentiate into cardiomyocytes in vivo after transplantation. The contracting myotubules do not operate in synchrony with the surrounding myocardium, which is most likely due to a lack of the gap-junction protein connexin 43 activity and lack of electrical coupling with the surrounding cardiac cells [20, 21]. First clinical studies were able to prove the feasibility and safety of skeletal myoblast implantation to the heart [19, 22, 23], yet the benefits seen were only marginal. Moreover these studies raised one considerable concern regarding the use of skeletal myoblasts for cardiac regeneration, which is their potential to create ventricular arrhythmias [24, 25]. However more recent clinical trials did not record an increase of arrhythmic events in vivo after intracardiac injection of skeletal myoblasts [26]. Preclinical studies have also shown that induced overexpression of connexin 43 might help to overcome this problem [27].

3.2 Bone Marrow Hematopoetic Progenitor/Stem Cells

Bone marrow derived haematopoietic stem cells (BMSC) or circulating peripheral blood progenitor cells have been shown to differentiate into cardiomyocytes in culture making them particulary interesting for the treatment of cardiac disease since they represent a well characterised and ample source of progenitor cells [28–30]. The isolation and the systemic delivery of bone marrow stem cells have been established before in the treatment of hematopoietic diseases [31]. Surface markers characterising hematopoietic stem cells in the adult human bone marrow include CD 133, CD 34, and CD 117 (c-kit). Besides the adult bone marrow stem cells expressing hematopoietic markers can also be obtained from the umbilical cord peripheral blood and the placenta. Although experimental data suggests that, for bone marrow derived stem cell with hematopoietic markers, cardiomyocyte differentiation is comparatively rare [32], transplantation of bone marrow derived hematopoietic stem cells is accompanied by myocardial regeneration, most likely due to neovascularisation and reduction of apoptosis [33, 34]. Vessel formation is shown as a result of both stem cell and resident cell action in the infarction area.

Thereby, rescue of hibernating myocardium in the infarction border zone by improved oxygen supply leads to functional improvement due to improved ventricular wall function. The largest of the randomised, placebo-controlled trials using bone marrow derived hematopoietic stem cells (The REPAIR-AMI-trial), performed in the field of interventional cardiology, and could proof the safety and feasibility of BMSC transplantation for the treatment of acute myocardial infarction and showed moreover a modest improvement of the left ventricular function and a significant reduction in subsequent cardiovascular events [35].

3.3 Mesenchymal Stem Cells from the Bone Marrow

Mesenchymal stem cells (MSC) [36] are a subset of stem cells that are located in the stroma of the bone marrow and can differentiate into osteoblasts, chondrocytes and adipocytes [7, 37]. They can be separated from haematopoietic stem cells by their ability to adhere to a culture dish [37]. MSC can also be induced to differentiate in vitro into cardiomyocytes [38]. MSC are potentially advantageous as they seem to be less immunogenic than other cell lines [39]. Preclinical animal studies with myocardial infarction models demonstrated improved left ventricular function and a reduction of infarct size [39, 40]. Difficulties however, may arise because of the broad differentiation capacity among MSC populations and consecutive low predictability of differentiation after intramyocardial transplantation. For example some studies reported MSC differentiation into osteoblasts after implantation into ventricular tissue [41]. This issue needs to be addressed prior to full scale therapy.

4 Stem Cell Isolation and Mode of Delivery

4.1 Stem Cell Isolation

Important conditions for clinical stem cell therapy are the precise and careful techniques of bone marrow cell preparation, availability of large cell concentrations within the area of interest (border zone of infarction), migration of stem cells into the apoptotic or necrotic myocardial area, and prevention of homing of transplanted cells to other extracardiac organs.

For stem cell transplantation in cardiac diseases, adult bone marrow (80–200 ml) is aspirated under local anaesthesia from the iliac crest. An alternative aspiration site is the sternal bone before median sternotomy. Although CABG, when performed "on-pump" using the heart–lung machine, requires systemic anticoagulation, up to now we had in our clinic no bleeding complications related to bone marrow aspiration [42]. Respective bone marrow stem cell populations then need to be isolated under good manufacturing practice (GMP) conditions. During cell preparation, viability needs to be determined several times and finally must reach around 95%. All microbiological tests of the clinically used cell preparations must prove negative for endogenous (HIV, HBV, HCV) or exogenous contamination. Also any hematologic disease compromising bone marrow cell quality has to be excluded prior to cell transplantation.

4.2 Mode of Cell Delivery

Surgical stem cell implantation is performed into well exposed ischemic areas, allowing for multiple injections within and principally around the infarct area with a thin needle. First clinical studies performed stem cell injection in combination with coronary artery bypass grafting (CABG). Once the graft-coronary artery anastomoses are completed the ischemic area is visualized and the cells are injected into the border zone of the infarcted area [4]. To date this method of direct injection represents the standard method of surgical stem cell therapy and has been applied successfully also during off-pump coronary artery bypass grafting as well as during "stand-alone" minimally invasive procedures where cell injection is performed without cardiac arrest via lateral minithoracotomy [43, 44]. Interventional therapy protocols utilizing stem cell transplantation for the treatment of acute myocardial infarction and chronic ischemic heart disease include intracoronary as well as endocardial stem cell injection. Homing of injected BMSC towards injured tissue is very complex and depends on the interplay of many factors, including stromal cell-derived factor-1 (SDF-1), the presence of an adequate inflammatory stimulus, adhesion molecules and a sufficient endothelial nitric oxide production. For this reason it may be hard to design an intracoronary treatment strategy, since local homing conditions depend a lot on the therapy timing. Therefore the amount of cells recruited is an obvious problem after intracoronary cell application. Local intramyocardial stem cell injection during surgery seems to overcome the problem linked to insufficient vascularisation, migration and homing of transplanted stem cells and is more promising than the attempts to influence stem cell migration processes in the vasculature. It results in a high stem cell persistence in the heart muscle [42]. In contrast to intracoronary application endocardial stem cell injection using the NOGA® injection catheter facilitates to deliver the stem cells directly into the target area of the myocardium without depending on sufficient cell migration across the endothelial barrier. First clinical studies were able to prove safety and feasibility of the transendocardial route in the setting of chronic ischemic heart disease [45] as well as for untreatable angina [46]. However orientation by electromechanical mapping is technically demanding and cell loss into the ventricle or misplacing of injection sites can occur.

Whatever route, whether epicardial (surgical), endocardial or intravascular, is used, the loss and high mortality of the injected cells is a problem. Therefore it seems to be of major importance always to deliver a maximum number of cells. Furthermore the advantageous oxygen and nutrient supply in the infarct border zone compared to the fibrotic scar should make the peri-infarct zone the preferred target for cell injection [47].

Regarding the safety of intramyocardial stem cell injection, potential ventricular arrhythmias caused by mechanical myocardial damage or intrinsic arrhythmogenicity of the injected cells have been a serious concern. Yet, bone marrow stem cell injection is reportedly safe. For example our group was able to maintain a high safety level while conducting Phase I and Phase II trials for evaluation of intramyocardial CD 133^+ stem cell injection during CABG [4]. Stem cell treated patients showed no signs of major therapy related complications.

5 Indications for Intramyocardial Stem Cell Transplantation

Until today there is preclinical as well as growing clinical evidence that stem cells might offer beneficial effects in subacute and chronic ischemic heart failure as the main surgical field and in the treatment of acute myocardial infarction as the domain of interventional therapy regimes by the cardiologists. There is only limited experience for the effects of bone marrow stem cell application in non-ischemic heart disease. However, via paracrine effects, gain in cardiac function in the absence of myocardial ischemia is reported [48].

The underlying inclusion criteria for cardiac cell therapy are reduction of the ventricular function of cardiac function with an akinetic, viable area which offers no target for standard revascularization procedures. Before cardiac cell transplantation heart failure symptoms, neurohumoral status and the myocardial function and viability should be assessed. Furthermore virus-free test should be performed. Selected patients should be in an early stage of heart failure since stem cell therapy is not an alternative to heart transplantation. The goal of this approach is to avoid or delay organ transplantation [47].

5.1 Ischemic Cardiomyopathy

Stem cell therapy for ischemic heart disease is indicated in patients presenting with impaired left ventricular ejection fraction between 20 and 40% due to myocardial infarction leading to symptoms of NYHA class II or III with or without angina. The underlying myocardial infarction should be of mild extension-approximately between 9 and 14 cm^2 with the presence of hibernating myocardium in the infarction border zone. Left ventricular wall thickness in echocardiographic evaluation should be greater than 4 mm in order to avoid extramyocardial injection and the risk of iatrogenic ventricular wall injury [47]. Early injection after infarction could be beneficial to prevent a large fibrotic scar. On the other hand, since myocardial infarction leads to severe impairment of heart function associated with rhythmic instability and poorer tolerance of additional treatment, including further ischemia during cardiac surgery, it might be reasonable to wait for the acute phase to pass until the infarction zone is consolidated. For surgical reasons myocardial consolidation is also preferable for any elective operative revascularisation procedure. Furthermore, cell transplantation should be more effective after the postischemic inflammatory reaction has subsided i.e. after days 8–12 following the acute attack [42, 47]. Transplantation of stem cells within the 'hot' phase post-myocardial infarction inflammation might lead them to take part in the inflammation cascade rather than in the formation of functional myocardium and vessels [49].

In our clinic we assign patients with impaired heart function after myocardial infarction and the presence of hibernating myocardium to stem cell treatment. The goal of intramyocardial stem cell transplantation in these patients is the improvement of myocardial function by augmentation of myocardial perfusion in the infarction border zone or regions of low contractility due to poor perfusion despite coronary revascularization. The treatment indication is additional to currently available revascularisation strategies Therefore, we select patients with the following characteristics [42]:

1. Presence of ischemic heart disease with need for extended revascularisation, e.g., presence of akinetic/hypokinetic left ventricular wall and hibernating myocardium; presence of an ischemic region without potential for direct revascularisation; presence of small vessel disease and myocardial hypoperfusion.
2. Reduced left ventricular ejection fraction (LVEF) below 40%.
3. Presence of refractory angina without target for revascularisation.

In the selected patients the treatment indication is tested by coronary angiogram, echocardiography, cardiac MRI, single photon emission computerized tomography (SPECT) perfusion scan, and Holter-ECG. For the orientation about the target region of stem cell injection in the operating room, a left ventricular topographic raster (Fig. 1) should be used by the diagnostic method characterizing the myocardium.

Fig. 1 Illustration of left ventricular topographic raster used for intramyocardial stem cell injection during coronary artery bypass surgery.
1 = anterior, 4 = posterior, 17 = apex

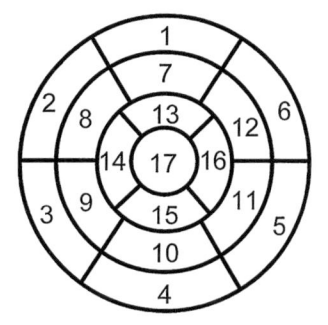

5.2 Non-ischemic Cardiomyopathy

Also patients suffering from cardiomyopathy of non-ischemic origin could benefit from cardiac stem cell therapy. In preclinical studies stem cell treatment has been performed successfully in small animal models, such as a canine model of idiopathic dilated cardiomyopathy and in doxorubicin induced heart failure [50]. These results initiated clinical trials in dilated cardiomyopathy. This first-in-man study of autologous bone marrow cells in dilated cardiomyopathy (Firstin-Man ABCD) investigated 44 patients and the Düsseldorf autologous bone marrow cells in dilated cardiomyopathy trial (Düsseldorfer ABCD Trial) investigated 20 patients. In both studies none of the patients had coronary disease (excluded by angiography) or myocarditis (excluded by endomyocardial biopsy). In both trials cell transplantation was performed via the intracoronary administration route in either coronary artery. New York Heart Association Functional Classification showed significant increase in treatment patients in both trials, furthermore ejection fraction improved by 5.4% in the First-in-Man ABCD trial and 8% in the Düsseldorf ABCD Trial [51, 52]. First results with paediatric patients suffering from idiopathic dilatative cardiomyopathy showed a clear improvement in left ventricular ejection fraction and left ventricular dimensions after intracoronary stem cell transplantation [53]. Transplanted cells seem to survive better in the host myocardium since in this pathology myocardial irrigation is not significantly impaired [51]. Data from the First-in-Man ABCD trial suggest that the benefit of stem cell therapy could be a paracrine effect with changes in vascularisation [51].

5.3 Diabetic Cardiomyopathy

It is well known that diabetic patients have an increased risk of developing heart failure due to diabetic cardiomyopathy, characterized by microvascular pathologies and interstitial fibrosis. Experimental studies have shown that mesenchymal stem cells can, when transplanted, prevent apoptosis of ischemic heart via up-regulation of Akt and endothelial nitric-oxide synthase. Furthermore they inhibit myocardial fibrosis of dilated cardiomyopathy by decreasing the expression of matrix metalloproteinase (MMP). Mesenchymal stem cell transplantation significantly increased myocardial arteriolar density and decreased the collagen volume in diabetic myocardium leading to an improved cardiac function [54]. The capability of mesenchymal stem cells to reduce apoptosis and remodelling processes was further improved after anoxic preconditioning [55]. Therefore mesenchymal stem cells could be a promising tool to attenuate cardiac remodeling and to improve cardiac function in the setting of diabetic cardiomyopathy most likely by attenuation of cardiac remodelling and angiogenesis.

5.4 Chagas Heart Disease

Chagas disease, also called American trypanosomiasis, is caused by the protozoan *Trypanosoma cruzi* and remains one of the major health problems in Latin America. In chronic cases 10–30% of the patients suffer from or will develop cardiomyopathy. Chagas disease is a progressive pathology and in its final stages there are to date no other treatment options than heart transplantation. Four main pathogenetic mechanisms characterize the Chagas heart disease: derangements of the autonomic nervous system, microvascular disturbance, parasite-dependent myocardial aggression and immune mediated myocardial injury [47]. In the long run patients develop severe cardiac arrhythmias, dilated cardiomyopathy and heart failure. Since the number of available organs for transplantation is very limited and furthermore late reactivation of the disease after transplantation due to isolated organ lesions is described, cardiac cell therapy is an important option for patients with secondary heart failure caused by Chagas disease. Selective intracoronary infusion of mononuclear stem cells has been performed and resulted in improved cardiac function and haemodynamics [56, 57]. When requiring additional coronary artery bypass grafting intramyocardial stem cell injection might be another option for these patients.

5.5 Stem Cells and Bridge-to-Recovery/Bridge-to-Transplant

One of the main unsolved problems in the treatment of end-stage heart failure is that the number of patients running out of therapeutic options and needing an urgent transplantation exceeds the availability of donor hearts. The promising results for the treatment of ischemic heart failure with bone marrow stem cells might set up a new therapeutic "bridge" for these patients. For example Yelda et al. report on successful treatment of heart-transplantation candidates with ischemic cardiomyopathy by intracoronary transplantation of autologous BMSC [58]. Especially the "stand alone" treatment mentioned above could- after further clinical efficacy studies- became a less invasive treatment option for patients waiting for a donor heart. There are also reports indicating that also pediatric heart failure could be a target for regenerative cardiac cell therapy, since it is known to have a good prognosis if causal therapy is possible. Olguntürk et al. report on paediatric patients with idiopathic dilatative cardiomyopathy (IDC) who showed a clear improvement in left ventricular ejection fraction and left ventricular dimensions after intracoronary stem cell transplantation [53]. Although heart transplantation will very likely remain the main therapeutic option for patients with IDC, stem cell transplantation could lessen the waiting list mortality. Until today half of children with symptomatic cardiomyopathy receive a transplant or die within 2 years.

Also in a bridge-to-recovery setting with heart failure patients managed by ventricular assist systems additional autologous stem cell therapy might augment the therapeutic effect. A very first clinical report showed signs of improved cardiac performance, angiogenesis, and reduced fibrosis [59].

6 Clinical Studies

To date several studies in humans with intramyocardial stem cell transplantation during coronary bypass surgery in patients with chronic ischemic heart disease have been performed. Although most of these studies have yielded encouraging results the extent to which stem cell transplantation can improve the patients outcome remains unclear since the amount and source of cells as well as inclusion criteria used have been quite heterogenous until today. The following chapter will try to give an overview about major surgical trials carried out so far.

Several clinical studies have revealed beneficial stem cell effects in subacute and chronic ischemic heart failure. Most of the surgical trials designed for this setting (Table 1) performed intramyocardial stem cell transplantation in combination with "on-pump" coronary artery bypass surgery. Surgical studies completed so far have been randomised [4, 43, 60, 61] as well as non randomised [62, 63]. All of these studies included a rather small number of patients weakening the validity of the results. While some investigators used a specified population of bone marrow stem cells, others injected a semi-enriched population of bone marrow mononuclear cells. Besides haematopoetic and mesenchymal stem cells this population contains also leucocytes making it difficult to attribute functional effects to a certain cell type. The improvement of cardiac function by intramyocardial cell transplantation cell has been described as an increase in left ventricular ejection fraction from 5 to 10% [4, 60, 63] as well as an improvement of wall motion caused by enhanced myocardial perfusion [61]. Moreover, a decrease of left ventricular dimensional parameters like LVEDV has been reported in some studies [4, 43], suggesting that intramyocardial cell therapy might contribute to reverse cardiac remodelling. Also studies combining stem cell transplantation with "off-pump" coronary surgery report similar results [43], implicating that cardiac arrest is not mandatory for safe and efficient stem cell transplantation. However, these results will always be difficult to interpret conclusively without consideration of revascularisation effects. Therefore recent reports about "stand alone stem cell treatment" for patients with ischemic heart failure are very interesting [64]. A recent, non- randomised study including reported not only a gain in cardiac function but also a clear improvement in quality of life for patients with chronic ischemic heart disease and refractory angina treated with "stand alone" bone marrow stem cell injection via lateral minithoracotomy [44]. Another trial conducted by Klein and co-workers reported a gain cardiac function in patients with ischemic heart disease after intramyocardial stem cell injection without concurrent CABG [64]. Yet, the number of patients included in these studies (five and ten) is too small to give a final judgement about the efficacy of "stand alone" therapy protocols.

Interventional studies using intracoronary or endocardial stem cell application have also been performed in the setting of chronic ischemic heart disease [45, 65, 66]. These studies report an improvement of left ventricular ejection fraction to

Table 1 Major surgical trials investigating the treatment of chronic ischemic heart disease by intramyocardial transplantation of autologous bone marrow stem cells

Reference	Sample size/design	Primary intervention	Co-intervention; mean stem cell dose (SD)	Route of injection	Follow-up duration
Hendrikx et al. [61]	20/RCT	CABG	CD 133 BMSC aspiration; 60.25 (31) × 10^6	IM	4 Months
Mocini et al. [62]	36/Cohort	CABG	BMC-MN aspiration; 292 (232) × 10^6	IM	3 Months
Klein et al. [64]	10/Cohort	"Stand alone" cell injection	CD 133 BMSC aspiration; 1.5–9.7 × 10^6 cells	IM	9 Months
Ahmadi et al. [63]	27/Cohort	CABG	CD 133 BMSC aspiration; 1.89 (0.03) × 10^6	IM	6 Months
Pompilio et al. [44]	5/Cohort	"Stand-alone" cell injection	CD 133 BMSC aspiration (3), BMSC mobilisation; (2); 7.6 × 10^6	IM	12 Months
Stamm et al. [4]	40/RCT	CABG	CD 133 BMSC aspiration; 5.80 × 10^6	IM	6 Months
Zhao et al. [60]	36/RCT	CABG	BMC-MN aspiration; 6.59 (5.12) × 10^8	IM	6 Months
Patel et al. [43]	20/RCT	CABG (off-pump)	CD 34 BMSC aspiration; 22 × 10^6	IM	6 Months

CABG coronary artery bypass grafting, *CD* cluster of differentiation, *BMSC* bone marrow stem cell, *BMC-MN* bone marrow mononuclear cell, *IM* intramyocardial, *RCT* randomised controlled trial, *cohort* cohort study, *SD* standard deviation, *SC* stem cells

similar extent to that in surgical trials. Furthermore, a significant decrease in infarction size and an improved overall myocardial oxygen uptake have been described [66].

7 Future Perspectives

Since the first description of bone marrow stem cell transplantation during coronary artery bypass surgery in 2001 several clinical studies have suggested intramyocardial stem cell injection to be beneficial for patients with chronic ischemic heart failure. Especially patients with a severe reduction of LV-EF (i.e. possible transplantation candidates) seem to profit from cell therapy in addition to revascularisation procedures. "Stand alone" stem cell therapy might offer a new bridging option for patients waiting for a donor heart. Still the underlying molecular mechanisms of stem cell related gain of cardiac function need to be further clarified. Besides myocardial regeneration due to enhanced neovascularisation, alternative modes of stem cell action like paracrine activity and immunomodulation have to be considered. For example the immunomodulatory properties of BMSC are promising to supplement the established immunosuppressive therapies in the setting of solid organ transplantation. On the other hand, especially the use of BMSC might also hold some pitfalls, like secondary infections due to uncontrolled immunosuppression and tumor growth. For establishing BMSC transplantation in the treatment of ischemic heart failure, randomized, double-blinded, multi-centre Phase III trails need to be accomplished as soon as possible. These trials should also be aimed to evaluate whether a significant gain in cardiac function is reached and to which this possible gain affects clinical outcome (NYHA score etc.) of stem cell treated patients on the long term. Furthermore attempts to create dynamic "multilinage" cardiac regeneration by combining cell therapy with tissue engineered scaffolds, heart valves or cardiac resynchronisation therapy [47, 67, 68] should be further supported since they offer a realistic perspective to come to an integrated regenerative approach. To reach these perspectives cooperative studies between clinical and preclinical research are mandatory.

References

1. Orlic, D., Kajstura, J., Chimenti, S., Limana, F., Jakoniuk, I., Quaini, F., et al.: Mobilized bone marrow cells repair the infarcted heart, improving function and survival. Proc. Natl. Acad. Sci. USA **98**(18), 10344–10349 (2001)
2. Kajstura, J., Rota, M., Whang, B., Cascapera, S., Hosoda, T., Bearzi, C., et al.: Bone marrow cells differentiate in cardiac cell lineages after infarction independently of cell fusion. Circ. Res. **96**(1), 127–137 (2005)
3. Strauer, B.E., Brehm, M., Zeus, T., Kostering, M., Hernandez, A., Sorg, R.V., et al.: Repair of infarcted myocardium by autologous intracoronary mononuclear bone marrow cell transplantation in humans. Circulation **106**(15), 1913–1918 (2002)

4. Stamm, C., Kleine, H.D., Choi, Y.H., Dunkelmann, S., Lauffs, J.A., Lorenzen, B., et al.: Intramyocardial delivery of CD133+ bone marrow cells and coronary artery bypass grafting for chronic ischemic heart disease: safety and efficacy studies. J. Thorac. Cardiovasc. Surg. **133**(3), 717–725 (2007)
5. Allgöwer, M.: The Cellular Basis of Wound Repair. Charles C. Thomas, Springfield (1956)
6. Krause, D.S., Theise, N.D., Collector, M.I., Henegariu, O., Hwang, S., Gardner, R., et al.: Multi-organ, multi-lineage engraftment by a single bone marrow-derived stem cell. Cell **105**(3), 369–377 (2001)
7. Jiang, Y., Jahagirdar, B.N., Reinhardt, R.L., Schwartz, R.E., Keene, C.D., Ortiz-Gonzalez, X.R., et al.: Pluripotency of mesenchymal stem cells derived from adult marrow. Nature **418**(6893), 41–49 (2002)
8. Quaini, F., Urbanek, K., Graiani, G., Lagrasta, C., Maestri, R., Monica, M., et al.: The regenerative potential of the human heart. Int. J. Cardiol. **95**(Suppl 1), S26–S28 (2004)
9. Goodell, M.A., Jackson, K.A., Majka, S.M., Mi, T., Wang, H., Pocius, J., et al.: Stem cell plasticity in muscle and bone marrow. Ann. N. Y. Acad. Sci. **938**, 208–218 (2001). Discussion 18–20
10. Pfeffer, M.A., Braunwald, E.: Ventricular remodeling after myocardial infarction. Experimental observations and clinical implications. Circulation **81**(4), 1161–1172 (1990)
11. Ren, G., Michael, L.H., Entman, M.L., Frangogiannis, N.G.: Morphological characteristics of the microvasculature in healing myocardial infarcts. J. Histochem. Cytochem. **50**(1), 71–79 (2002)
12. Orlic, D., Kajstura, J., Chimenti, S., Jakoniuk, I., Anderson, S.M., Li, B., et al.: Bone marrow cells regenerate infarcted myocardium. Nature **410**(6829), 701–705 (2001)
13. Murry, C.E., Soonpaa, M.H., Reinecke, H., Nakajima, H., Nakajima, H.O., Rubart, M., et al.: Haematopoietic stem cells do not transdifferentiate into cardiac myocytes in myocardial infarcts. Nature **428**(6983), 664–668 (2004)
14. Segers, V.F., Lee, R.T.: Stem-cell therapy for cardiac disease. Nature **451**(7181), 937–942 (2008)
15. Stamm, C., Westphal, B., Kleine, H.D., Petzsch, M., Kittner, C., Klinge, H., et al.: Autologous bone-marrow stem-cell transplantation for myocardial regeneration. Lancet **361**(9351), 45–46 (2003)
16. Bergmann, O., Bhardwaj, R.D., Bernard, S., Zdunek, S., Barnabe-Heider, F., Walsh, S., et al.: Evidence for cardiomyocyte renewal in humans. Science (New York, NY) **324**(5923), 98–102 (2009)
17. Buckingham, M., Montarras, D.: Skeletal muscle stem cells. Curr. Opin. Genet. Dev. **18**(4), 330–336 (2008)
18. Arsic, N., Mamaeva, D., Lamb, N.J., Fernandez, A.: Muscle-derived stem cells isolated as non-adherent population give rise to cardiac, skeletal muscle and neural lineages. Exp. Cell Res. **314**(6), 1266–1280 (2008)
19. Pagani, F.D., DerSimonian, H., Zawadzka, A., Wetzel, K., Edge, A.S., Jacoby, D.B., et al.: Autologous skeletal myoblasts transplanted to ischemia-damaged myocardium in humans. Histological analysis of cell survival and differentiation. J. Am. Coll. Cardiol. **41**(5), 879–888 (2003)
20. Leobon, B., Garcin, I., Menasche, P., Vilquin, J.T., Audinat, E., Charpak, S.: Myoblasts transplanted into rat infarcted myocardium are functionally isolated from their host. Proc. Natl. Acad. Sci. U S A **100**(13), 7808–7811 (2003)
21. Reinecke, H., Poppa, V., Murry, C.E.: Skeletal muscle stem cells do not transdifferentiate into cardiomyocytes after cardiac grafting. J. Mol. Cell. Cardiol. **34**(2), 241–249 (2002)
22. Menasche, P., Hagege, A.A., Vilquin, J.T., Desnos, M., Abergel, E., Pouzet, B., et al.: Autologous skeletal myoblast transplantation for severe postinfarction left ventricular dysfunction. J. Am. Coll. Cardiol. **41**(7), 1078–1083 (2003)
23. Herreros, J., Prosper, F., Perez, A., Gavira, J.J., Garcia-Velloso, M.J., Barba, J., et al.: Autologous intramyocardial injection of cultured skeletal muscle-derived stem cells in patients with non-acute myocardial infarction. Eur. Heart J. **24**(22), 2012–2020 (2003)

24. Fouts, K., Fernandes, B., Mal, N., Liu, J., Laurita, K.R.: Electrophysiological consequence of skeletal myoblast transplantation in normal and infarcted canine myocardium. Heart Rhythm **3**(4), 452–461 (2006)
25. Itabashi, Y., Miyoshi, S., Yuasa, S., Fujita, J., Shimizu, T., Okano, T., et al.: Analysis of the electrophysiological properties and arrhythmias in directly contacted skeletal and cardiac muscle cell sheets. Cardiovasc. Res. **67**(3), 561–570 (2005)
26. Menasche, P., Alfieri, O., Janssens, S., McKenna, W., Reichenspurner, H., Trinquart, L., et al.: The Myoblast Autologous Grafting in Ischemic Cardiomyopathy (MAGIC) trial: first randomized placebo-controlled study of myoblast transplantation. Circulation **117**(9), 1189–1200 (2008)
27. Abraham, M.R., Henrikson, C.A., Tung, L., Chang, M.G., Aon, M., Xue, T., et al.: Antiarrhythmic engineering of skeletal myoblasts for cardiac transplantation. Circ. Res. **97**(2), 159–167 (2005)
28. Yeh, E.T., Zhang, S., Wu, H.D., Korbling, M., Willerson, J.T., Estrov, Z.: Transdifferentiation of human peripheral blood CD34 + -enriched cell population into cardiomyocytes, endothelial cells, and smooth muscle cells in vivo. Circulation **108**(17), 2070–2073 (2003)
29. Bedada, F.B., Braun, T.: Partial induction of the myogenic program in noncommitted adult stem cells. Cells Tissues Organs **188**(1–2), 189–201 (2008)
30. Koyanagi, M., Bushoven, P., Iwasaki, M., Urbich, C., Zeiher, A.M., Dimmeler, S.: Notch signaling contributes to the expression of cardiac markers in human circulating progenitor cells. Circ. Res. **101**(11), 1139–1145 (2007)
31. Hock, H.: Some hematopoietic stem cells are more equal than others. J. Exp. Med. **207**(6), 1127–1130 (2010)
32. Nygren, J.M., Jovinge, S., Breitbach, M., Sawen, P., Roll, W., Hescheler, J., et al.: Bone marrow-derived hematopoietic cells generate cardiomyocytes at a low frequency through cell fusion, but not transdifferentiation. Nat. Med. **10**(5), 494–501 (2004)
33. Tse, H.F., Yiu, K.H., Lau, C.P.: Bone marrow stem cell therapy for myocardial angiogenesis. Curr. Vasc. Pharmacol. **5**(2), 103–112 (2007)
34. Ma, N., Ladilov, Y., Kaminski, A., Piechaczek, C., Choi, Y.H., Li, W., et al.: Umbilical cord blood cell transplantation for myocardial regeneration. Transplant. Proc. **38**(3), 771–773 (2006)
35. Schachinger, V., Erbs, S., Elsasser, A., Haberbosch, W., Hambrecht, R., Holschermann, H., et al.: Improved clinical outcome after intracoronary administration of bone-marrow-derived progenitor cells in acute myocardial infarction: final 1-year results of the REPAIR-AMI trial. Eur. Heart J. **27**(23), 2775–2783 (2006)
36. Dobert, N., Britten, M., Assmus, B., Berner, U., Menzel, C., Lehmann, R., et al.: Transplantation of progenitor cells after reperfused acute myocardial infarction: evaluation of perfusion and myocardial viability with FDG-PET and thallium SPECT. Eur. J. Nucl. Med. Mol. Imaging **31**(8), 1146–1151 (2004)
37. Alhadlaq, A., Mao, J.J.: Mesenchymal stem cells: isolation and therapeutics. Stem Cells Dev. **13**(4), 436–448 (2004)
38. Makino, S., Fukuda, K., Miyoshi, S., Konishi, F., Kodama, H., Pan, J., et al.: Cardiomyocytes can be generated from marrow stromal cells in vitro. J. Clin. Invest. **103**(5), 697–705 (1999)
39. Dai, W., Hale, S.L., Martin, B.J., Kuang, J.Q., Dow, J.S., Wold, L.E., et al.: Allogeneic mesenchymal stem cell transplantation in postinfarcted rat myocardium: short- and long-term effects. Circulation **112**(2), 214–223 (2005)
40. Amado, L.C., Saliaris, A.P., Schuleri, K.H., St John, M., Xie, J.S., Cattaneo, S., et al.: Cardiac repair with intramyocardial injection of allogeneic mesenchymal stem cells after myocardial infarction. Proc. Natl. Acad. Sci. USA **102**(32), 11474–11479 (2005)
41. Yoon, Y.S., Park, J.S., Tkebuchava, T., Luedeman, C., Losordo, D.W.: Unexpected severe calcification after transplantation of bone marrow cells in acute myocardial infarction. Circulation **109**(25), 3154–3157 (2004)
42. Kaminski, A., Steinhoff, G.: Current status of intramyocardial bone marrow stem cell transplantation. Semin. Thorac. Cardiovasc. Surg. **20**(2), 119–125 (2008)

43. Patel, A.N., Geffner, L., Vina, R.F., Saslavsky, J., Urschel Jr., H.C., Kormos, R., et al.: Surgical treatment for congestive heart failure with autologous adult stem cell transplantation: a prospective randomized study. J. Thorac. Cardiovasc. Surg. **130**(6), 1631–1638 (2005)
44. Pompilio, G., Steinhoff, G., Liebold, A., Pesce, M., Alamanni, F., Capogrossi, M.C., et al.: Direct minimally invasive intramyocardial injection of bone marrow-derived AC133+ stem cells in patients with refractory ischemia: preliminary results. Thorac. Cardiovasc. Surg. **56**(2), 71–76 (2008)
45. Perin, E.C., Dohmann, H.F., Borojevic, R., Silva, S.A., Sousa, A.L., Mesquita, C.T., et al.: Transendocardial, autologous bone marrow cell transplantation for severe, chronic ischemic heart failure. Circulation **107**(18), 2294–2302 (2003)
46. Losordo, D.W., Schatz, R.A., White, C.J., Udelson, J.E., Veereshwarayya, V., Durgin, M., et al.: Intramyocardial transplantation of autologous CD34 + stem cells for intractable angina: a phase I/IIa double-blind, randomized controlled trial. Circulation **115**(25), 3165–3172 (2007)
47. Chachques, J.C.: Cellular cardiac regenerative therapy in which patients? Expert Rev. Cardiovasc. Ther. **7**(8), 911–919 (2009)
48. Ohnishi, S., Yanagawa, B., Tanaka, K., Miyahara, Y., Obata, H., Kataoka, M., et al.: Transplantation of mesenchymal stem cells attenuates myocardial injury and dysfunction in a rat model of acute myocarditis. J. Mol. Cell. Cardiol. **42**(1), 88–97 (2007)
49. Soeki, T., Tamura, Y., Shinohara, H., Tanaka, H., Bando, K., Fukuda, N.: Serial changes in serum VEGF and HGF in patients with acute myocardial infarction. Cardiology **93**(3), 168–174 (2000)
50. Dhein, S., Garbade, J., Rouabah, D., Abraham, G., Ungemach, F.R., Schneider, K., et al.: Effects of autologous bone marrow stem cell transplantation. Journal of Cardiothoracic Surgery **1**, 17 (2006)
51. Seth, S., Narang, R., Bhargava, B., Ray, R., Mohanty, S., Gulati, G., et al.: Percutaneous intracoronary cellular cardiomyoplasty for nonischemic cardiomyopathy: clinical and histopathological results: the first-in-man ABCD (Autologous Bone Marrow Cells in Dilated Cardiomyopathy) trial. J. Am. Coll. Cardiol. **48**(11), 2350–2351 (2006)
52. Seth, S., Bhargava, B., Narang, R., Ray, R., Mohanty, S., Gulati, G., et al.: The ABCD (Autologous Bone Marrow Cells in Dilated Cardiomyopathy) trial a long-term follow-up study. J. Am. Coll. Cardiol. **55**(15), 1643–1644 (2010)
53. Olgunturk, R., Kula, S., Sucak, G.T., Ozdogan, M.E., Erer, D., Saygili, A.: Peripheric stem cell transplantation in children with dilated cardiomyopathy: preliminary report of first two cases. Pediatric Transplantation **14**(2), 257–260 (2009)
54. Zhang, N., Li, J., Luo, R., Jiang, J., Wang, J.A.: Bone marrow mesenchymal stem cells induce angiogenesis and attenuate the remodeling of diabetic cardiomyopathy. Exp. Clin. Endocrinol. Diabetes **116**(2), 104–111 (2008)
55. Li, J.H., Zhang, N., Wang, J.A.: Improved anti-apoptotic and anti-remodeling potency of bone marrow mesenchymal stem cells by anoxic pre-conditioning in diabetic cardiomyopathy. J. Endocrinol. Invest. **31**(2), 103–110 (2008)
56. Vilas-Boas, F., Feitosa, G.S., Soares, M.B., Mota, A., Pinho-Filho, J.A., Almeida, A.J., et al.: Early results of bone marrow cell transplantation to the myocardium of patients with heart failure due to Chagas disease. Arquivos brasileiros de cardiologia **87**(2), 159–166 (2006)
57. Soares, M.B., Garcia, S., Campos de Carvalho, A.C., Ribeiro dos Santos, R.: Cellular therapy in Chagas' disease: potential applications in patients with chronic cardiomyopathy. Regenerative Med. **2**(3), 257–264 (2007)
58. Yelda, T., Berrin, U., Murat, S., Aytac, O., Sevgi, B., Yasemin, S., et al.: Intracoronary stem cell infusion in heart transplant candidates. Tohoku J. Exp. Med. **213**(2), 113–120 (2007)
59. Miyagawa, S., Matsumiya, G., Funatsu, T., Yoshitatsu, M., Sekiya, N., Fukui, S., et al.: Combined autologous cellular cardiomyoplasty using skeletal myoblasts and bone marrow cells for human ischemic cardiomyopathy with left ventricular assist system implantation: report of a case. Surg. Today **39**(2), 133–136 (2009)

60. Zhao, Q., Sun, Y., Xia, L., Chen, A., Wang, Z.: Randomized study of mononuclear bone marrow cell transplantation in patients with coronary surgery. Ann. Thorac. Surg. **86**(6), 1833–1840 (2008)
61. Hendrikx, M., Hensen, K., Clijsters, C., Jongen, H., Koninckx, R., Bijnens, E., et al.: Recovery of regional but not global contractile function by the direct intramyocardial autologous bone marrow transplantation: results from a randomized controlled clinical trial. Circulation **114**(1 Suppl), I101–I107 (2006)
62. Mocini, D., Staibano, M., Mele, L., Giannantoni, P., Menichella, G., Colivicchi, F., et al.: Autologous bone marrow mononuclear cell transplantation in patients undergoing coronary artery bypass grafting. Am. Heart J. **151**(1), 192–197 (2006)
63. Ahmadi, H., Baharvand, H., Ashtiani, S.K., Soleimani, M., Sadeghian, H., Ardekani, J.M., et al.: Safety analysis and improved cardiac function following local autologous transplantation of CD133(+) enriched bone marrow cells after myocardial infarction. Curr. Neurovas. Res. **4**(3), 153–160 (2007)
64. Klein, H.M., Ghodsizad, A., Marktanner, R., Poll, L., Voelkel, T., Mohammad Hasani, M.R., et al.: Intramyocardial implantation of CD133+ stem cells improved cardiac function without bypass surgery. Heart Surg. Forum **10**(1), E66–E69 (2007)
65. Erbs, S., Linke, A., Adams, V., Lenk, K., Thiele, H., Diederich, K.W., et al.: Transplantation of blood-derived progenitor cells after recanalization of chronic coronary artery occlusion: first randomized and placebo-controlled study. Circ. Res. **97**(8), 756–762 (2005)
66. Strauer, B.E., Brehm, M., Zeus, T., Bartsch, T., Schannwell, C., Antke, C., et al.: Regeneration of human infarcted heart muscle by intracoronary autologous bone marrow cell transplantation in chronic coronary artery disease: the IACT Study. J. Am. Coll. Cardiol. **46**(9), 1651–1658 (2005)
67. Shafy, A., Lavergne, T., Latremouille, C., Cortes-Morichetti, M., Carpentier, A., Chachques, J.C.: Association of electrostimulation with cell transplantation in ischemic heart disease. J. Thorac. Cardiovasc. Surg. **138**(4), 994–1001 (2009)
68. Sodian, R., Schaefermeier, P., Abegg-Zips, S., Kuebler, W.M., Shakibaei, M., Daebritz, S., et al.: Use of human umbilical cord blood-derived progenitor cells for tissue-engineered heart valves. Ann. Thorac. Surg. **89**(3), 819–828 (2010)

Tissue Engineered Myocardium

Wolfram-Hubertus Zimmermann

Abstract Myocardial tissue engineering is equally attractive for basic and translational cardiovascular research as it may ultimately provide "realistic" in vitro heart muscle models and therapeutic myocardial substitutes. A prerequisite for successful cardiac muscle engineering is simulation of natural cardiomyogenesis in vitro to yield true myocardial structures with appropriate macro- and micro-morphology as well as function. This requires an assembly of the various cellular and extracellular components of the living heart under so called biomimetic culture conditions. This chapter will give an introduction into different tissue engineering modalities and discuss essential cellular and extracellular components as well as other biomimetic factors, controlling myocardial assembly in vitro. Finally, potential in vitro and in vivo applications such as modeling of heart muscle development, applications in functional genomics and disease modeling, drug development and safety assessment as well as cardiac repair will be reviewed.

1 Introduction

Tissue engineering has been identified as a key technology for the future with a chance to have an important impact on economy and health in the twenty first century [1, 2]. This prediction is based on the anticipation that tissue engineered products will reach the clinic as biological substitutes or find a broad application in the pharmaceutical industry, mainly as test-beds for substance development and

W.-H. Zimmermann (✉)
Department of Pharmacology, Heart Research Center Göttingen, Georg-August University Göttingen, Robert-Koch-Str. 40, 37075 Göttingen, Germany
e-mail: w.zimmermann@med.uni-goettingen.de

safety assessment. In cardiac muscle engineering, several biological and regulatory issues remain to be addressed on the way to the envisioned applications. These include: (1) the utilization of human cells with an unrestricted capacity to generate myocardium; (2) development of tissue engineering modalities which support not only muscle formation but also organotypic maturation; (3) generation of tissue with electrophysiological and contractile properties of native myocardium to enable drug testing and development in a realistic—but simplified—heart-like test bed; (4) scalability and risk free applicability in case of a clinical exploitation. The first paragraphs of this chapter summarize different in vitro macro-tissue engineering formats; interested readers are referred to more comprehensive reviews for a detailed discussion of the specifics on the mentioned tissue engineering modalities [3–6]. This chapter will then focus on essential general concepts that must be considered in all cardiac muscle engineering approaches and finally speculate on potential in vitro and in vivo applications of tissue engineered myocardium.

2 Evolution of Cardiac Muscle Engineering

First reports on functional cardiomyocyte cultures stem from the beginning of the twentieth century [7]. Subsequently, the propensity of myocytes to form spherical re-aggregates in rotating high density cultures [8] and the utility of the resulting "mini hearts" in for example assessment of cardiac toxicology have been identified [9]. Cellular re-aggregation is until today utilized to engineer myocardial micro-tissue [10] and likely the underlying mechanism of tissue formation also in macro-tissue engineering formats [11]. In macro-tissue engineering tissue size and geometry are normally defined by scaffold material or casting molds, which serve either as a predefined growth substrate or concentrate cells in a defined space to facilitate self-organization into multi-cellular aggregates. Alternative tissue engineering modalities employ stacks of cell monolayers [12] or re-seeding of decellularized cadaveric organs [13]. Recently, in vivo cardiac tissue engineering has been established as an alternative method to generate vascularized cardiac muscle [14, 15]. Importantly, either of these approaches yields three-dimensional tissue equivalents and it appears fair to state that, despite its slow start approximately 50 years ago [8], enormous advances especially during the past 15 years have led to the development of enabling technologies that will likely provide true myocardial models and substitutes in the foreseeable future. The following paragraphs provide an overview of the conceptually different approaches in macro-tissue engineering (Fig. 1):

2.1 Engineering of Myocardium

Tissue engineering, as a technology to facilitate three-dimensional cell growth on chemically engineered matrices, was initially proposed by Langer and Vacanti [16].

Tissue Engineered Myocardium

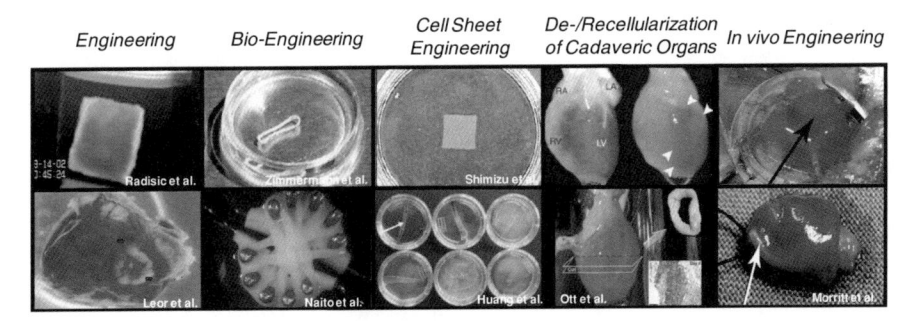

Fig. 1 Overview of conceptually different cardiac muscle engineering approaches (adapted from [102] and [15])

This particular engineering approach has been adopted by several groups to generate myocardial tissue constructs [17–25]. Most widely used biologically degradable synthetic or natural materials include polylactic acid (PLA), polyglycolic acid (PGA), polyglycerol-sebacate (PGS; also known as biorubber), alginate from seaweed, and collagen (for example as Gelfoam®). The unique feature of the "classical" engineering approach is the use of spatially and structurally (pre-) defined scaffolds as growth substrates. The challenge is to identify scaffold materials, compositions, and design algorithms that instruct cardiomyocytes to assemble structurally and functionally mature syncytia.

2.2 Bio-Engineering of Myocardium

Bio-engineering of myocardium can be distinguished from the classical engineering approach as it primarily takes advantage of the propensity of immature cardiomyocytes to re-aggregate [9]. Matrix composition and design are of subordinate importance. The re-aggregation process can, however, be spatially restricted and physically controlled by suspending cells for example in hanging drop cultures [10] or by concentrating cells in unstructured hydrogels [26]. Here collagen, matrigel, fibrin or mixtures thereof [27–30] have been widely utilized. The unique feature of the bio-engineering technology is that, and this is in clear contrast to the above mentioned engineering approach, the employed matrix does not serve as cell attachment substrate, but simply entraps cells and thereby supports the inherent propensity of dispersed cells to self-assemble into multi-cellular aggregates. Accordingly, the challenge of the bio-engineering approach is to control tissue size and geometry as well as to identify exogenous stimuli that may be exploited to facilitate myocardial tissue formation.

2.3 Cell Sheets Approach

An essentially scaffold free approach in myocardial tissue engineering employs temperature-controlled release of cell monolayers from culture dishes for subsequent stacking into three-dimensional tissue "sandwiches" [12]. Alternatively, spontaneous detachment of cell-dense monolayers and rolling of cardiomyocyte monolayers into cylindrical tissues has been attempted ([31, 32] and own unpublished data). The unique strength of the cell sheet approach is that cell layers are not enzymatically dispersed during the tissue construction process and thus maintain basal lamina proteins and extracellular domains of transmembraneous proteins. This may in fact explain the rapid reconstitution of electrical communications between adjacent monolayers through already established junctional protein complexes [33]. Bottlenecks of the cell sheet engineering approach appear to be low tensile strength and structural heterogeneity of cell monolayers.

2.4 Repopulation of Decellularized Hearts

Decellularization has been applied widely in heart valve and vessel engineering [34]. Only recently, decellularization of whole organs, including the heart, became feasible [13]. A unique feature of this technology is the possibility to generate human organ-sized extracellular matrix structures for example by decellularization of pig hearts. Moreover, depletion of cells is expected to reduce immunogenicity of the "biological" scaffold. After the process of decellularization, recellularization has to be performed to re-muscularize the extracellular matrix skeleton of the heart. One of the associated key challenges is to identify cells capable of reconstituting the complex cellular tissue architecture of the heart, including muscle, conduction system and vasculature. First experiments demonstrate that partial remuscularization of the matrix and reendothelialization of denuded vasculature within the myocardial matrix skeleton are possible [13].

2.5 In Vivo Tissue Engineering

Establishing oxygen and nutrient supply through adequate vasculature is a key challenge in cardiac muscle engineering. Despite the presence of capillaries in different tissue engineering formats [24, 26], it is likely that more intense vascularization [35] would be necessary to generate and maintain thick myocardial structures. To circumvent the difficulties associated with in vitro vascularization, the concept of in vivo tissue engineering has been developed [14, 15]. Here the natural capacity of vessel in-growth into tissue implants is utilized either in preformed chambers or sequentially stacked cell sheets. As an alternative approach, intramyocardial injection of cell-free or cell-containing hydrogels has been suggested [36–38]. This approach does not particularly aim at re-muscularization of diseased myocardium, but attempts to stabilize scar tissue, prevent further dilation, and increase thickness of the ventricular wall by either increasing extracellular matrix or non-myocyte (i.e. myofibroblast) infiltration. Any thickening of the ventricular wall will ultimately reduce ventricular wall tension according to La Place's law (tension = pressure x radius/wall thickness) and thereby stress and oxygen consumption.

3 Cardiac Muscle Engineering Essentials

Cardiomyocyte re-aggregation and formation of structurally as well as functionally anisotropic myocardium is the desired goal. To successfully engineer proper heart muscle, a deep understanding of cardiac developmental biology as well as cell physiology of the heart is fundamentally important and following issues need particular attention: (1) tissue engineering of myocardium does not only rely on the re-aggregation of cardiomyocytes, but depends crucially on the contribution of non-myocytes as they appear to control cardiogenesis and cardiac homeostasis not only in vivo, but also in vitro; (2) extracellular matrix in tissue engineered myocardium should ideally contain all components and exhibit functionality of the extracellular matrix in native myocardium to offer structural support, but also instructive stimuli (e.g. through growth factor reservoirs and relay of physical stimuli) to facilitate proper growth; (3) engineered myocardium has to be subjected to physiological physical stimuli, including mechanical load and electrical activity not only to maintain functionality and structure, but likely also as a prerequisite for organotypic assembly of heart cells into a functional syncytium in vitro.

Taken together, the provision of proper cellular, extracellular, and physical micro-milieus in addition to sufficient oxygen and nutrient supply is essential to establish a biomimetic culture environment [39] and thereby facilitate myocardial tissue assembly in vitro (Fig. 2). The following paragraphs discuss these essential components, which are equally important to all myocardial tissue engineering concepts:

3.1 Biomimetic Cultures—Cells and Cell Composition

The heart is, as any other organ, composed of multiple cell types with distinct functions. Cardiomyocytes make up approximately 70% of the total heart mass, but account for only 30% of the total cell number [40, 41]. Fibroblasts are the most abundant cell type within the non-myocyte fraction [42, 43]. Smooth muscle cells, endothelial cells, macrophages, and neurons are also present, but collectively make up only a small fraction of the cell number in the heart (10–20%). In addition,

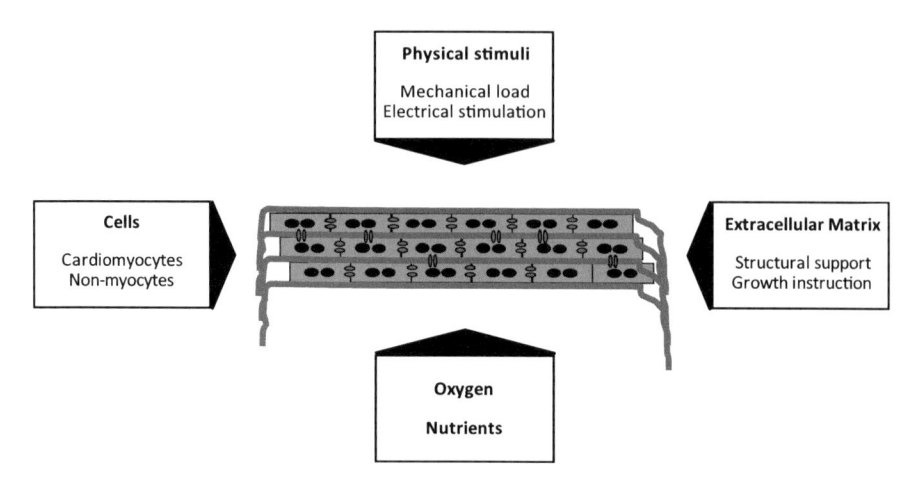

Fig. 2 Summary of factors controlling the in vitro assembly of heart cells into a functional syncytium

it may be important to consider that alterations in cell composition have been observed during development and that cell composition appears to vary in different species, with higher fibroblast numbers in larger animals [43].

The beneficial role of non-myocytes in cardiac muscle engineering was observed earlier [11] and recently identified to be likely fibroblast-related [42, 44]. In agreement with these findings, the role of fibroblasts for in vivo cardiogenesis has recently been established [45–47]. While myocyte proliferation can be influenced by fibroblasts in the embryo, it is likely that their role differs in postnatal hearts. Here cardiac fibroblasts are the main regulators of extracellular matrix synthesis and may in addition elicit paracrine actions on surrounding cells, a mechanism that may also physiologically contribute to hypertrophic growth of cardiomyocytes [45].

Other cell types, which are in absolute numbers not as prominent as fibroblasts and cardiomyocytes, include endothelial cells, smooth muscle cells, leukocytes, and neurons. Endothelial cells and smooth muscle cells are necessary for vessel formation in vivo. Tissue engineered myocardium can be supplemented with either of these cell types and appears to benefit at least structurally, e.g. through formation of primitive capillaries in vitro or better vascularization after implantation in vivo [24, 26, 42, 44, 48]. Leukocytes, and especially macrophages, may also play a similar role in capillarization of tissue engineered myocardium as observed in vivo, i.e. through "drilling holes into tissue" which are subsequently lined with endothelial cells [49, 50]. In addition, macrophages may fulfill an important cell debris clearing function in tissue engineered myocardium [26]. Largely underestimated is the contribution of neurons to regular myocardial performance and development. Here it is important to note that the heart is extensively innervated and that especially sympathetic neuron dysfunction may on the one hand facilitate arrhythmias and could on the other hand have an impact on myocardial growth [51, 52].

Whether addition of putatively important cell-derived factors to the culture medium is sufficient to substitute for local paracrine activity is questionable, as accumulating evidence points to a high diversity of signal quality and quantity in defined extracellular micro-milieus or cardiogenic niches within the heart.

Taken together, myocytes are naturally the key for the generation of functional myocardium as they provide the appropriate contractile machinery. Yet, there is clear evidence that myocytes alone cannot develop into myocardium and that the process of cell assembly into tissue is essentially supported by non-myoctyes, and in particular fibroblasts (own unpublished data). It is presently rather poorly defined how non-myocytes elicit their cardiomyogenic effects. Possible mechanisms include the provision of unique growth-controlling extracellular micro-milieus, composed of defined sets of paracrine factors and extracellular matrix.

3.2 Biomimetic Cultures—Extracellular Matrix

The extracellular matrix of the heart can be grossly categorized in interstitial and perivascular matrix as well as the basal membrane. These components have several unique functions: one is to maintain organ architecture others are to provide a proper growth milieu or growth factor reservoir and relay mechanical signals. Alterations in matrix composition or abundance can lead to disease, but are also necessary to enable adaptation to physiological growth and mechanical load [53]. The importance of proper cell suspension within the extracellular matrix skeleton can be readily observed in cell culture, here isolated non-attached cells regularly undergo anoikis [54].

Collagen type I and III are the most abundant components of the cardiac extracellular matrix and make up most of the interstitial and perivascular matrix. Collagen type V, laminin, nidogen, and proteoglycans are key constituents of the basal membrane, which anchors stationary cells within the interstitial extracellular matrix. Additional components of the extracellular matrix include fibronectin, periostin, osteopontin, thrombospondin, and tenescin-c. In addition to its structural role and function as local growth factor reservoir [55], extracellular matrix relays physical stimuli and "injects" them into cells via specific integrins [56, 57]. Considering these properties, it appears important to simulate the multi-facetted properties of natural extracellular matrix in cardiac muscle engineering. One attempt is to create scaffold materials with all biological and structural properties of natural extracellular matrix, e.g. by modeling extracellular matrix compliance and integration of specific integrin-binding motifs, such as arginine-glycine-aspartate [RGD]-peptides, or by binding growth factors to matrix materials in order to create an artificial reservoir of paracrine factors [58]. An alternative approach is to stimulate cells in tissue engineered myocardium to produce extracellular matrix. In this context, utilization of fibroblasts in cardiac muscle engineering appears crucial. The role of extracellular matrix is further exemplified by the finding that embryonic stem cells with deficient integrin signaling demonstrate a retardation of cardiogenesis [59].

Given the biological complexity of natural extracellular matrix it appears reasonable to argue that chemical engineering its diverse and at the same time unique properties would be an insurmountable challenge. Thus utilizing cells to provide matrix may ultimately be more realistic, but the challenge will be to stimulate the natural process of extracellular matrix formation in vitro. This will on the one hand require extracellular matrix-producing cells, but likely also additional growth factor/cytokine support to trigger the extracellular matrix formation process. Complete "endogenous" extracellular matrix production, while employing autologous cells to engineer patient-specific tissue, would in fact be favorable to avoid immunological complication which may arise from the use of cells and extracellular matrix from allogenic or even xenogenic sources.

3.3 Biomimetic Culture—Physical Stimuli

Hypertrophic growth of the heart is mainly a result of adaptation to physical stimuli, i.e. for example load [60–62]. Lack of hemodynamic load will lead to atrophy, while increased load causes hypertrophy [63]. In addition, differences in beating frequency may have an impact on sarcomere constitution, such as myosin heavy chain isoform composition and thus force development [64]. Providing developmental stage specific physical stimuli is likely important to induce organotypic maturation and tissue formation.

Available experimental data clearly supports the role of loading and electrical stimulation in myocardial tissue engineering [23, 28, 65]. Interestingly, loading and electrical stimulation appear to elicit similar effects on cardiomyocyte morphology [23, 26]. Preliminary data suggests, however, that a combination of mechanical and electrical stimulation can yield engineered heart tissue with superior contractile performance (own unpublished data).

Timing as well as quality and quantity of external stimuli have to be properly controlled to achieve optimal results. In engineered heart tissue cultures for example, external load can only be established once tissues have properly solidified, i.e. after 1–7 culture days depending on the non-myocyte content. Premature overloading will lead to tissue rupture. Ideally, load has to be adapted to intrinsic beating properties for example by subjecting engineered myocardial tissues to flexible load to facilitate auxotonic contractions [66].

Endogenous beating frequency of tissue engineered myocardium depends on the input cells (enriched ventricular cell population vs. mixture of ventricular and atrial cells including pacemaker cells) and species (intrinsic beating frequency: human < rat < mouse). External stimulation has to be adjusted accordingly to avoid over-stimulation and subsequent tissue failure. Electrical stimulation using external electrodes is complicated by electrolysis-related culture medium acidification which needs to be controlled by frequent medium exchanges or adequate buffers to achieve optimal results.

Additional essential physical/chemical factors include oxygenation and supply with nutrients in the absence of functional vascularization. Accordingly, utilization of scaffold materials with microchannels and perfusion bioreactor systems have been suggested [67, 68]. Despite the fact that matrix material may impose an oxygen diffusion barrier in cardiac muscle engineering, little experimental data is available on its extent and relevance in hydrogel-based bio-engineering, cell sheet engineering, and organ recellularization. Here it is also important to identify an appropriate technique to assess and quantify potentially harmful hypoxia. Cell distribution within tissue engineered myocardium appears to be a good indicator of oxygen and nutrient supply. Detection of alterations in hypoxia-inducible factor 1 alpha protein and prolyl-hydroxylase 2 and 3 mRNA would be even more sensitive to identify acute and chronic hypoxia, respectively [69]. In scaffold-based bio-engineering with preformed PGS- or collagen-sponges clear evidence for hypoxia related "core necrosis" has been provided; and approaches to improve oxygen allocation by means of medium supplementation with oxygen carriers, such as perfluorocarbon, or perfusion through micro-channels have been successfully applied [68, 70, 71]. In cell sheet engineering, core necrosis develops if more than 3 cell sheets (diameter >80 μm) are stacked [14]. In contrast, engineered heart tissue preparations from neonatal rat cardiomyocytes (diameter ~1 mm) do not display core necrosis or other evidence for apparent hypoxia ([26]; Tiburcy et al., own unpublished data). This surprising finding may be explained by the particular morphology of engineered heart tissue, which is the formation of a network of thin muscle bundles with variable diameter within relatively loose extracellular matrix. Here individual muscle bundles may have an impressive volume, but appear to not extend over 100 μm at least in one dimension. This unique "in vitro anatomy" suggests that (1) the extracellular matrix of engineered heart tissue does not constitute a major obstacle for adequate oxygenation as well as nutrition and (2) individual muscle strands in engineered heart tissue are capable of forming energetically optimal dimensions, enabling proper diffusion of oxygen into the core of individual muscle strands within a more complex muscle network. This unique morphology may also offer an explanation for the survival of engineered heart muscle grafts in the absence of immediate blood perfusion [66, 72].

4 In Vitro Applications

The anticipated in vitro utility of tissue engineered myocardium is far reaching and includes applications in developmental biology, functional genomics, disease modeling, and drug development. A potential advantage of tissue engineered myocardium in any of these applications is the possibility to define cell and matrix composition as well as grow and study tissue samples under well controlled in vitro conditions. Morphological and functional resemblance with native myocardium

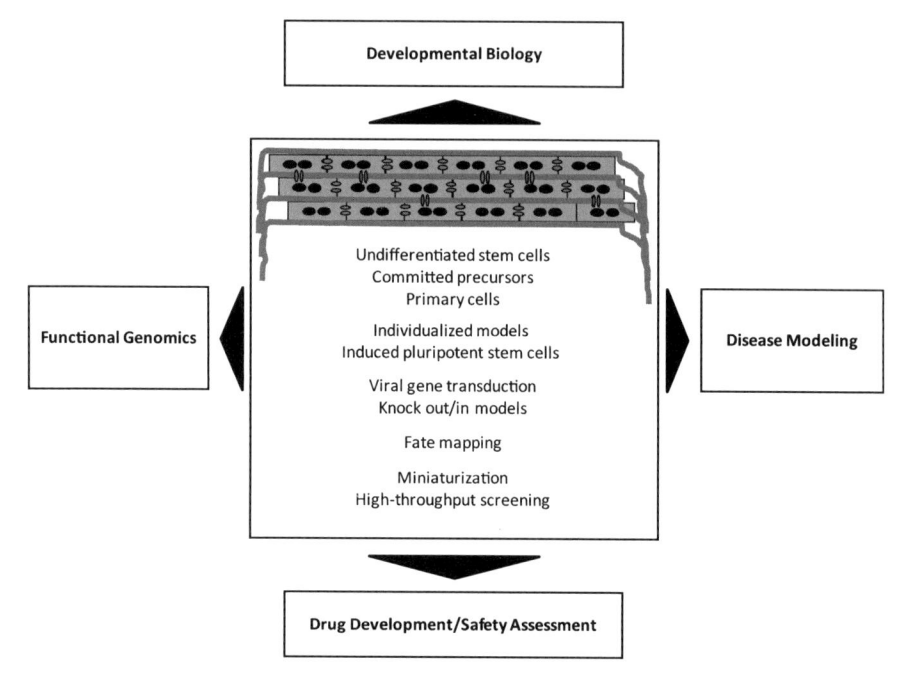

Fig. 3 Overview of potential in vitro applications of tissue engineered myocardium

is a prerequisite. Depending on the desired application primary heart cells or stem cell-derived cardiomyocytes and non-myocytes or mixtures thereof may be used (Fig. 3):

4.1 Heart Muscle Development

Cardiac specification and cardiac maturation are biologically distinguishable phases in heart development. Deciphering the molecular programs that govern cardiac specification and maturation would not only broaden our understanding of cardiac developmental biology, but may also lead to new insight into disease mechanisms. Tissue engineered myocardium may be used in this context as a well controllable model of in vitro heart development as it recapitulates many aspects of natural cardiogenesis, such as formation of functional tissue from single cells.

Cardiac specification encompasses the early steps in embryonic development from undifferentiated or mesodermally (pre-)committed stem cells into distinct heart muscle cell populations [73]. Subsequently, maturation of individual cells within a functional syncytium is of key importance to generate "heart muscle" with properties of native myocardium. Most of the so far published myocardial tissue engineering studies investigated myocardial maturation as they utilized either primary [12, 13, 17–20, 22, 27, 28, 31] or stem cell-derived [24, 74] cardiomyocytes.

Cardiac maturation is associated with terminal differentiation (i.e. withdrawal from the cell cycle) and development of the characteristic myocardial cytoskeleton (i.e. registered sarcomeres in rod-shaped myocytes with one or two centrally localized nuclei). Studies of cardiac specification are based on the assumption that cardiogenic niches exist and that these can be engineered in vitro [75]. Potential components of cardiogenic niches encompass paracrine and matrix factors [76–78] as well as physical stimuli [79, 80]. Identification of niche factors and subsequent validation of their functionality is expected to be facilitated by the use of uncommitted stem cells with genetic markers, enabling for example cell fate tracking by optical imaging technologies [81–83]. Disadvantages due to the non-physiologic nature of tissue engineered myocardium may be outweighed by full experimental control of cell and matrix composition as well as culture conditions, making tissue engineered myocardium an attractive supplementary tool to in vivo models.

4.2 Functional Genomics/Disease Modeling

After the successful decryption of the genetic code it is essential to make use of the available information and validate the function of specific gene product in reliable models and establish links to particular disease conditions. In vivo mouse models are clearly the state-of-the-art technology in functional genomics and there are several examples where human mutations and associated diseases could be simulated in genetically modified mice [84, 85]. It is, however, unrealistic to effectively screen all genes of interest in a reasonable time in animal models. In addition, the complexity of experimental animals may complicate the interpretation of the resulting data as they have to be viewed in the context of potential systemic confounders and sometimes even strain differences.

As an alternative simple, yet more complex model than monolayer cultures tissue engineered myocardium can be employed by making use of readily available gene transduction or novel stem cell technologies. The feasibility of adeno- and lentivirus-mediated gene transduction in tissue engineered myocardium has been previously established [86, 87]. Alternatively, genetically manipulated stem cells or induced pluripotent stem cells from individuals with defined genetic defects [88] may also be used as starting material for tissue engineered myocardium and disease modeling. Respective human heart disease models may ultimately not only serve as in vitro systems to define disease-underlying molecular mechanisms, but potentially also as in vitro test-beds for perspective "gene therapy" approaches.

4.3 Drug Development and Safety Assessment

The majority of experimental drugs fail during preclinical and clinical development due to organ toxicity. Liver and cardiac toxicity are the most common obstacles. Cardiac toxicity presents mainly as arrhythmias (as for example in response to anti-histamine administration; [89]) or structural myocardial damage (as for example in response to the administration of tyrosine kinase inhibitors; [90]). Detection of relevant cardiac toxicity during pre-clinical stages of drug development would help to markedly reduce costs in industrial research and development programs and eventually improve safety of clinical trials.

Tissue engineered myocardium may turn out as an attractive screening platform to identify drug toxicity [4]. To be applicable in this context, it is important to provide tissue that closely resembles natural myocardium in terms of function, structure, and maturity. In addition, tissue engineered myocardium should be generated from human cells [91]. The availability of human pluripotent stem cells [92, 93] and the possibility to establish cardiac derivatives from the latter [94, 95] were pivotal to advance myocardial tissue engineering technologies [24]. Key to a successful application in drug development is, however, demonstration of reliability and predictability of tissue engineered myocardium in drug testing and risk assessment as well as scalability to enable high through-put screening applications [96]. Moreover, tissue engineered myocardium will have to demonstrate its superiority to already established human embryonic stem cell-based cell assays [97, 98]. An exciting prospect is to use patient specific induced pluripotent stem cells to test drugs within the context of defined genetic predispositions, such as for example long QT-syndrome.

5 In Vivo Applications

Repairing a failing heart with biological substitutes is one of the penultimate dreams in regenerative medicine and it appears as if tissue engineered myocardium has the potential to become useful in this area. Biological repair would address a so far unmet need for reconstructive therapy especially in disease conditions which cannot be treated successfully with classical pharmacological agents. The heart with its at best minimal endogenous repair activity [99] and the lack of adequate organ donors [100] provide the rational for therapeutic tissue engineering in patients with heart failure. The alternative in severe heart failure is implantation of left ventricular assist devices (LVADs) either as a bridge to transplantation or destination therapy [101]. Here a median survival of 408 days post LVAD implantation, in contrast to a median survival of 150 days in the medical-therapy group, has been reported (data from the REMATCH-trial; [101]).

Clinical scenarios in which tissue engineered myocardium could be helpful include end-stage heart failure as a consequence of a severe myocardial infarction and substantial scarring [66] and congenital malformation with the necessity for reconstructive surgery or myocardial augmentation [102]. In pediatric applications, growth potential of tissue engineered myocardium would be an important prerequisite as it would potentially circumvent the necessity for repeated surgeries during childhood and adolescence. This would naturally not only reduce morbidity, but also mortality and procedural costs. While anatomically defined defects may enable precise replacement of dysfunctional myocardium, it appears more complex to address the needs in global heart failure. Regardless of the disease scenario, potential benefits of a therapeutic application of tissue engineered myocardium have to compare favorably to foreseeable risks, which include transplant failure with potentially dramatic consequences, arrhythmia induction, and in case of stem cell-based tissue engineering tumor formation.

Costs are often used as an argument against the clinical application of tissue engineered myocardium. Here one has to consider that the ultimate goal in cardiac tissue engineering is to provide immunologically compatible substitutes with the potential to seamlessly integrate into the recipient myocardium. This ideal scenario would require minimal follow-up costs, which can make up a significant share in patients with heart transplantation or LVAD implantation. The average costs of an LVAD insertion has for example been calculated at $200.000 with additional costs of $200.000 per annum in follow-up costs [103]. If tissue engineering is to be a successful alternative, it has to not only elicit a significant and sustained therapeutic effect but also be cheaper than presently available treatment modalities.

All envisioned therapeutic applications of tissue engineered myocardium eventually require the application of vast amounts of immunologically compatible cells [104]. Allocating sufficient cell quantities to tissue engineering appears to be only feasible through the exploitation of stem cell technologies [105]. Whether patient-specific, for example through reprogramming of patient-specific somatic cells [93], or immune-matched cells [106] can eventually find an application remains to be determined. The following paragraphs will discuss the feasibility of therapeutic tissue engineering in different disease conditions (Fig. 4):

5.1 Repair Post Myocardial Infarction

Myocardial infarction is associated with substantial irreversible loss of cardiomyocytes [104]. Endogenous repair mechanisms are insufficient to functionally replace the affected tissue [99]. Instead of re-muscularization, scar formation serves as biological repair mechanism and stiffens the defective myocardium. This ultimately prevents myocardial rupture, but also disrupts the functional integrity of the myocardium and leads to increased wall stress and initially compensatory hypertrophy of the remaining myocardium. Chronic overload will eventually potentiate myocardial damage and cardiomyocyte death. In case of localized defects, aneurysm formation with hemodynamic consequences may occur. Here surgical repair by aneurysm dissection is an option to improve myocardial performance. It will, however, not result in proper re-muscularization of the heart. In this scenario, integration of biological myocardial substitutes instead of for example non-vital Dacron patches [107] appears highly attractive. Available tissue

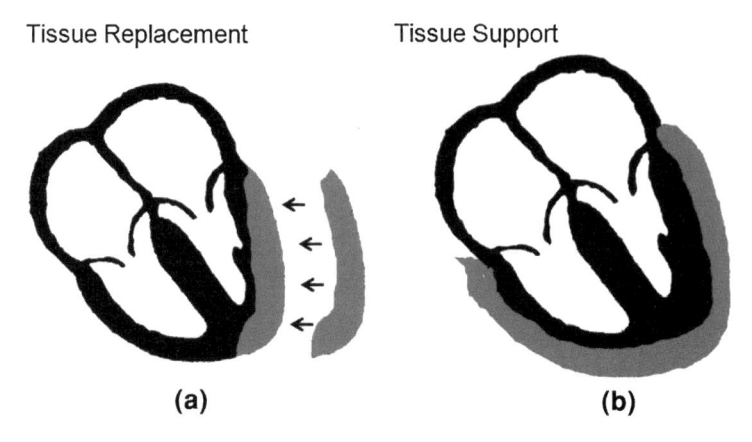

Fig. 4 Potential therapeutic application of tissue engineered myocardium in (**a**) tissue reconstruction (e.g. replacement of scarred myocardium or augmentation of hypoplastic hearts) and (**b**) tissue support (e.g. in dilated cardiomyopathy)

engineering technologies may already today provide the needed myocardial substitutes. Proof-of-concept for tissue engineering-based cardiac re-muscularization is readily available [66], but awaits further confirmation in large animal models. In addition, refinement of the proposed technology is necessary to enable transmural wall repair [108, 109] instead of epicardial augmentation.

5.2 Congenital Heart Disease

One in hundred newborns can be diagnosed with a congenital heart defect (CHD) [102]. The severity of CHD ranges from defects which may remain undetected throughout life (for example minor septum defects) to defects that are incompatible with postnatal life (for example complex outflow tract obstruction). Accordingly, surgical correction may either not be necessary or life saving if performed shortly after birth or even in utero. Especially in tissue reconstruction or augmentation, tissue engineered myocardium may find an application. Under ideal conditions, tissue engineered myocardium would be seamlessly integrated, cause no immune reaction, and grow with the postnatally developing heart. Another challenge will be to adjust physical properties of tissue engineered myocardium to achieve active and passive mechanical properties of native myocardium and establish blood perfusion. The lack of appropriate therapy for severe myocardial defects such as hypoplastic left heart syndrome (HLHS) and double inlet left or outlet right ventricles (DILV/DORV) may ultimately facilitate an exploitation of tissue engineered myocardium in pediatric patients.

5.3 Treatment of Congestive Heart Failure

Congestive heart failure develops classically after myocardial damage induced by myocardial infarctions, chronic hypertension, myocarditis, genetic predisposition, or drug toxicity (e.g. in patients under anthracyclines or tyrosine kinase inhibitors). It is typically associated with loss in contractile performance and dilation. Thus, tissue engineered myocardium should actively improve systolic performance and passively stabilize diastolic function. This may be achieved by implantation of heart embracing engineered myocardial pouches. These should, in contrast to the Acorn-Cor-Cap devices [110, 111], be fully biological and not cause pericardial constriction. To this end, so called "Biological Ventricular Assist Devices (BioVADs)" have been developed and first experimental data has been acquired in support of the general applicability of BioVADs in vivo [112]. Additional, studies will have to demonstrate whether or not BioVAD-implantation can indeed provide contractile support to a failing heart and at the same time prevent further dilation of the ventricles.

6 Summary and Perspective

It appears fair to state that tissue engineered myocardium with some key properties of native myocardium can be generated in vitro by different means. Each of the above mentioned technologies has unique strengths and weaknesses and it remains open whether or not and if so which approach will ultimately be most suitable for the different envisioned in vitro and in vivo applications. Most likely appear to be applications in drug development and safety assessment as well as target validation and disease modeling. However, substantial progress in the past 15 years has also opened the door for therapeutic applications, and early tissue engineering approaches are presently under clinical investigation (MAGNUM-trial; [113]). Any perspective in vitro and in vivo application of tissue engineered myocardium will require the use of human cells, which will likely be provided through stem cell technologies. To finally develop a tissue engineered product for commercial use, it appears essential to not only further advance the related science, but also optimize the manufacturing process and achieve full process control and automation; only then translation into an FDA approved therapeutic or diagnostic products will be realistic [114].

Acknowledgments WHZ is supported by the German Research Council (Zi 708/7–1, 8–1, 10–1), the Federal Ministry of Science and Education (DLR FKZ: 01GN0827 and 01GN0957), and the European Union (EU FP7 CARE-MI).

References

1. Cohen, S., Leor, J.: Rebuilding broken hearts. Biologists and engineers working together in the fledgling field of tissue engineering are within reach of one of their greatest goals: Constructing a living human heart patch. Sci. Am. **291**(5), 44–51 (2004)

2. Khademhosseini, A., Vacanti, J.P., Langer, R.: Progress in tissue engineering. Sci. Am. **300**(5), 64–71 (2009)
3. Zimmermann, W.H., Melnychenko, I., Eschenhagen, T.: Engineered heart tissue for regeneration of diseased hearts. Biomaterials **25**(9), 1639–1647 (2004)
4. Eschenhagen, T., Zimmermann, W.H.: Engineering myocardial tissue. Circ. Res. **97**(12), 1220–1231 (2005)
5. Jawad, H., Lyon, A.R., Harding, S.E., Ali, N.N., Boccaccini, A.R.: Myocardial tissue engineering. Br. Med. Bull. **87**, 31–47 (2008)
6. Zimmermann, W.H.: Remuscularizing failing hearts with tissue engineered myocardium. Antioxid. Redox Signal. **11**(8), 2011–2023 (2009)
7. Burrows, M.T.: Rhythmical activity of isolated heart muscle cells in vitro. Science **36**(916), 90–92 (1912)
8. Moscona, A.: Rotation-mediated histogenetic aggregation of dissociated cells. A quantifiable approach to cell interactions in vitro. Exp. Cell Res. **22**, 455–475 (1961)
9. Sperelakis, N.: Cultured heart cell reaggregate model for studying cardiac toxicology. Environ. Health Perspect. **26**, 243–267 (1978)
10. Kelm, J.M., Ehler, E., Nielsen, L.K., Schlatter, S., Perriard, J.C., Fussenegger, M.: Design of artificial myocardial microtissues. Tissue Eng. **10**(1–2), 201–214 (2004)
11. Zimmermann, W.H., Eschenhagen, T.: Cardiac tissue engineering for replacement therapy. Heart Fail Rev. **8**(3), 259–269 (2003)
12. Shimizu, T., Yamato, M., Isoi, Y., Akutsu, T., Setomaru, T., Abe, K., Kikuchi, A., Umezu, M., Okano, T.: Fabrication of pulsatile cardiac tissue grafts using a novel 3-dimensional cell sheet manipulation technique and temperature-responsive cell culture surfaces. Circ. Res. **90**(3), e40 (2002)
13. Ott, H.C., Matthiesen, T.S., Goh, S.K., Black, L.D., Kren, S.M., Netoff, T.I., Taylor, D.A.: Perfusion-decellularized matrix: Using nature's platform to engineer a bioartificial heart. Nat. Med. **14**(2), 213–221 (2008)
14. Shimizu, T., Sekine, H., Yang, J., Isoi, Y., Yamato, M., Kikuchi, A., Kobayashi, E., Okano, T.: Polysurgery of cell sheet grafts overcomes diffusion limits to produce thick, vascularized myocardial tissues. FASEB J. **20**(6), 708–710 (2006)
15. Morritt, A.N., Bortolotto, S.K., Dilley, R.J., Han, X., Kompa, A.R., McCombe, D., Wright, C.E., Itescu, S., Angus, J.A., Morrison, W.A.: Cardiac tissue engineering in an in vivo vascularized chamber. Circulation **115**(3), 353–360 (2007)
16. Langer, R., Vacanti, J.P.: Tissue engineering. Science **260**(5110), 920–926 (1993)
17. Bursac, N., Papadaki, M., Cohen, R.J., Schoen, F.J., Eisenberg, S.R., Carrier, R., Vunjak-Novakovic, G., Freed, L.E.: Cardiac muscle tissue engineering: toward an in vitro model for electrophysiological studies. Am. J. Physiol. **277**(2 Pt 2), H433–H444 (1999)
18. Carrier, R.L., Papadaki, M., Rupnick, M., Schoen, F.J., Bursac, N., Langer, R., Freed, L.E., Vunjak-Novakovic, G.: Cardiac tissue engineering: Cell seeding, cultivation parameters, and tissue construct characterization. Biotechnol. Bioeng. **64**(5), 580–589 (1999)
19. Leor, J., Aboulafia-Etzion, S., Dar, A., Shapiro, L., Barbash, I.M., Battler, A., Granot, Y., Cohen, S.: Bioengineered cardiac grafts: A new approach to repair the infarcted myocardium? Circulation **102**(19 Suppl 3), III56–III61 (2000)
20. Sakai, T., Li, R.K., Weisel, R.D., Mickle, D.A., Kim, E.T., Jia, Z.Q., Yau, T.M.: The fate of a tissue-engineered cardiac graft in the right ventricular outflow tract of the rat. J. Thorac. Cardiovasc. Surg. **121**(5), 932–942 (2001)
21. Kofidis, T., Akhyari, P., Boublik, J., Theodorou, P., Martin, U., Ruhparwar, A., Fischer, S., Eschenhagen, T., Kubis, H.P., Kraft, T., Leyh, R., Haverich, A.: In vitro engineering of heart muscle: Artificial myocardial tissue. J. Thorac. Cardiovasc. Surg. **124**(1), 63–69 (2002)
22. van Luyn, M.J., Tio, R.A., Gallego y van Seijen, X.J., Plantinga, J.A., de Leij, L.F., DeJongste, M.J., van Wachem, P.B.: Cardiac tissue engineering: characteristics of in unison contracting two- and three-dimensional neonatal rat ventricle cell (co)-cultures. Biomaterials **23**(24), 4793–4801 (2002)

23. Radisic, M., Park, H., Shing, H., Consi, T., Schoen, F.J., Langer, R., Freed, L.E., Vunjak-Novakovic, G.: Functional assembly of engineered myocardium by electrical stimulation of cardiac myocytes cultured on scaffolds. Proc. Natl. Acad. Sci. USA **101**(52), 18129–18134 (2004)
24. Caspi, O., Lesman, A., Basevitch, Y., Gepstein, A., Arbel, G., Habib, I.H., Gepstein, L., Levenberg, S.: Tissue engineering of vascularized cardiac muscle from human embryonic stem cells. Circ. Res. **100**(2), 263–272 (2007)
25. Engelmayr Jr., G.C., Cheng, M., Bettinger, C.J., Borenstein, J.T., Langer, R., Freed, L.E.: Accordion-like honeycombs for tissue engineering of cardiac anisotropy. Nat. Mater. **7**(12), 1003–1010 (2008)
26. Zimmermann, W.H., Schneiderbanger, K., Schubert, P., Didie, M., Munzel, F., Heubach, J.F., Kostin, S., Neuhuber, W.L., Eschenhagen, T.: Tissue engineering of a differentiated cardiac muscle construct. Circ. Res. **90**(2), 223–230 (2002)
27. Eschenhagen, T., Fink, C., Remmers, U., Scholz, H., Wattchow, J., Weil, J., Zimmermann, W., Dohmen, H.H., Schafer, H., Bishopric, N., Wakatsuki, T., Elson, E.L.: Three-dimensional reconstitution of embryonic cardiomyocytes in a collagen matrix: A new heart muscle model system. FASEB J. **11**(8), 683–694 (1997)
28. Zimmermann, W.H., Fink, C., Kralisch, D., Remmers, U., Weil, J., Eschenhagen, T.: Three-dimensional engineered heart tissue from neonatal rat cardiac myocytes. Biotechnol. Bioeng. **68**(1), 106–114 (2000)
29. Birla, R.K., Borschel, G.H., Dennis, R.G., Brown, D.L.: Myocardial engineering in vivo: Formation and characterization of contractile, vascularized three-dimensional cardiac tissue. Tissue Eng. **11**(5–6), 803–813 (2005)
30. Bakunts, K., Gillum, N., Karabekian, Z., Sarvazyan, N.: Formation of cardiac fibers in matrigel matrix. Biotechniques **44**(3), 341–348 (2008)
31. Baar, K., Birla, R., Boluyt, M.O., Borschel, G.H., Arruda, E.M., Dennis, R.G.: Self-organization of rat cardiac cells into contractile 3-d cardiac tissue. FASEB J. **19**(2), 275–277 (2005)
32. Huang, Y.C., Khait, L., Birla, R.K.: Contractile three-dimensional bioengineered heart muscle for myocardial regeneration. J. Biomed. Mater. Res. A **80**(3), 719–731 (2007)
33. Haraguchi, Y., Shimizu, T., Yamato, M., Kikuchi, A., Okano, T.: Electrical coupling of cardiomyocyte sheets occurs rapidly via functional gap junction formation. Biomaterials **27**(27), 4765–4774 (2006)
34. Schmidt, C.E., Baier, J.M.: Acellular vascular tissues: Natural biomaterials for tissue repair and tissue engineering. Biomaterials **21**(22), 2215–2231 (2000)
35. Korecky, B., Hai, C.M., Rakusan, K.: Functional capillary density in normal and transplanted rat hearts. Can. J. Physiol. Pharmacol. **60**(1), 23–32 (1982)
36. Kofidis, T., Lebl, D.R., Martinez, E.C., Hoyt, G., Tanaka, M., Robbins, R.C.: Novel injectable bioartificial tissue facilitates targeted, less invasive, large-scale tissue restoration on the beating heart after myocardial injury. Circulation **112**(9 Suppl), I173–I177 (2005)
37. Landa, N., Miller, L., Feinberg, M.S., Holbova, R., Shachar, M., Freeman, I., Cohen, S., Leor, J.: Effect of injectable alginate implant on cardiac remodeling and function after recent and old infarcts in rat. Circulation **117**(11), 1388–1396 (2008)
38. Leor, J., Tuvia, S., Guetta, V., Manczur, F., Castel, D., Willenz, U., Petnehazy, O., Landa, N., Feinberg, M.S., Konen, E., Goitein, O., Tsur-Gang, O., Shaul, M., Klapper, L., Cohen, S.: Intracoronary injection of in situ forming alginate hydrogel reverses left ventricular remodeling after myocardial infarction in swine. J. Am. Coll. Cardiol. **54**(11), 1014–1023 (2009)
39. Radisic, M., Park, H., Gerecht, S., Cannizzaro, C., Langer, R., Vunjak-Novakovic, G.: Biomimetic approach to cardiac tissue engineering. Philos. Trans. R. Soc. Lond. B Biol. Sci. **362**(1484), 1357–1368 (2007)
40. Zak, R.: Development and proliferative capacity of cardiac muscle cells. Circ. Res. **35**((2) suppl II), 17–26 (1974)

41. Nag, A.C., Zak, R.: Dissociation of adult mammalian heart into single cell suspension: An ultrastructural study. J. Anat. **129**(Pt 3), 541–559 (1979)
42. Naito, H., Melnychenko, I., Didie, M., Schneiderbanger, K., Schubert, P., Rosenkranz, S., Eschenhagen, T., Zimmermann, W.H.: Optimizing engineered heart tissue for therapeutic applications as surrogate heart muscle. Circulation **114**(1 Suppl), I72–I78 (2006)
43. Banerjee, I., Fuseler, J.W., Price, R.L., Borg, T.K., Baudino, T.A.: Determination of cell types and numbers during cardiac development in the neonatal and adult rat and mouse. Am. J. Physiol. Heart Circ. Physiol. **293**(3), H1883–H1891 (2007)
44. Radisic, M., Park, H., Martens, T.P., Salazar-Lazaro, J.E., Geng, W., Wang, Y., Langer, R., Freed, L.E., Vunjak-Novakovic, G.: Pre-treatment of synthetic elastomeric scaffolds by cardiac fibroblasts improves engineered heart tissue. J. Biomed. Mater. Res. A **86**(3), 713–724 (2008)
45. Kakkar, R., Lee, R.T.: Intramyocardial fibroblast myocyte communication. Circ. Res. **106**(1), 47–57 (2010)
46. Ieda, M., Tsuchihashi, T., Ivey, K.N., Ross, R.S., Hong, T.T., Shaw, R.M., Srivastava, D.: Cardiac fibroblasts regulate myocardial proliferation through beta1 integrin signaling. Dev. Cell. **16**(2), 233–244 (2009)
47. Souders, C.A., Bowers, S.L., Baudino, T.A.: Cardiac fibroblast: the renaissance cell. Circ. Res. **105**(12), 1164–1176 (2009)
48. Sekine, H., Shimizu, T., Hobo, K., Sekiya, S., Yang, J., Yamato, M., Kurosawa, H., Kobayashi, E., Okano, T.: Endothelial cell coculture within tissue-engineered cardiomyocyte sheets enhances neovascularization and improves cardiac function of ischemic hearts. Circulation **118**(14 Suppl), S145–S152 (2008)
49. Moldovan, N.I., Goldschmidt-Clermont, P.J., Parker-Thornburg, J., Shapiro, S.D., Kolattukudy, P.E.: Contribution of monocytes/macrophages to compensatory neovascularization: The drilling of metalloelastase-positive tunnels in ischemic myocardium. Circ. Res. **87**(5), 378–384 (2000)
50. Leor, J., Rozen, L., Zuloff-Shani, A., Feinberg, M.S., Amsalem, Y., Barbash, I.M., Kachel, E., Holbova, R., Mardor, Y., Daniels, D., Ocherashvilli, A., Orenstein, A., Danon, D.: Ex vivo activated human macrophages improve healing, remodeling, and function of the infarcted heart. Circulation **114**(1 Suppl), I94–I100 (2006)
51. Ieda, M., Kanazawa, H., Kimura, K., Hattori, F., Ieda, Y., Taniguchi, M., Lee, J.K., Matsumura, K., Tomita, Y., Miyoshi, S., Shimoda, K., Makino, S., Sano, M., Kodama, I., Ogawa, S., Fukuda, K.: Sema3a maintains normal heart rhythm through sympathetic innervation patterning. Nat. Med. **13**(5), 604–612 (2007)
52. Ieda, M., Fukuda, K.: New aspects for the treatment of cardiac diseases based on the diversity of functional controls on cardiac muscles: The regulatory mechanisms of cardiac innervation and their critical roles in cardiac performance. J. Pharmacol. Sci. **109**(3), 348–353 (2009)
53. Bowers, S.L., Banerjee, I., Baudino, T.A.: The extracellular matrix: at the center of it all. J. Mol. Cell. Cardiol. **48**(3), 474–482 (2010)
54. Michel, J.B.: Anoikis in the cardiovascular system: Known and unknown extracellular mediators. Arterioscler Thromb. Vasc. Biol. **23**(12), 2146–2154 (2003)
55. Corda, S., Samuel, J.L., Rappaport, L.: Extracellular matrix and growth factors during heart growth. Heart Fail Rev. **5**(2), 119–130 (2000)
56. Barczyk, M., Carracedo, S., Gullberg, D.: Integrins. Cell Tissue Res. **339**(1), 269–280
57. Ross, R.S., Borg, T.K.: Integrins and the myocardium. Circ. Res. **88**(11), 1112–1119 (2001)
58. von der Mark, K., Park, J., Bauer, S., Schmuki, P.: Nanoscale engineering of biomimetic surfaces: Cues from the extracellular matrix. Cell Tissue Res. **339**(1), 131–153 (2010)
59. Fassler, R., Rohwedel, J., Maltsev, V., Bloch, W., Lentini, S., Guan, K., Gullberg, D., Hescheler, J., Addicks, K., Wobus, A.M.: Differentiation and integrity of cardiac muscle cells are impaired in the absence of beta 1 integrin. J. Cell Sci. **109**(Pt 13), 2989–2999 (1996)

60. Manner, J., Wessel, A., Yelbuz, T.M.: How does the tubular embryonic heart work? Looking for the physical mechanism generating unidirectional blood flow in the valveless embryonic heart tube. Dev. Dyn. **239**(4), 1035–1046 (2010)
61. Opie, L.H., Commerford, P.J., Gersh, B.J., Pfeffer, M.A.: Controversies in ventricular remodelling. Lancet **367**(9507), 356–367 (2006)
62. Keller, B.B., Liu, L.J., Tinney, J.P., Tobita, K.: Cardiovascular developmental insights from embryos. Ann. N. Y. Acad. Sci. **1101**, 377–388 (2007)
63. Depre, C., Shipley, G.L., Chen, W., Han, Q., Doenst, T., Moore, M.L., Stepkowski, S., Davies, P.J., Taegtmeyer, H.: Unloaded heart in vivo replicates fetal gene expression of cardiac hypertrophy. Nat. Med. **4**(11), 1269–1275 (1998)
64. Korte, F.S., Herron, T.J., Rovetto, M.J., McDonald, K.S.: Power output is linearly related to myhc content in rat skinned myocytes and isolated working hearts. Am. J. Physiol. Heart Circ. Physiol. **289**(2), H801–H812 (2005)
65. Fink, C., Ergun, S., Kralisch, D., Remmers, U., Weil, J., Eschenhagen, T.: Chronic stretch of engineered heart tissue induces hypertrophy and functional improvement. FASEB J. **14**(5), 669–679 (2000)
66. Zimmermann, W.H., Melnychenko, I., Wasmeier, G., Didie, M., Naito, H., Nixdorff, U., Hess, A., Budinsky, L., Brune, K., Michaelis, B., Dhein, S., Schwoerer, A., Ehmke, H., Eschenhagen, T.: Engineered heart tissue grafts improve systolic and diastolic function in infarcted rat hearts. Nat. Med. **12**(4), 452–458 (2006)
67. Radisic, M., Park, H., Chen, F., Salazar-Lazzaro, J.E., Wang, Y., Dennis, R., Langer, R., Freed, L.E., Vunjak-Novakovic, G.: Biomimetic approach to cardiac tissue engineering: Oxygen carriers and channeled scaffolds. Tissue Eng. **12**(8), 2077–2091 (2006)
68. Radisic, M., Marsano, A., Maidhof, R., Wang, Y., Vunjak-Novakovic, G.: Cardiac tissue engineering using perfusion bioreactor systems. Nat. Protoc. **3**(4), 719–738 (2008)
69. Katschinski, D.M.: In vivo functions of the prolyl-4-hydroxylase domain oxygen sensors: Direct route to the treatment of anaemia and the protection of ischaemic tissues. Acta Physiol. (Oxf) **195**(4), 407–414 (2009)
70. Radisic, M., Deen, W., Langer, R., Vunjak-Novakovic, G.: Mathematical model of oxygen distribution in engineered cardiac tissue with parallel channel array perfused with culture medium containing oxygen carriers. Am. J. Physiol. Heart Circ. Physiol. **288**(3), H1278–H1289 (2005)
71. Radisic, M., Malda, J., Epping, E., Geng, W., Langer, R., Vunjak-Novakovic, G.: Oxygen gradients correlate with cell density and cell viability in engineered cardiac tissue. Biotechnol. Bioeng. **93**(2), 332–343 (2006)
72. Zimmermann, W.H., Didie, M., Wasmeier, G.H., Nixdorff, U., Hess, A., Melnychenko, I., Boy, O., Neuhuber, W.L., Weyand, M., Eschenhagen, T.: Cardiac grafting of engineered heart tissue in syngenic rats. Circulation **106**(12 Suppl 1), I151–I157 (2002)
73. Kattman, S.J., Adler, E.D., Keller, G.M.: Specification of multipotential cardiovascular progenitor cells during embryonic stem cell differentiation and embryonic development. Trends Cardiovasc. Med. **17**(7), 240–246 (2007)
74. Guo, X.M., Zhao, Y.S., Chang, H.X., Wang CY, E.L.L., Zhang, X.A., Duan, C.M., Dong, L.Z., Jiang, H., Li, J., Song, Y., Yang, X.J.: Creation of engineered cardiac tissue in vitro from mouse embryonic stem cells. Circulation **113**(18), 2229–2237 (2006)
75. Peerani, R., Zandstra, P.W.: Enabling stem cell therapies through synthetic stem cell-niche engineering. J. Clin. Invest. **120**(1), 60–70 (2010)
76. Mummery, C., Ward-van Oostwaard, D., Doevendans, P., Spijker, R., van den Brink, S., Hassink, R., van der Heyden, M., Opthof, T., Pera, M., de la Riviere, A.B., Passier, R., Tertoolen, L.: Differentiation of human embryonic stem cells to cardiomyocytes: Role of coculture with visceral endoderm-like cells. Circulation **107**(21), 2733–2740 (2003)
77. Kattman, S.J., Huber, T.L., Keller, G.M.: Multipotent flk-1+ cardiovascular progenitor cells give rise to the cardiomyocyte, endothelial, and vascular smooth muscle lineages. Dev. Cell **11**(5), 723–732 (2006)

78. Yang, L., Soonpaa, M.H., Adler, E.D., Roepke, T.K., Kattman, S.J., Kennedy, M., Henckaerts, E., Bonham, K., Abbott, G.W., Linden, R.M., Field, L.J., Keller, G.M.: Human cardiovascular progenitor cells develop from a kdr+ embryonic-stem-cell-derived population. Nature **453**(7194), 524–528 (2008)
79. Bhana, B., Iyer, R.K., Chen, W.L., Zhao, R., Sider, K.L., Likhitpanichkul, M., Simmons, C.A., Radisic, M.: Influence of substrate stiffness on the phenotype of heart cells. Biotechnol. Bioeng. **105**(6), 1148–1160 (2010)
80. Engler, A.J., Sen, S., Sweeney, H.L., Discher, D.E.: Matrix elasticity directs stem cell lineage specification. Cell **126**(4), 677–689 (2006)
81. Song, H., Yoon, C., Kattman, S.J., Dengler, J., Masse, S., Thavaratnam, T., Gewarges, M., Nanthakumar, K., Rubart, M., Keller, G.M., Radisic, M., Zandstra, P.W.: Interrogating functional integration between injected pluripotent stem cell-derived cells and surrogate cardiac tissue. Proc. Natl. Acad. Sci. USA **107**(8), 3329–3334 (2010)
82. Moretti, A., Caron, L., Nakano, A., Lam, J.T., Bernshausen, A., Chen, Y., Qyang, Y., Bu, L., Sasaki, M., Martin-Puig, S., Sun, Y., Evans, S.M., Laugwitz, K.L., Chien, K.R.: Multipotent embryonic isl1+ progenitor cells lead to cardiac, smooth muscle, and endothelial cell diversification. Cell **127**(6), 1151–1165 (2006)
83. Domian, I.J., Chiravuri, M., van der Meer, P., Feinberg, A.W., Shi, X., Shao, Y., Wu, S.M., Parker, K.K., Chien, K.R.: Generation of functional ventricular heart muscle from mouse ventricular progenitor cells. Science **326**(5951), 426–429 (2009)
84. Knoll, R., Kostin, S., Klede, S., Savvatis, K., Klinge, L., Stehle, I., Gunkel, S., Kotter, S., Babicz, K., Sohns, M., Miocic, S., Didie, M., Knoll, G., Zimmermann, W.H., Thelen, P., Bickeboller, H., Maier, L.S., Schaper, W., Schaper, J., Kraft, T., Tschope, C., Linke, W.A., Chien, K.R.: A common mlp (muscle lim protein) variant is associated with cardiomyopathy. Circ. Res. **106**(4), 695–704 (2010)
85. Wehrens, X.H., Lehnart, S.E., Huang, F., Vest, J.A., Reiken, S.R., Mohler, P.J., Sun, J., Guatimosim, S., Song, L.S., Rosemblit, N., D'Armiento, J.M., Napolitano, C., Memmi, M., Priori, S.G., Lederer, W.J., Marks, A.R.: Fkbp12.6 deficiency and defective calcium release channel (ryanodine receptor) function linked to exercise-induced sudden cardiac death. Cell **113**(7), 829–840 (2003)
86. El-Armouche, A., Rau, T., Zolk, O., Ditz, D., Pamminger, T., Zimmermann, W.H., Jackel, E., Harding, S.E., Boknik, P., Neumann, J., Eschenhagen, T.: Evidence for protein phosphatase inhibitor-1 playing an amplifier role in beta-adrenergic signaling in cardiac myocytes. FASEB J. **17**(3), 437–439 (2003)
87. El-Armouche, A., Singh, J., Naito, H., Wittkopper, K., Didie, M., Laatsch, A., Zimmermann, W.H., Eschenhagen, T.: Adenovirus-delivered short hairpin rna targeting pkcalpha improves contractile function in reconstituted heart tissue. J. Mol. Cell Cardiol. **43**(3), 371–376 (2007)
88. Carvajal-Vergara, X., Sevilla, A., D'Souza, S.L., Ang, Y.S., Schaniel, C., Lee, D.F., Yang, L., Kaplan, A.D., Adler, E.D., Rozov, R., Ge, Y., Cohen, N., Edelmann, L.J., Chang, B., Waghray, A., Su, J., Pardo, S., Lichtenbelt, K.D., Tartaglia, M., Gelb, B.D., Lemischka, I.R.: Patient-specific induced pluripotent stem-cell-derived models of leopard syndrome. Nature **465**(7299), 808–812 (2010)
89. Finlayson, K., Witchel, H.J., McCulloch, J., Sharkey, J.: Acquired qt interval prolongation and herg: Implications for drug discovery and development. Eur. J. Pharmacol. **500**(1–3), 129–142 (2004)
90. Force, T., Krause, D.S., Van Etten, R.A.: Molecular mechanisms of cardiotoxicity of tyrosine kinase inhibition. Nat. Rev. Cancer **7**(5), 332–344 (2007)
91. Zimmermann, W.H., Kehat, I., Boy, O., Gepstein, A., Neuhuber, W.L., Gepstein, L.: Three-dimensional culture induces advanced differentiation of primary rat and human embryonic stem cell derived cardiomyocytes: Implications for cardiac tissue engineering. In: Scientific Sessions of the American Heart Association, 2003. Circulation, pp IV-243 Abtract

92. Thomson, J.A., Itskovitz-Eldor, J., Shapiro, S.S., Waknitz, M.A., Swiergiel, J.J., Marshall, V.S., Jones, J.M.: Embryonic stem cell lines derived from human blastocysts. Science **282**(5391), 1145–1147 (1998)
93. Takahashi, K., Tanabe, K., Ohnuki, M., Narita, M., Ichisaka, T., Tomoda, K., Yamanaka, S.: Induction of pluripotent stem cells from adult human fibroblasts by defined factors. Cell **131**(5), 861–872 (2007)
94. Kehat, I., Kenyagin-Karsenti, D., Snir, M., Segev, H., Amit, M., Gepstein, A., Livne, E., Binah, O., Itskovitz-Eldor, J., Gepstein, L.: Human embryonic stem cells can differentiate into myocytes with structural and functional properties of cardiomyocytes. J. Clin. Invest. **108**(3), 407–414 (2001)
95. Zwi, L., Caspi, O., Arbel, G., Huber, I., Gepstein, A., Park, I.H., Gepstein, L.: Cardiomyocyte differentiation of human induced pluripotent stem cells. Circulation **120**(15), 1513–1523 (2009)
96. Hansen, A., Eder, A., Bonstrup, M., Flato, M., Mewe, M., Schaaf, S., Aksehirlioglu, B., Schworer, A., Uebeler, J., Eschenhagen, T.: Development of a drug screening platform based on engineered heart tissue. Circ Res
97. Jonsson, M.K., Duker, G., Tropp, C., Andersson, B., Sartipy, P., Vos, M.A., van Veen, T.A.: Quantified proarrhythmic potential of selected human embryonic stem cell-derived cardiomyocytes. Stem Cell Res. **4**(3), 189–200 (2010)
98. Caspi, O., Itzhaki, I., Kehat, I., Gepstein, A., Arbel, G., Huber, I., Satin, J., Gepstein, L.: In vitro electrophysiological drug testing using human embryonic stem cell derived cardiomyocytes. Stem Cells Dev. **18**(1), 161–172 (2009)
99. Bergmann, O., Bhardwaj, R.D., Bernard, S., Zdunek, S., Barnabe-Heider, F., Walsh, S., Zupicich, J., Alkass, K., Buchholz, B.A., Druid, H., Jovinge, S., Frisen, J.: Evidence for cardiomyocyte renewal in humans. Science **324**(5923), 98–102 (2009)
100. Sharples, L.D., Cafferty, F., Demitis, N., Freeman, C., Dyer, M., Banner, N., Birks, E.J., Khaghani, A., Large, S.R., Tsui, S., Caine, N., Buxton, M.: Evaluation of the clinical effectiveness of the ventricular assist device program in the United Kingdom (evad UK). J. Heart Lung Transplant. **26**(1), 9–15 (2007)
101. Rose, E.A., Gelijns, A.C., Moskowitz, A.J., Heitjan, D.F., Stevenson, L.W., Dembitsky, W., Long, J.W., Ascheim, D.D., Tierney, A.R., Levitan, R.G., Watson, J.T., Meier, P., Ronan, N.S., Shapiro, P.A., Lazar, R.M., Miller, L.W., Gupta, L., Frazier, O.H., Desvigne-Nickens, P., Oz, M.C., Poirier, V.L.: Long-term mechanical left ventricular assistance for end-stage heart failure. N. Engl. J. Med. **345**(20), 1435–1443 (2001)
102. Zimmermann, W.H., Cesnjevar, R.: Cardiac tissue engineering: Implications for pediatric heart surgery. Pediatr. Cardiol. **30**(5), 716–723 (2009)
103. Oz, M.C., Gelijns, A.C., Miller, L., Wang, C., Nickens, P., Arons, R., Aaronson, K., Richenbacher, W., van Meter, C., Nelson, K., Weinberg, A., Watson, J., Rose, E.A., Moskowitz, A.J.: Left ventricular assist devices as permanent heart failure therapy: the price of progress. Ann. Surg. **238**(4), 577–583 (2003). discussion 583–575
104. Gepstein, L.: Derivation and potential applications of human embryonic stem cells. Circ. Res. **91**(10), 866–876 (2002)
105. Zandstra, P.W., Bauwens, C., Yin, T., Liu, Q., Schiller, H., Zweigerdt, R., Pasumarthi, K.B., Field, L.J.: Scalable production of embryonic stem cell-derived cardiomyocytes. Tissue Eng. **9**(4), 767–778 (2003)
106. Kim, K., Lerou, P., Yabuuchi, A., Lengerke, C., Ng, K., West, J., Kirby, A., Daly, M.J., Daley, G.Q.: Histocompatible embryonic stem cells by parthenogenesis. Science **315**(5811), 482–486 (2007)
107. Athanasuleas, C.L., Stanley Jr., A.W., Buckberg, G.D., Dor, V., DiDonato, M., Blackstone, E.H.: Surgical anterior ventricular endocardial restoration (saver) in the dilated remodeled ventricle after anterior myocardial infarction. Restore group. Reconstructive endoventricular surgery, returning torsion original radius elliptical shape to the lv. J. Am. Coll. Cardiol. **37**(5), 1199–1209 (2001)

108. Matsubayashi, K., Fedak, P.W., Mickle, D.A., Weisel, R.D., Ozawa, T., Li, R.K.: Improved left ventricular aneurysm repair with bioengineered vascular smooth muscle grafts. Circulation **108**(Suppl 1), II219–II225 (2003)
109. Ozawa, T., Mickle, D.A., Weisel, R.D., Matsubayashi, K., Fujii, T., Fedak, P.W., Koyama, N., Ikada, Y., Li, R.K.: Tissue-engineered grafts matured in the right ventricular outflow tract. Cell Transplant. **13**(2), 169–177 (2004)
110. Bredin, F., Franco-Cereceda, A., Midterm results of passive containment surgery using the acorn cor cap cardiac support device in dilated cardiomyopathy. J Card Surg **25**(1):107–112 (2010)
111. Walsh, R.G.: Design and features of the acorn corcap cardiac support device: The concept of passive mechanical diastolic support. Heart Fail Rev. **10**(2), 101–107 (2005)
112. Yildirim, Y., Naito, H., Didie, M., Karikkineth, B.C., Biermann, D., Eschenhagen, T., Zimmermann, W.H.: Development of a biological ventricular assist device: Preliminary data from a small animal model. Circulation **116**(11 Suppl), I16–I23 (2007)
113. Chachques, J.C., Trainini, J.C., Lago, N., Masoli, O.H., Barisani, J.L., Cortes-Morichetti, M., Schussler, O., Carpentier, A.: Myocardial assistance by grafting a new bioartificial upgraded myocardium (magnum clinical trial): one year follow-up. Cell Transplant. **16**(9), 927–934 (2007)
114. Archer, R., Williams, D.J.: Why tissue engineering needs process engineering. Nat. Biotechnol. **23**(11), 1353–1355 (2005)

Injectable Materials for Myocardial Tissue Engineering

Jennifer M. Singelyn and Karen L. Christman

Abstract Injectable materials have gained recent focus as therapeutic alternatives to treat and prevent heart failure post-myocardial infarction. These materials offer the potential to treat the damaged region of the heart through minimally invasive catheter delivery. A variety of naturally derived and inspired materials, as well as synthetic materials have been explored as potential extracellular matrix replacement scaffolds to prevent a decline in cardiac function and/or improve cell transplant survival. Most recently, decellularized matrices have been suggested, to provide a cardiac-specific biomimetic replacement. This chapter will review the variety of materials that have been explored as injectable therapies for cardiac repair, with a particular focus on decellularized matrices. Additionally, this chapter will review the injection systems currently available, and the design criteria materials must meet for compatibility with minimally invasive catheter delivery.

1 Introduction

Biomaterials have become a recent area of research for the treatment of heart failure (HF) post-myocardial infarction (MI) and other cardiovascular diseases. The use of biomaterials for cardiac repair can be divided into three areas of focus:

J. M. Singelyn and K. L. Christman (✉)
Department of Bioengineering, University of California, San Diego,
9500 Gilman Dr. MC 0412, La Jolla, CA 92093, USA
e-mail: kchristm@ucsd.edu

J. M. Singelyn
e-mail: jsingely@ucsd.edu

external wall supports, cardiac patches, and injectable materials. External wall supports, or left ventricular (LV) restraint devices, are mesh materials that serve to support the ventricles, to prevent dilation and thus preserve function. Materials such as polypropylene and polyester have been used as restraint devices [1]. Cardiac patches (in vitro tissue engineering approach) are cellular or acellular patches that are placed over the damaged region to aid in repair. A variety of materials, including alginate, collagen, and small intestinal submucosa (SIS) have been explored with or without cells for myocardial tissue engineering approaches [1–7]. These materials, placed on the epicardial surface of the myocardium allow for wall support, as well as promote neovascularization into the patch region. While LV restraints and cardiac patches have demonstrated advancement in the field of biomaterials for myocardial tissue engineering, both techniques require an invasive surgical procedure for implantation. Injectable materials offer the unique potential advantage of being delivered through minimally invasive, catheter-based approaches.

Injectable materials can be used on their own, as acellular scaffolds, or they can be delivered with cells. On their own, injectable materials offer the advantage of eliminating the complications of cell-based therapies, such as immunogenicity, culture time and the issues of cell source [8]. Initially, cell transplantation was performed by injecting cells in culture medium or saline directly into the damaged myocardium, a technique termed cellular cardiomyoplasty [9–15]. In large animal studies, as well as in pre-clinical and clinical trials, cells are injected through a catheter, allowing for minimally invasive delivery [9–11, 15, 16]. Despite numerous clinical trials, cellular transplantation techniques are limited by poor cell retention, survival, and integration with the host myocardium. These limitations are due in part to the lack of an appropriate microenvironment to allow for and guide cell development [17]. Thus, the initial focus of injectable biomaterials to increase cell survival during cellular cardiomyoplasty remains a priority today [18, 19]. With or without cells, injectable materials can be injected intramyocardially, rather than just the epicardial surface, to allow for direct repair within the region of damage. Injectable materials thus offer the potential to mitigate the negative LV remodeling process by thickening the LV free wall and reducing wall stress, thereby preventing HF (Fig. 1). As mentioned, these beneficial effects of increased cell survival and/or preserved cardiac function can be obtained through a potentially minimally invasive injection procedure, rather than an open-chest surgery. Less invasive procedures, such as injectable materials, offer a decreased risk to patients, as well as decreased hospitalization time.

A variety of injectable materials, both synthetic and naturally derived or inspired, are being investigated as injectable scaffolds, with or without cells, for myocardial tissue engineering (Fig. 2). This chapter will serve to summarize the success and limitations of each of the materials that have been explored as injectable scaffolds for myocardial tissue engineering, as well as comment on the different intramyocardial delivery methods available.

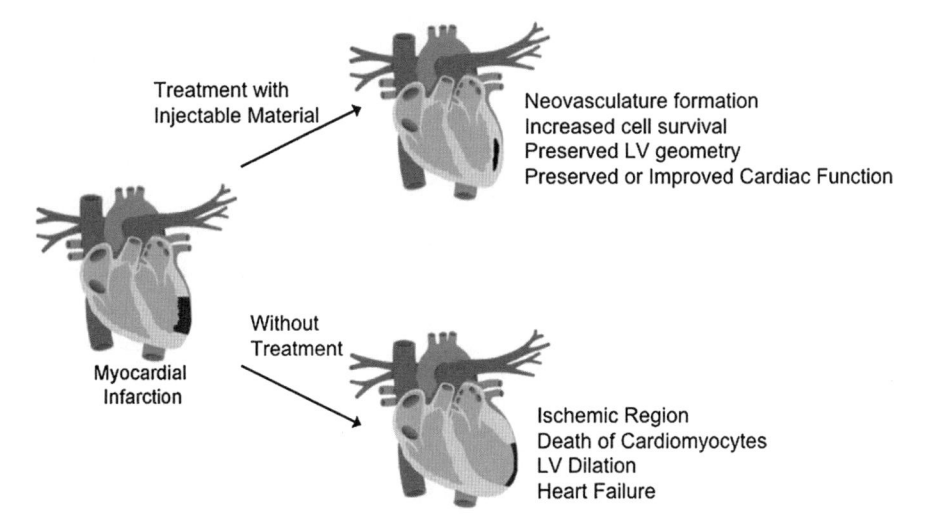

Fig. 1 Schematic of heart failure post-MI and the potential benefits of treating the heart with an injectable material

2 Natural Injectable Materials

Naturally derived or inspired materials, including proteins, peptides, and polysaccharides, are often the first types of materials to be explored for tissue engineering applications, as they provide components of the native extracellular matrix (ECM) to be replaced or repaired. The body thus recognizes these materials and is able to easily break them down with hydrolytic or proteolytic activity into degradation products which are inherently safe [20]. While there are many benefits to using natural materials, it can be difficult to guarantee purity and consistency of the material. To date, a majority of the materials explored as injectable treatment post-MI have been naturally derived or inspired materials, including collagen [21–24], fibrin [18, 25–27], Matrigel [28], self assembling peptide nanofibers [29], alginate [30, 31], chitosan [32], and decellularized matrices [33, 34] (Table 1).

2.1 Collagen

Collagen is a natural choice for tissue engineering, as it is the main protein of the ECM. Collagen has been widely explored for in vitro tissue engineering [5], as an implantable patch [4, 35], and as an injectable material for myocardial tissue engineering. In the latter, collagen has been explored for use on its own [21, 22], with cells [23], or in combination with other materials [24].

Two groups have evaluated the use of collagen as an injectable material within the damaged myocardium. Dai et al. [21] assessed the use of a commercially

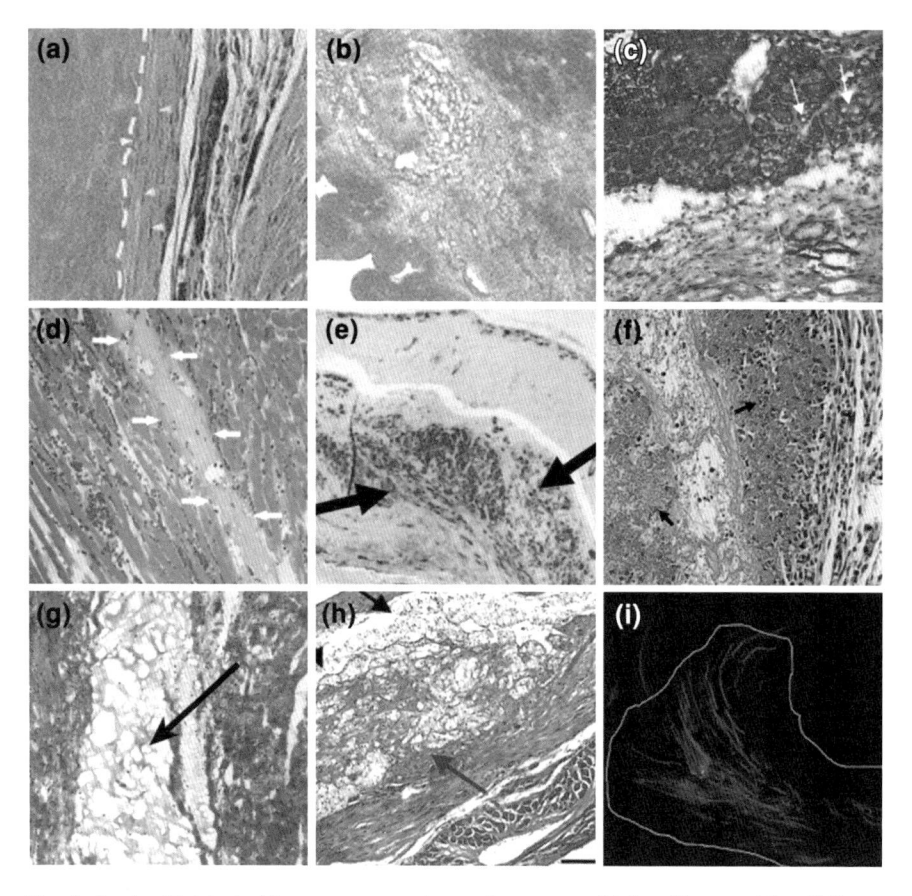

Fig. 2 In vivo histological images representing each category of injectable material. **a** Collagen [21], **b** fibrin glue [18], **c** Matrigel [74], **d** nanofibers (RAD16-II) [29], **e** alginate, [30], **f** chitosan [32], **g** decellularized myocardial matrix [33], **h** PNIPAAm [70], **i** PEG [72]

available collagen product, Zyderm, for its ability to preserve cardiac function post-MI. The Zyderm II product used was 65 mg/mL purified collagen (95% Collagen I and 5% Collagen III) in liquid form, ready for injection. Although the product was commercially available for dermal soft tissue repair, it is no longer on the market. In this study, 100 μL of the Zyderm product was injected into a total occlusion rat model at one-week post-MI, and assessed 6 weeks post-injection, as compared to saline injection controls. Evaluation at 6 weeks post-injection showed that collagen increased scar thickness, stroke volume, and ejection fraction (EF), as compared to saline controls. In addition, it was stated that collagen was able to prevent paradoxical systolic bulging, evident when the end-systolic lumen was larger than the end-diastolic lumen, and measured by total LV diastolic circumference and bulging circumferential length. Hemodynamic evaluation did not demonstrate improved diastolic or systolic blood pressure. It was also reported that there was no cell infiltration into the injection region at 6 weeks post-injection,

Injectable Materials for Myocardial Tissue Engineering 137

Table 1 Naturally derived or inspired injectable materials used for myocardial tissue engineering

Material	Model	Time post-MI	Injection type	Cells	Vessels	Geometry	Cardiac function	Ref.
Collagen	Rat; ligation	1 wk	Alone		–	=	+	[21]
	Rat; 30 min	1 wk	Alone		+			[22]
	Porcine; healthy	n/a	Alone; w/BM	+				[23]
Fibrin	Rat; 17 min	1 wk	Alone; w/SkM	+	+		+	[18, 26]
	Rat; 17 min	1 wk	Alone		+			[22]
	Rat; cryoinjury	3 wks	w/BM MNCs		+			[27]
	Rat; 17 min	1 wk	w/PTN		+ (w/PTN)			[25]
	Sheep; ligation	1 wk	w/EC		+ (w/cells)		+ (w/cells)	[39]
Matrigel	Mouse; ligation	0 d	w/ESC				+ (w/cells)	[28]
	Rat; 17 min	1 wk	Alone		+			[22]
	Rat; healthy heart	n/a	Alone		–			[29]
	Mouse; 60 min	4 d	w/hESC	=				[44]
Nanofibers	Mouse; healthy	n/a	Alone; w/nCM		+			[29]
	Rat; healthy, ligation	0 d	Alone; w/nCM + IGF-1	+		= (w/GF + cells)	+ (w/GF + cells)	[48]
	Rat; ligation	0 d	w/IGF-1; w/CPC + IGF-1		+		= (w/GF + cells)	[49]
	Rat; ligation, 60 min	0 d	Alone; w/PDGF		–		+ (w/GF)	[45, 50]
	Rat; ligation	2 wks	Alone; w/PDGF, w/SkM	=	+	–	–	[51]
Alginate	Rat; ligation	1 wk; 60 d	Alone			= ED	=	[30]
	Rat; ligation	1 wk	Alone; w/GF		+ (w/GF)	= ED	=	[54]
	Rat; 25 min	5 wks	Alone; w/RGD		+ (w/RGD)	= ED	=	[55]
	Porcine; 90 min	3–4 d	Alone			=	=	[31]

(continued)

Table 1 (continued)

Material	Model	Time post-MI	Injection type	Cells	Vessels	Geometry	Cardiac function	Ref.
Chitosan	Rat; ligation	1 wk	Alone; w/mESC	+	+ (w/cells)	+ (w/cells)	+ (w/cells)	[32]
Myocardial Matrix	Rat; healthy	n/a	Alone		=		=	[33]
	Rat; 25 min	2 wks	Alone					–
	Porcine; healthy	n/a	Alone					–
Pericardial Matrix	Rat; healthy	n/a	Alone		=			[34]
Collagen-Matrigel	Rat; ligation	3 wks	Alone; w/nCM				=, + (w/cells)	[24]
Fibrin-Alginate	Porcine; ligation	1 wk	Alone		=	+ ES	=	[67]

BM bone marrow cells, *SkM* skeletal myoblasts, *BM MNC* bone marrow mononuclear cells, *PTN* pleiotrophin, *EC* endothelial cells, *ESC* embryonic stem cells, *hESC* human embryonic stem cells, *nCM* neonatal cardiomyocytes, *mESC* mouse embryonic stem cells, *GF* growth factors, *IGF*-1 insulin like growth factor-1, *CPC* cardiac progenitor cells, *PDGF* platelet derived growth factor

Key: cell survival: "+": increased cell survival, "=": no improvement in cell survival; neovasculature: "+": increased vessel density over control, "=": vessel formation present (either not improved or no control), "–": lack of vessel formation; LV geometry: symbols apply to both ED and ES values, unless otherwise indicated: "+": improved geometry (decreased value) as compared to other groups, or via pairwise comparison pre- and post-treatment, "=": preserved geometry pre- and post-treatment, or the same as other groups, "–": worsened geometry (dilation, increase pre- and post-treatment); cardiac function: "+": improved cardiac function, "=": preserved cardiac function, "–": no functional benefit including an absence of vascular cells and vessel formation. Thus, the authors claim that this provides insight into the mechanism of functional improvement. Since other materials that demonstrate preserved cardiac function, such as fibrin glue [26], show increased wall thickness and neovasculature formation, the mechanism of functional improvement is not clear. Here, the authors demonstrate that cardiac function can be preserved, as a result of increased wall thickness, which suggests that vascularization is not essential to preserve function.

Interestingly, a study by Huang et al. [22], does show vascular cell infiltration into a collagen injected region at 5 weeks post-injection. While these results seem contradictory, it should be noted that there are several differences in the experimental model, as well as the material used. Huang et al. set out to compare the angiogenic potential of different injectable materials: fibrin glue, collagen, and Matrigel. In this rat model, the left anterior descending artery (LAD) was ligated for 30 min and allowed to reperfuse. At one-week post-MI, 50 µL of fibrin glue, collagen, Matrigel, or saline was injected into the damaged region of the heart. The collagen used in this study was rat-tail Collagen I (BD Biosciences) at 1 mg/mL. Five weeks post-injection, hearts were removed and stained with anti-CD31 to visualize capillaries and anti-α-smooth muscle actin to identify myofibroblasts and smooth muscle cells. Quantification revealed a significant increase in both capillary density and myofibroblast infiltration into collagen-injected scars, as compared to scars injected with saline. The variation in infarct model or the type of collagen could be responsible for the difference seen from the previous study. Huang used an ischemia reperfusion model, as opposed to a total occlusion model of the previous study, which would more easily facilitate neovascularization.

In another study, intended to test the safety and feasibility of a new percutaneous injection method in a porcine model, collagen was injected with cells obtained from bone marrow aspirate [23]. In this study, a type I collagen product, Cellagen (ICN Biomedical) was used. Autologous bone marrow cells were cultured and reconstituted so that $\sim 2 \times 10^7$ cells were suspended in 1 cc of 0.3% collagen biogel and injected into healthy swine hearts via trans(coronary) venous injection. A collagen biogel control injection was also performed in one animal. A total of 88 punctures were made, averaging 14.6 per animal throughout the left ventricular septal wall and free wall. The presence of transplanted cells at 28 days was confirmed, however there was not a control to test the ability of collagen to increase cell survival. No further testing of collagen in a large animal has been reported.

2.2 Fibrin

Fibrin is a fibrillar protein important in the coagulation cascade and wound healing [36], and is known to promote angiogenesis [37, 38], making it an attractive option for tissue engineering applications. Several commercial products are available, which consist of two-component systems of fibrinogen and thrombin. When injected together, through a double-barreled injector, the thrombin enzymatically cleaves fibrinogen, forming fibrin, which self-assembles into a fibrous gel. The ease of use and appealing properties of fibrin have lead to its use as an injectable material for cardiac repair, which is able to improve cell survival [18], as well as induce angiogenesis and preserve cardiac function both with cells and as an acellular scaffold [18, 22, 26, 39].

Christman et al. were the first to demonstrate increased cell survival using a biomaterial [18], as well as demonstrate that the material alone could preserve cardiac function [26]. For both studies, the tisseel fibrin glue product from Baxter was used. In the first experiment, four groups of animals were injected one-week post-MI and assessed at 5 weeks post-injection [18]. One week after a 17 min ligation of the left anterior descending coronary artery (LAD) of female rats, 50 μL of fibrin glue alone, saline plus bovine serum albumin (BSA), skeletal myoblasts in fibrin glue, or skeletal myoblasts in saline (plus BSA) was injected into the damaged myocardium. After 5 weeks, immunohistochemical staining revealed a significant increase in the arteriole density within the infarct region of injections involving fibrin. Additionally, there was a significant increase in the percentage of cells within the infarct region when myoblasts were injected with fibrin glue, as compared to saline. Thus, fibrin glue was shown to improve arteriole formation as well as cell survival. In a second study, evaluating the same four groups, myoblasts alone, myoblasts with fibrin glue, and fibrin glue alone all showed preserved cardiac function at 5 weeks post-injection, as compared to pre-injection fraction shortening (FS) by echocardiographic (Echo) analysis, while the saline control group showed deteriorating cardiac function [26], indicating cell transplantation may not be critical. Wall thickness was also preserved in all groups except the control. It was thus hypothesized that increased wall thickness aids in the attenuation of left ventricular remodeling which preserves cardiac function.

To further validate the angiogenic potential of fibrin glue within an ischemic environment, increased capillary density, within the ischemic scar (created by a 30 min occlusion), was observed 5 weeks post-injection, using the same tisseel system. However, fibrin did not show an increased concentration of myofibroblasts within the infarct post-injection, as collagen had in the same study. In another study, bone marrow mononuclear cells (BMMNCs) were injected within a fibrin matrix, Greenplast (GreencrossPD Co.) [27]. A cryoinjury model was used to create an ischemic region on the LV wall in a rat model, and at 3 weeks post-injury, BMMNC suspended in cell culture medium, medium alone, or BMMNC in fibrin were injected. Fibrinogen and thrombin solutions were mixed in a 1:1 ratio for two 100 μL injections into the damaged region with or without cells. At 8 weeks post-injection, cells were detected in hearts that had BMMNCs injected with or without fibrin, yet quantification was not reported. Further assessment revealed an increased arteriole density within the infarcted region of hearts injected with BMMNCs within fibrin, as compared to BMMNCs in medium or in medium alone. This study corroborates the increased angiogenic potential as presented by Christman et al. [18, 26]. Additionally, it has been shown in a rat model that fibrin matrix enhances the neovascularization within an infarct, upon injection with a plasmid encoding pleiotrophin (PTN), an angiogenic growth factor, as compared to fibrin alone or PTN plasmid in saline [25].

One large animal study, utilizing a fibrin matrix as an injectable material to treat damaged myocardial tissue, has been reported. A sheep model was used to test the injection of autologous endothelial cells in a fibrin matrix, as compared to saline alone [39]. Transepicardial injection of cells within fibrin was performed at one-week post-ligation, and hearts were evaluated at 8 weeks post-injection. The group containing cells with fibrin showed improved ventricular function, as well as increased blood flow and neovascularization. While interesting, it is not clear from this study if the cells, the fibrin, or a combination was responsible for the improved function and neovasculature formation.

2.3 Matrigel

Matrigel is a naturally derived basement membrane matrix, derived from Engelbreth-Holm-Swarm (EHS) mouse sarcoma that is commercially available from BD Biosciences. This material is commonly used for its ability to offer a range of components found in natural ECM, including laminin, collagen IV, and heparan sulfate proteoglycans [40, 41]. Matrigel has been utilized as a platform for enhanced cell growth in vitro, as well as for promoting angiogenesis in vivo. Although Matrigel has been used for cardiac tissue engineering as a potential injectable material [22, 28], and the benefits of being naturally derived are appealing, clinical translation may be hindered by reports of the basement membrane matrix supporting tumor growth [42, 43].

Matrigel has been explored by Kofidis et al. in combination with mouse embryonic stem cells (ESCs), for injection into damaged myocardium of a BALB/c mouse model, following complete LAD ligation [28]. At 4 weeks post-surgery, hearts were assessed using Echo and histology. Injections were 50 μL each of Matrigel with ESCs, Matrigel alone, or ESCs alone, with non-injected hearts as controls. At 4 weeks, Matrigel with ESCs showed preserved function, as assessed by FS. All groups had similar wall thickness, which was increased over control animals with no injection. Additionally, in a study exploring the angiogenic potential of collagen, fibrin, and Matrigel within an ischemic region, Matrigel showed increased capillary density over saline control injections [22]. However, Matrigel was also used as a control material in another study, and shown not to induce cell migration in healthy myocardium at 7 or 28 days post-injection [29].

Matrigel was also tested with human embryonic stem cell (hESC) derived cardiomyocytes (CMs), and as part of a prosurvival cocktail (PSC) created to increase cell survival and engraftment [44]. The goal of this study was to improve the survival of hESC derived CMs, through the prevention of cell death by a variety of potential mechanisms. Matrigel, one component in the PSC, was included to prevent anoikis. Cells alone, cells in Matrigel alone, or cells with PSC (which included Matrigel) were injected into an infarcted nude rat heart, 4 days after the induction of a 60 min ischemia–reperfusion. Histological analysis at one-week post-injection revealed that Matrigel alone was not enough to increase cell survival, over injection of cells on their own. The data did demonstrate that hESC derived CMs injected with the PSC had increased survival at one week. Further functional analysis was performed, but cells with Matrigel alone were not assessed. Interestingly, when baseline (pre-injection) Echo data, including LV end diastolic dimension (LVEDD), LV end systolic dimension (LVESD), and FS measurements were compared to post-injection data, collected at 4 weeks, all groups showed dilated LVEDD and LVESD, whereas, hESC derived CM + PSC was the only group to show an improved FS. When groups were compared to one another at 4 weeks post-injection, hESC derived CMs with PSC showed lower LVESD, and higher FS than the groups of PSC alone, non-cardiac cells with PSC, or media alone.

2.4 Nanofibers

The use of self-assembling peptide nanofibers for myocardial tissue engineering has been recently explored [29, 45]. These self-assembling peptides are composed of short peptides that assemble at physiologic pH into a fibrous 3D structure. The transition at physiologic pH makes this approach particularly appealing as an injectable material since it can remain a liquid until injection into the myocardial tissue, where it can assemble to form a 3D scaffold for repair. Two peptide sequences have been explored in vivo: RAD16-I, termed Puramatrix (peptide sequence AcN-RADARADARADARADA-CNH2), and RAD16-II (peptide sequence AcN-RARADADARARADADA-CNH2).

Previous work involving these peptides demonstrated survival of both myocytes and endothelial cells within the peptide microenvironments in vitro [46, 47]. In the first in vivo test by Davis et al. [29], C57BL/6 mice were used, and 10 μL of the RAD16-II or Matrigel was injected into healthy hearts. Endothelial cell and myocyte infiltration was assessed at 3 h post-injection as well as 7, 14, and 28 days post-injections. Isolection staining for endothelial cells showed a peak number of cells at 7 days, followed by lower values at 14 and 28 days. Cardiomyocytes continued to increase within the peptide injected region, as quantified by cardiac specific α-sarcomeric actin staining. Cellular infiltration was compared to Matrigel, which showed few to no nuclei within the injection region at 7 days or 28 days (the only time points assessed). A further experiment compared cell infiltration into the region of injection of the RAD16-II peptide on its own or with GFP labeled neonatal CMs. The cell peptide combination yielded an increased number of non-GFP labeled cardiac cells infiltrating at 7 and 14 days. By 28 days however, both groups showed a similar number of cardiac cells. A 14 day time point was also assessed for hESCs injected with the peptide gel. The hESCs were labeled with eGFP, which was under the control of a α-myosin heavy chain promoter, and thus only fluoresced once differentiated into CMs. At 14 days post-injection, eGFP fluorescence was present, indicating that the hESC had differentiated to CMs. However, since there was no control group of hESCs without the RAD16-II peptides, it is not clear if injection with the peptides caused the transition to CMs in vivo. Further studies would be necessary to determine if the RAD16-II peptide gel was responsible for the cardiomyogenesis of hESC.

Further work using these peptides, with the incorporation of insulin-like growth factor-1 (IGF-1) [48, 49] and platelet derived growth factor (PDGF) [45, 50] post-MI have yielded interesting results. Shortly after the first in vivo results using RAD16-II, Davis reported the incorporation of IGF-1 tethered to the nanofibers to aid in cell survival and improve cardiac function [48]. First, an experiment in healthy rats showed that IGF-1 tethered to the peptides allowed for a sustained release, with higher levels of IGF-1 remaining in vivo at 28 days, as well as increased activation of survival factor Akt, as compared to free IGF-1 and un-tethered IGF-1 with peptides. A second experiment, in healthy rats, examined the injection of neonatal cardiomyocytes with peptide, with IGF-1 and peptide (un-tethered), and with IGF-1 tethered to the peptides. At 14 days post-injection, tethered IGF-1 showed a 25% increase in cell survival, as well as a decrease in the expression of cleaved caspase 3, an apoptosis marker. Cross-sectional area of individual transplanted cells was also quantified to be significantly greater when IGF-1 was tethered vs. untethered in healthy myocardium. A third experiment tested cells alone, in addition to the previous groups, injected immediately post-MI. At 21 days post-injection, the group with cells and tethered IGF-1 showed improved FS, attenuated LV dilation, as compared to cells with untethered IGF-1. Furthermore, an additional group of cells infected with a hemagglutinin-tagged dominant negative Akt (dnAkt) adenovirus with tethered IGF-1 was shown to block the therapeutic effects of improved function. A smaller study showed increased FS of cells with tethered IGF-1 vs. cells with untethered IGF-1 in a nude rat model, to demonstrate that the same effects were seen in an immune compromised situation. The IGF-1 tethered peptide was also tested with injection of cardiac progenitor cells and shown to preserved EF, diastolic wall thickness, +dP/dt, and –dP/dt one month after injection immediately post-MI [49]. Similarly, controlled release of PDGF-BB from RAD16-II nanofibers was demonstrated in healthy rats [45]. The growth factor with material demonstrated increased FS 14 days and 4 months following immediate injection post-MI, while nanofibers alone did not improve function [45, 50]. All of the previously described experiments involved 80 μL injections.

Despite the recent success, both the RAD16-I and RAD16-II peptide sequences with or without PDGF failed to improve skeletal myoblast survival and cardiac function post-MI [51]. A permanent ligation of the LAD was performed in Lewis rats, and 200 μL injections into 3 or 4 sites of the infarct and border zone were made 2 weeks post-MI. In one experiment the injection groups were: medium, RAD16-I, skeletal myoblasts, and skeletal myoblasts with RAD16-I. In a second experiment, groups were: medium, skeletal myoblasts, RAD16-II with PDGF, and skeletal myoblasts with RAD16-II and PDGF. In both experiments, hearts were assessed by Echo prior to injection and again at one-month post-injection.

No functional improvement was observed in any group, suggesting that 2 weeks post-MI is too late to allow for functional improvement. All groups involving nanofiber injections showed increased angiogenesis, yet skeletal myoblast survival was not improved by RAD16-I or by RAD16-II with PDGF. These data may seem to contradict the previous reports that RAD16-II with IGF-1 improved function and increased cardiomyocyte survival. However, the authors highlight that the cardiomyocyte injections took place immediately post-MI, rather than 2 weeks post-MI, which could account for survival differences. They also point out that the cell types were different and stress the potential importance of finding the appropriate cell-specific biomaterial scaffold for improved cell survival.

2.5 Alginate

The material that has gained the most momentum over the past few years and shown clinical potential has been alginate. Collagen, fibrin glue, and Matrigel were explored with enthusiasm from 2004–2007, but publications on advancements using these materials have been limited since. The use of alginate, however, has continued to be explored in both small and large animal models [30, 31]. Alginate, a polysaccharide derived from brown seaweed, is able to cross-link to form a gel in the presence of divalent ions, such as calcium [52]. It is a commonly used biomaterial for tissue engineering applications, despite its limited cellular interaction and integration [53].

Landa et al. [30] first set out to study the distribution and degradation properties of alginate as an acellular injectable scaffold. MI was induced by permanent occlusion in a rat model, and 100–150 μL of biotin labeled calcium-crosslinked alginate was injected one-week post-MI. Animals were euthanized at 1 h, 1 week, 4 weeks, and 6 weeks, revealing that very little alginate remained in the myocardial tissue at 6 weeks. To investigate the usefulness of alginate to preserve cardiac function post-MI, rats were injected with 100–150 μL of alginate, 1×10^6 neonatal CMs in alginate, or saline at one-week post-MI. Echo, performed pre-injection, and 8 weeks post-injection was used to assess the impact of alginate injections on LV geometry and function. Pairwise comparison of pre- and post-injection data revealed that alginate was able to increase wall thickness at ED and ES, as well as preserve LVEDD, FS, and fractional area change (FAC). Animals injected with CMs in saline also showed preserved FS, indicating that both the biomaterial and cellular injection are able to preserve cardiac function when injected one-week post-MI. Doppler interrogation of mitral inflow confirmed that material injection did not cause an alteration of diastolic function. Additionally, an influx of myofibroblast cells in biomaterial injected hearts was observed, through α-actin staining. The final experiment in this work compared the injection of alginate or saline control into hearts of rats considered to have old infarcts, meaning that injections were 60 days post-MI. Hearts were assessed at 8 weeks post-injection, and as with recent infarcts, alginate injection significantly thickened the anterior wall at ED and ES. Alginate was also shown to preserve FS, thus attenuating LV dysfunction.

Additionally, alginate had been explored as a scaffold for growth factor (GF) delivery [54]. The effects of sequential release of VEGF-A_{165} and PDGF-BB from an injected alginate gel were studied in a total occlusion rat model. Briefly, 3 μg of VEGF, 3 μg of PDGF-BB, or 3 μg of both proteins together were combined in 100 μL of alginate gel and injected into the infarct border zone one-week post-MI. PBS and alginate alone served as control injections. Sequential release of both GFs lead to significantly increased capillary density, as compared to all other groups, while arteriole density was increased in all injections containing GFs. Alginate alone showed no increase in vessel density over the PBS control. Assessment of EF at 4 weeks post-treatment showed that although incorporation of GFs trended towards improved cardiac function, when compared to alginate alone, there was no significant difference in LVDD or EF among groups.

In another study, alginate was modified by conjugation of RGD peptide and tested within a rat model for effects on LV function and angiogenesis [55]. Briefly, rats underwent a 25-min ischemia reperfusion event, as caused by temporary ligation of the LAD. Five weeks post-MI, 0.5% BSA in saline, non-modified alginate, or RGD peptide-modified alginate was injected. Echo was performed prior to the injection surgery (5 weeks post-MI), 2 days post-injection and 5 weeks post-injection. At just 2 days post-injection, the alginate group showed increased wall thickness and FS, as compared to baseline. At 5 weeks post-injection, PBS injected hearts continued to decline, with increased dilatation and decreased FS, while alginate injection (modified and unmodified) preserved LV internal dimensions (LVID) and function. As RGD is a peptide known to promote cell adhesion, angiogenesis was assessed as a way to quantify increased cell–matrix interactions. Through α-smooth muscle actin staining, quantification revealed that the alginate group modified with RGD showed a significantly higher arteriole density as compared to non-modified alginate. The results indicate that RGD conjugated alginate improves the angiogenic potential of the material.

This past year, it is notable that alginate was tested in a swine model, through transcoronary catheter delivery [31]. Catheter access to the left main coronary artery was achieved, and injections were made in a similar manner to injections of contrast agent during angiography. A balloon over the catheter wire was inflated, allowing the material to infuse into the myocardium and form a gel upon reaction with the calcium present in the tissue. A pilot experiment tested the bounds of injection volume and found that a total injection volume of <10 mL would be tolerated via catheter delivery to the heart. For safety and feasibility assessment, healthy and infarcted animals were studied. A transient balloon occlusion of the LAD was performed to create an MI, and alginate or control saline was injected into healthy hearts or 3–4 days post-MI. Injection of alginate through transcoronary delivery was found to be safe and feasible, without causing negative ECG changes, or any type of arrhythmias, and without detrimental distribution to satellite organs. A functional study sought to determine the optimal volume of injection to prevent LV remodeling and preserve cardiac function. Thus, 1, 2 or 4 mL of alginate or

2 mL saline was injected into the damaged tissue 3–4 days after a 90-min artery occlusion, and Echo was used to assess LV diastolic area (LVDA), LV systolic area (LVSA), LV mass, and FS prior to injection (3 days post-MI), as well as at 30 and 60 days post-MI. Alginate injection into the infarct preserved LV geometry as well as cardiac function, while saline injected animals showed a decline in LV geometry, as demonstrated by increased LVDA, LVSA, LV mass. It was concluded from these data that 2 or 4 mL are appropriate volumes of alginate injection to mitigate negative LV remodeling. In addition, histological analysis revealed that alginate injected hearts have increased wall thickness, as compared to saline controls, as well as an accumulation of myofibroblasts. This important study demonstrates that an acellular, material scaffold is able to preserve cardiac function in a large animal model, through minimally invasive catheter delivery.

2.6 Chitosan

Chitosan, a commonly used biopolymer in other tissue engineering applications, is derived from chitin, the main structural component in crustacean shells. Chitosan is a linear polysaccharide of glucosamine and N-acetyl glucosamine units, created by removing acetyl groups from chitin. The degree of deacytylation controls the material degradation properties, which can be beneficial for tissue engineering applications [56]. In addition, chitosan can undergo a temperature-phase transition, thus allowing it to gel in situ, upon injection, at physiologic temperature [32]. While there are many advantages to using chitosan, potential barriers include the lack of solubility in neutral solutions and limited or inconsistent cellular attachment [53].

Lu et al. has proposed the use of chitosan to improve cell survival [32]. The authors argue that fibrin glue degrades too rapidly in vivo to be effective, and highlight that Matrigel does not support cell infiltration. Thus, they sought to test the effects of chitosan as a scaffold material for injection with mouse embryonic stem cells (mESCs). In order to demonstrate feasibility of thermally responsive chitosan as an injectable scaffold to increase cell retention, and engraftment, as well as and preserve cardiac function, injections into ischemic border zone were made one-week post-MI. MI was induced using an ischemia–reperfusion model and three groups were examined: mESCs in phosphate buffered saline (PBS), mESCs in chitosan, chitosan alone, or PBS alone. Cell retention at 24 h and cell engraftment at 4 weeks were improved by injection with chitosan, compared to saline. Echo of the groups chitosan, PBS + mESC, and chitosan + mESC showed improved LVESD, LVEDD, EF, and FS as compared to PBS controls at 4 weeks. Additionally, animals receiving chitosan + mESCs showed improved cardiac function over PBS + mESCs. The chitosan + mESCs also showed significant increase in function over injection of PBS + mESC or chitosan alone. The angiogenic potential of this cell material combination was also tested, through quantification of vessels with a lumen diameter 10–100 μm. Chitosan + mESCs showed statistically increased microvessel density as compared to PBS alone, chitosan alone, or PBS + mESC. Thus, this work has shown that chitosan is able to increase cell retention and engraftment, and when injected with cells is an effective material to preserve LV geometry and cardiac function, and promote neovascularization within an infarct.

2.7 Decellularized Matrices

Decellularized matrices, scaffolds derived from the removal of a tissue's cellular components, have been used clinically, primarily in sheet form, derived from skin, pericardium, or small intestinal submucosa (SIS) [57]. For cardiac repair, SIS and urinary bladder matrix (UBM) have been utilized as cardiac patch therapies [6, 7], yet do not provide a cardiac specific material, and require an invasive surgical procedure for implantation. Recently, new techniques have allowed for the decellularization of more complex organs, such as a whole heart [58, 59], which allows for the generation of tissue engineering scaffolds containing the native ECM [60]. Additionally, decellularized porcine ventricular ECM and decellularized pericardial ECM have been developed as injectable materials for cardiac tissue engineering [33, 34]. These ECMs can be processed into a powder and solubilized through enzymatic digestion to remain a liquid at room temperature, but self-assemble to form a gel at physiologic pH and temperature.

Singelyn et al. first demonstrated feasibility of an injectable version of decellularized porcine ventricular ECM, termed myocardial matrix, for use as a scaffold in a rat model [33]. Characterization of the myocardial matrix revealed a complexity of proteins, peptides, and glycosaminoglycans, as well as a nanofibrous structure upon gelation. Retention of the biochemical complexity is important to provide appropriate cues that mimic the native microenvironment [8, 20, 61]. An in vitro migration assay showed that myocardial matrix was the preferred chemoattractant for rat aortic smooth muscle cells (RASMC) (as compared to collagen, fetal bovine serum (FBS), and pepsin), and also encouraged migration of human coronary artery endothelial cells (HCAEC). In vivo assessment of the matrix upon injection into healthy rat hearts showed vascular cell infiltration, including arteriole formation at 11 days post-injection. The myocardial matrix was also recently tested in a rat infarct model (data unpublished). Briefly, 75 μL of the myocardial matrix or saline control was injected 2 weeks following a 25 min ischemia–reperfusion event. Baseline EF was calculated from magnetic resonance (MR) images at one-week post-MI, and post-treatment measurements were evaluated 4 weeks post-injection (6 weeks post-MI). Myocardial matrix was shown to preserve cardiac function, as assessed through paired comparisons of pre- and post-treatment EF, while saline injected hearts showed continued decline in EF. Additionally, the myocardial matrix showed preserved LVEDV and LVESV, while saline controls had dilatation of the LV.

The myocardial matrix was also recently tested in a porcine model to assess potential minimally-invasive, percutaneous delivery (data unpublished).

25 successful injections of 0.2 mL each were made via transendocardial catheter delivery throughout the LV free wall and septal wall, under the guidance of an electomechanical voltage (NOGA) map. It is notable that the viscosity of the material was such to allow consistent resistance over multiple injections, and that no premature gelation occurred within the catheter. Two hours post-injection, the heart was removed and histological analysis revealed gelation of the matrix within the ventricular tissue. Additionally, the brain, kidneys, lungs, liver, and spleen were removed to confirm that the material did not spread to satellite organs. To our knowledge, this is the first successful material injection via transendocardial delivery.

Initial feasibility of an injectable form of decellularized pericardium as a potentially autologous treatment was also recently demonstrated [34]. Decellularized, intact pericardium is an FDA approved material for use in a variety of applications including soft tissue repair and prosthetic valves [62]. Porcine and human pericardium was recently decellularized and processed into a liquid using a similar preparation as the injectable myocardial matrix [33]. Likewise, these formed gels at physiologic temperature and pH. In vitro, the porcine pericardial gel was shown to be the preferred chemoattractant for RASMC, HCAEC, and rat epicardial cells. Within normal myocardium, porcine and human pericardial gels showed similar arteriole formation at 2 weeks post-injection. In addition, it was noted that there was a presence of c-kit$^+$ cells within the injection region of both materials, indicating that the material may have the potential to recruit stem cells into the injected region. Since the pericardium is routinely removed with no adverse effects, [63] a patient's own pericardium could be removed and processed, as has been done for other autologous treatments [64, 65]. Additionally, a patient's own pepsin (which is necessary to solubilized the material) can be isolated [66], thereby making the entire material from autologous sources. While the pericardial gel does not provide the exact cues present in the myocardium, it has the distinct advantage of potentially providing a completely autologous injectable treatment for patients in end-stage HF.

2.8 Combination Materials

In addition to the individual materials that have been explored as injectable scaffolds, a few groups have considered combining materials, to take advantage of the beneficial effects of each material. One such material, a collagen-Matrigel hydrogel has been studied in vitro with CMs, and was recently tested in vivo, as an injectable material [24]. The collagen-Matrigel combination with neonatal CMs was injected 3 weeks post-MI into a rat total occlusion model, as compared to cells in DMEM, material alone, or DMEM alone. 2–3 million cells were injected in 150 µL of DMEM or material. At 4 weeks post-injection, the authors state that the cells in hydrogel mixture showed improved FS from baseline Echo ($p = 0.049$), while all other groups maintained function pre- and post-injection. The DMEM control did not show a decline in cardiac function as would be expected. In addition, ANOVA revealed significant differences among the groups, with the mixture injection showing the highest FS, as well as lowest LVESD among the groups. While this study shows that cells in combination with material has enhanced therapeutic effects, single material controls were not performed to determine which material provides the benefit.

Another study using a combination material utilized a fibrin-alginate (fib-alg) biocomposite for injection in a porcine model [67]. A ligation model was used to create the MI, and injections of either saline or the fibrin-alginate composite were performed one-week post-MI. Epicardial injections were made, through a double-barreled injector system, requiring an invasive open chest procedure. 25 0.2 mL injections were made within a 2 × 2 cm region. A map grid was drawn on the heart, with markings every 0.5 cm to indicate the location sites for each of the 25 injections. Echo was used to assess LV geometry and EF at 7, 14, 21, and 28 days post-MI, and compared to baseline measurements taken prior to MI. Statistical analysis compared values at each time point, as compared to baseline (pre-MI), and as compared to 7 days post-MI. Although ES wall thickness was higher in fib-alg injected hearts, as compared to saline, both groups showed preserved wall thickness as compared to 7 days post-MI (pre-injection). LVEDV continued to increase in both groups, showing a significant difference from 7 days post-MI and from baseline. The fib-alg composite also preserved infarct expansion and EF, while saline injected hearts continued to show an increased infarct expansion value, and a decreased EF. Thus, the material injection was able to preserve cardiac function, as well as prevent continued infarct expansion. As compared to remote myocardium within the same heart, there was a decrease in capillary density within the infarcted region of saline and fib-alg injected hearts. However, the capillary density in the MI border zone in material injected hearts was similar to the remote myocardium. Without the controls of each material on its own, one cannot conclude if the composite material is necessary, or if one of the materials would have been sufficient. Interestingly, this study chose to also evaluate matrix metalloproteinase-9 (MMP-9) and MMP-2 levels within the MI region, border zone, and remote myocardium. While both MMPs were upregulated within the infarct, active MMP-2 levels were lower in the fibrin-alginate MI region than in the saline injected MI region. The other main finding from this study is that injection of the composite material reduced the amount of soluble collagen in the MI region, suggesting a prevention of collagen degradation, and thus a potential mechanism for preventing further infarct expansion.

3 Synthetic Injectable Materials

Although most of the work on injectable materials for myocardial tissue engineering has been focused on naturally derived or inspired biomaterials, synthetic materials offer the distinct advantages of tunability and control [20]. A variety of synthetic polymers have been explored in vitro or in vivo as patch materials [1, 68], yet only two polymer classes have made been tested as an injectable treatment. Variations of poly(N-isopropylacrylamide) and poly(ethylene glycol) have both been injected and evaluated within infarct models (Table 2).

3.1 Poly(N-isopropylacrylamide)

Poly(N-isopropylacrylamide) (PNIPAAm) is a thermoresponsive hydrogel that undergoes a phase transition at its lower critical solution temperature (LCST), which is just below physiologic temperature. This material has been explored in a variety of fields, but is applicable as an injectable scaffold because of its gelation capabilities at body temperature. Variations of PNIPAAm are currently being explored as injectable materials for cardiac tissue engineering [69, 70].

Wang et al. have explored the use of a biodegradable dextran (Dex) grafted poly(E-caprolactone)-2-hydroxyethyl methacrylate (PCL-HEMA)/PNIPAAm system (Dex-PCL-HEMA/PNIPAAm) as an injectable hydrogel [69]. The material is liquid at room temperature, but undergoes a sol–gel, reversible phase transition to form a gel at 37°C. The hydrogel or saline was injected into the infarcted myocardial tissue of rabbits, 4 days after MI was induced by LAD ligation. Several animals underwent a sham operation, where the vessel was not occluded. Injections consisted of four individual injections of 50 µL each, for a total of 200 µL injected into each heart. M-mode Echo was used to compare LV geometry and function 30 days post-injection. Both saline and hydrogel injected hearts showed a significant increase in LVEDD and a significant decline in EF as compared to LVEDD and EF of sham operation animals. While the hydrogel injected hearts had an increased LVEDD and decreased EF, as compared to sham, LVEDD was significantly lower for hydrogel hearts than saline hearts. Likewise, EF was significantly higher in hydrogel-injected hearts vs. saline. Thus, although cardiac function was decreased compared to sham, the hydrogel was likely able to slow the progression of HF, as demonstrated by an increased EF compared to saline control. Hemodynamic data, LV systolic pressure and dp/dt, also showed a decrease from sham values for both groups, while the hydrogel had significantly increased values as compared to saline injection. Hydrogel injection also caused significantly reduced infarct size, and increased scar thickness, as assessed by histology. The authors argue that this Dex-PCL-HEMA/PNIPAAm material is better than the others available (fibrin, collagen, alginate, and self-assembling peptides) because it offers tunable mechanical properties, is relatively inert, is more uniform (because it is synthetic), and would be easier and less expensive to manufacture.

Another variation of PNIPAAm has been developed for use as an injectable material by Fujimoto et al. [70]. Copolymerization of NIPAAm, acrylic acid (AAc), and hydroxyethyl methacrylate-poly(trimethylene carbonate) (HEMA-PTMC) allows for the development of a biodegradable, thermoresponsive poly(NIPAAm-co-AAc-co-HEMAPTMC) hydrogel. Injection of the PNIPAAm

Table 2 Synthetic injectable materials used for myocardial tissue engineering

Material	Model	Time post-MI	Injection type	Cells	Vessels	Geometry	Cardiac function	Ref.
Dex-PCL-HEMA/PNIPAAm	Rabbit; ligation	4 d	Alone			+	+	[69]
poly(NIPAAm-co-AAc-co-HEMAPTMC)	Rat; ligation	2 wks	Alone			= ED	=	[70]
PEG	Rat; ligation	0 d	Alone			−	−	[72]
PEGylated fibrin	Mouse; ligation	0 d	Alone; w/HGF, w/BMNC	+ (w/GF)		+ (w/ cells + GF)	+ (w/ cells + GF)	[19]

PEG poly(ethylene glycol), *Dex-PCL-HEMA/PNIPAAm* biodegradable dextran (Dex) grafted poly(E-caprolactone)-2-hydroxylethyl methacrylate (PCL-HEMA)/PNIPAAm system, *poly(NIPAAm-co-AAc-co-HEMAPTMC)* copolymerization of NIPAAm, acrylic acid (AAc), and hydroxyethyl methacrylate-poly(trimethylene carbonate) (HEMAPTMC), *HGF* hepatocyte growth factor, *BMNC* bone marrow derived mononuclear cells polymer in PBS was performed 2 weeks after ligation in a rat model, with PBS alone as a control. Five injections of 100 µL each were made in each heart in the apical, proximal, lateral, and septal wall border zones of the infarct. The authors reported that the wall thickness in PNIPAAm injected hearts was significantly greater than in saline injected hearts. In addition, what appeared to be a muscle-like layer was observed in hydrogel injected hearts as indicated by positive αSMA staining along with co-staining for proteins associated with contractile function within the hydrogel injected region. Baseline Echo was performed prior to injection, and again at 4 and 8 weeks post-injection. Echo showed preserved LV end diastolic area (EDA) and FAC at 4 and 8 weeks, while saline injected hearts continued to expand and show decreased function.

3.2 Poly(ethylene glycol)

Poly(ethylene glycol) (PEG) is a simple polymer of repeating ethylene oxide units that is well known to resist protein and cell adhesion, but can be modified to support cell binding [71]. Dobner et al. recently sought to determine the effects of an inert, non-degradable PEG on damaged hearts [72]. A rat permanent ligation model was used to create a MI, and 100 µL of PEG-vinyl sulfone (PEG-VS) or saline was injected (via multiple injections) immediately following MI. In addition, sham MI surgeries were performed, followed by saline or PEG injection. Initial retention experiments confirmed that the amount of PEG present remained constant at 4 and 13 weeks post-injection. Analysis with Echo at 2 and 4 weeks post-injection showed that the PEG hydrogel injection was able to attenuate the dilatation that occurred in saline injected hearts, by preventing an increase in LVEDD. However, by 13 weeks, LVEDD was the same in both groups. Additionally, no difference in FS or LVESD was seen between saline and PEG injected infarcted hearts at any time point. Thus, future studies will be necessary to elucidate the potential use of PEG as an injectable material to provide mechanical reinforcement as a means to prevent the progression of HF.

3.3 Combination Materials

Finally, a PEGylated fibrin biomatrix has been developed and tested as an injectable therapy for cardiac repair [19]. This is the only instance (to date) of a natural/synthetic hybrid material being injected into the myocardium. The biomatrix, requiring integration of fibrinogen, thrombin, and PEG was used to test cell survival of BM MNCs with and without hepatocyte growth factor (HGF), a GF recently shown to be cardioprotective [73]. A PEGylated fibrin biomatrix offers the ability to covalently bind growth factors, as well as control their release. To test the material, a permanent ligation was performed in Balb/c mice and injections were made within the infarct and in the 'periscar' (border zone) region. Six different injection groups were analyzed, including a saline control, biomatrix alone, cells alone, biomatrix + cells, biomatrix + HGF, and biomatix + HGF + cells. At 4 weeks, each group containing cells was assessed for cell survival. The biomatrix alone did not statistically increase cell survival, although biomatrix + HGF with cells did show an increase. Thus, it is hypothesized that HGF was necessary to encourage cell engraftment. Cell fate was also assessed through a variety of antibodies, but no differentiation was identified. All groups demonstrated increased vascular density as compared to saline controls in the periscar zone. The LVEDV of biomatrix + HGF and biomatrix + HGF + cells groups, as assessed by Echo, each showed significant difference from saline at 2 weeks, while LVESV of cells alone and biomatrix + HGF + cells showed a significant difference from saline at 2 weeks. At 4 weeks post-treatment, only the biomatrix + HGF + cells group maintained the difference from saline in LVEDV and LVESV. FS was assessed and shown to be statistically different from saline in all treatment groups, while only biomatrix + HGF + cells showed a difference in EF at 4 weeks. This paper demonstrates the effectiveness of incorporating a GF with a material for increased cell survival and preserved cardiac function, although it is not clear if the combination biomatrix is necessary, since the individual materials were not evaluated.

4 Injection Delivery Methods

A variety of injection systems have been explored to deliver cells, growth factors, or materials to the myocardium. Initial delivery of growth factors or small molecules was achieved through intravenous injection. However, intramyocardial delivery allows for targeted delivery and treatment that is focused on the region of interest, with limited systemic effects. To achieve intramyocardial delivery, both epicardial or endocardial injection techniques, as well as intracoronary injection methods have been developed (Fig. 3). Direct epicardial injection was the first avenue explored clinically, as it can be performed during already scheduled open chest procedures, such as coronary artery bypass graft (CABG) surgery or LV assist device (LVAD) implantation [14, 19], and is easily achieved with a standard syringe and needle. Furthermore, this method is also used for intramyocardial injections in small animal models, as more complex delivery systems would be difficult to scale down for a mouse, rat, or rabbit. While direct epicardial injection is the most common method explored for feasibility or pre-clinical studies in small animals [18, 21, 26, 30, 32, 74], a minimally invasive delivery option would be more clinically desirable to avoid an open-chest procedure if the patient does not require one.

To achieve percutaneous delivery to the myocardial tissue, transcoronary and transendocardial catheter based delivery systems have been developed and utilized for cell transplantation and growth factor delivery in large animal models and

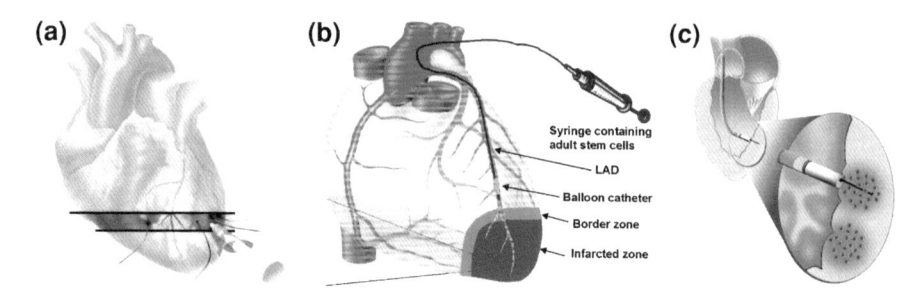

Fig. 3 Illustrations of each injection delivery method. **a** Direct epicardial injection: injection through a standard needle and syringe into the myocardial tissue; **b** transcoronary injection: injection via catheter access to the coronary vessels; **c** transendocardial injection: injection into the myocardial wall via catheter access into the lumen of the LV. Modified from: [15, 89, 90]

humans [9–11, 15, 16, 31]. Transendocardial delivery (or endoventricular delivery) involves the insertion of an injection catheter through a sheath that has already been passed through the left or right femoral artery into the left ventricle. A retractable needle can then be inserted into the endocardial wall for injection. The Cordis MyoStar catheter [9, 11, 75, 76], a Guidant custom made catheter [77], and the recently patented "Cell-Fix" catheter [78] are three catheters reported for injection that fit in an 8 F sheath, each with a 27 G retractable needle. In addition BioCardia's Helix Biotherapuetic Delivery catheter systems, including one version designed to insulate thermally sensitive materials, and the Bioheart MyoCath and MyoCell catheter systems offer additional endoventricular delivery options. For each system, once the injection catheter is in place, multiple injections are typically made throughout the infarct region, or within the border zone. Thus, a potential limitation of this technique is the challenge of delivery into an often very thin myocardial wall associated with infarction; but needle length can be adjusted to avoid puncturing through the entire wall. Injection location is extremely important to ensure treatment of the damaged tissue. Thus, endocardial injection is often coupled with electromechanical mapping of the heart, to identify the infarct and border zones. A commonly used mapping system is the NOGA mapping system, which creates a color coded electromechanical map of the heart, and allows for directed injection [9–11, 79].

Transcoronary delivery involves the use of catheter access to the coronary vessels to deliver treatment. Typically, an over-the-wire angiographic catheter is used to place a balloon within a coronary artery [31, 78, 80]. The balloon is then deployed, blocking blood flow to prevent the injection from entering the blood stream. This is repeated several times with wait periods in between injections to limit prolonged ischemia. Another intracoronary injection system has also been tested in a porcine model [23]. In this approach, access is gained through a coronary vein using a TransAccess tip, guided by ultrasound. Injection is then achieved through an IntraLume catheter via a needle that can advance into the myocardium. The disadvantage to coronary delivery, is the need for access to the coronary vessels, which may not be possible in all patients post-MI. Transcoronary delivery also has a risk of microembolism [78]. With both endocardial and intracoronary methods there is risk that the injected treatment will leak into the LV lumen or coronary arteries, and thus travel systemically, although recent studies have shown this risk to be minimal with a transendocardial approach [11].

Several studies have explored the benefits and effectiveness of transcoronary and transendocardial catheter delivery, as compared to one another and as compared to epicardial injection. Two early studies evaluating the intracoronary and endocardial catheter delivery systems concluded that the methods were comparable [77, 78], while recently it was declared that transendocardial delivery is the preferred method for cell injection [79]. Laham et al. set out to first compare intracoronary vs. intravenous delivery, followed by a second study to compare endocardial vs. epicardial delivery, using fibroblast growth factor-2. These studies concluded that distribution and retention of the growth factor was similar for endocardial and epicardial injection techniques, but was much improved over intravenous or intracoronary delivery [81, 82]. As most of the materials explored as injectable therapies have been tested in small animal models, relying on epicardial injection, it is unclear if they will successfully translate to large animals and eventually to humans. Thus far, the recently developed myocardial matrix, derived from decellularized ventricular ECM, is the only material to have been tested via transendocardial delivery in a large animal model.

While most injectable materials are currently being tested in small animal models, through epicardial injection, it is important to consider that the goal for clinical translation for an injectable scaffold is catheter-based delivery. Transcoronary and transendocardial catheter based injectable systems offer the unique advantage of minimally invasive delivery, eliminating the need for a risky open chest surgical procedure.

5 Conclusion

Given the recent promise shown by a variety of materials, it is important to look ahead to clinically relevant delivery methods and design criteria necessary for clinical translation. In any aspect of tissue engineering, it is important to consider the requirements and guidelines that would make treatment clinically relevant. Broadly, the goal of myocardial tissue engineering is to preserve cardiac function and prevent heart failure. To best achieve this goal, with minimal risk to the patient, minimally invasive delivery of treatment is essential. Thus, translation of a material for myocardial tissue engineering will be dependent on its feasibility for catheter delivery. To meet this requirement, a material must remain liquid until it reaches the myocardial tissue to prevent clogging of the catheter during a multiple injection procedure, making gelation kinetics an important consideration in material development. In most small animal studies the injection is a fairly quick procedure involving a single injection through a standard small diameter needle and syringe. For humans and large animals, the material must remain a liquid for a much longer period of time, while multiple injections are made through a catheter. Many materials that have been developed for in vitro experimentation or small animal models would fail minimally-invasive clinical translation, because of rapid gelation times that would create a risk of clogging the catheter. Additionally, several therapies require a double-barreled injection system [18, 26, 67, 74], for which there is not currently a catheter equivalent. Thus, only the materials with proper viscosity and gelation kinetics have the potential to translate to the clinic for minimally invasive delivery.

From a tissue engineering perspective, a well-designed scaffold is one that will provide cells with the proper cell–matrix interaction cues for adhesion, proliferation, and maturation [8, 20, 61]. To accomplish this, recent tissue engineering strategies have focused on materials that are biomimetic of the structure, morphology, and biochemical cues of the native microenvironment they are attempting to replace [8, 20, 61]. This approach is of particular importance in myocardial tissue engineering. Post-MI, cell death occurs in the ischemic region, and the ECM is degraded by MMPs. The ECM, which is degraded post-MI, is a combination of collagen, laminin, and fibronectin, and other proteins, as well as proteoglycans. The complex combination serves to guide cellular attachment, survival, migration, proliferation, and differentiation [8, 20, 83–86]. Thus, tissue engineering strategies focus on replacement of the native ECM, to provide cells with the proper environment to develop into mature tissue [17]. Of the materials explored as injectable scaffolds for myocardial tissue engineering, Matrigel, injectable myocardial matrix, and injectable pericardial matrix provide the best mimic of native myocardial ECM, as they are a complex combination of ECM components. Yet, the recently developed myocardial matrix is the only material that offers a tissue specific matrix, as it is derived from ventricular tissue. Matrigel is derived from a sarcoma cell line, and thus does not provide the correct microenvironment. Synthetic materials and other non-mammalian or single protein naturally derived or inspired materials also do not offer the complexity of the native ECM.

Of particular interest in myocardial tissue engineering is the restoration of a blood supply to the ischemic region. The ability of a scaffold to support and promote vascular infiltration is considered one of the most important requirements in most types of tissue engineering [20, 61, 87]. Thus, it is important that a material or signals from degradation products should encourage the recruitment of vascular cells [20, 61] and promote neovascularization within the infarcted tissue. Matrigel and collagen have each been reported to both induce neovascularization or to fail at allowing cellular infiltration. Synthetic materials likely do not provide specific biological cues to promote cell or vessel infiltration, but can allow for angiogenesis through their porous structures [88], or can be biochemically modified. Several of the naturally derived and inspired materials explored as injectable scaffolds have shown capillary and arteriole formation, including fibrin [18, 22], nanofibers [29], and decellularized matrices [33, 34]. Thus, the most attractive candidates for injection therapy are those with the ability to replenish the lacking blood supply in the ischemic tissue.

The results of the studies included in this chapter, using naturally derived or inspired materials and synthetic materials as injectable therapies for the treatment of myocardial tissue engineering have been an area of great excitement in the field. Injectable therapies offer the potential advantage of minimally invasive delivery, which could alleviate the need for open chest surgical procedures. Of particular interest is the use of materials as injectable scaffolds to provide the appropriate microenvironment to increase cell survival during cellular cardiomyoplasty. Fibrin glue [18] and chitosan [32] were both reported to increase cell survival on their own, and PEGylated fibrin was shown to increase cell survival when HGF was conjugated [19], demonstrating the potential of materials to provide cells with the needed matrix replacement. Interestingly, self-assembling nanofibers have been shown to increase the survival of CM [48], but have no effect on the survival of skeletal myoblasts [51]. It was thus suggested that a biomaterial may be cell-specific, highlighting that the appropriate selection of biomaterial and/or cell type is an important consideration [51].

To be an effective treatment to prevent HF post-MI, it is advantageous that a therapy mitigate negative LV remodeling to thus prevent or improve cardiac function through preserved LV geometry. A variety of materials have shown preserved or improved cardiac function with or without cells in small and large animal models [21, 26, 30–32]. Additional results have demonstrated that the injection of collagen [21], fibrin glue [26], chitosan [32], alginate [30, 31], decellularized myocardial matrix, and PNIPAAm [70], without cells can preserve LV geometry and cardiac function, suggesting that an acellular treatment could stand alone. The distinct advantage of an acellular therapy is the potential to be an off the shelf type of treatment, avoiding the complications associated with cell therapies, including the selection of an appropriate cell type and source, the need for expansion, or potential disease transmission.

The studies reviewed here demonstrate the potential of injectable materials and the substantial progress that has been made in this field since its beginning, less than a decade ago. While each material has its advantages, there is not yet an obvious choice as to the best solution. Fibrin glue is able to induce neovasculature formation [18, 22, 27], enhance cell survival [18], and preserve cardiac function with or without cells [26], making it the most successful material in small animal models. However, fibrin is likely not able to meet the criteria for catheter delivery, as its gelation is quite rapid and it requires a double-barreled injection system. The only large animal study involving fibrin utilized direct epicardial injection [39]. Alginate is an attractive acellular therapy option, as it is able to preserve cardiac function in small and large animal models [30, 31], and meets the clinical translation criteria of transcoronary delivery [31]. The recently developed decellularized myocardial matrix is another attractive acellular option shown to allow for vessel infiltration [33], preserve cardiac function in small animals, and be compatible with successful transendocardial delivery (data unpublished). Although neither alginate nor the myocardial matrix materials have been tested for their ability to improve cell survival, the myocardial matrix has been shown to retain biochemical cues of the adult ventricular ECM that would potentially provide injected cells with the appropriate microenvironment.

As the field continues to develop and move toward clinical application, it will be important to keep in mind critical design criteria: having appropriate properties to allow for catheter delivery and in vivo gelation, providing the appropriate cell–matrix interactions, offering angiogenic capabilities, and preserving cardiac function. Studies in large animal models will be necessary to demonstrate catheter injectability, as well as safety and efficacy for clinical translation. The recent advances of transendocardial delivery of the myocardial matrix and the demonstration of preserved cardiac function upon transcoronary injection of alginate [31], have brought injectable materials one step closer to clinical translation. These exciting advancements have the potential to revolutionize cardiac care, as injectable materials offer the option of minimally invasive treatment post-MI.

References

1. Christman, K.L., Lee, R.J.: Biomaterials for the treatment of myocardial infarction. J. Am. Coll. Cardiol. **48**(5), 907–913 (2006)
2. Kellar, R.S., Shepherd, B.R., Larson, D.F., Naughton, G.K., Williams, S.K.: Cardiac patch constructed from human fibroblasts attenuates reduction in cardiac function after acute infarct. Tissue Eng. **11**(11–12), 1678–1687 (2005)
3. Leor, J., Aboulafia-Etzion, S., Dar, A., Shapiro, L., Barbash, I.M., Battler, A., et al.: Bioengineered cardiac grafts: a new approach to repair the infarcted myocardium? Circulation **102**(19 Suppl 3), III56–III61 (2000)
4. Zimmermann, W.H., Didie, M., Wasmeier, G.H., Nixdorff, U., Hess, A., Melnychenko, I., et al.: Cardiac grafting of engineered heart tissue in syngenic rats. Circulation **106**(12 Suppl 1), I151–I157 (2002)
5. Kofidis, T., Akhyari, P., Boublik, J., Theodorou, P., Martin, U., Ruhparwar, A., et al.: In vitro engineering of heart muscle: artificial myocardial tissue. J. Thorac. Cardiovasc. Surg. **124**(1), 63–69 (2002)
6. Robinson, K.A., Li, J., Mathison, M., Redkar, A., Cui, J., Chronos, N.A.F., et al.: Extracellular matrix scaffold for cardiac repair. Circulation **112**(suppl I), I-135–I-143 (2005)
7. Badylak, S.F., Obermiller, J., Geddes, L., Matheny, R.: Extracellular matrix for myocardial repair. The Heart Surgery Forum **6**(2), E20–E26 (2002)
8. Leor, J., Amsalem, Y., Cohen, S.: Cells, scaffolds, and molecules for myocardial tissue engineering. Pharmacol. Ther. **105**(2), 151–163 (2005)
9. Smits, P.C., van Geuns, R.J., Poldermans, D., Bountioukos, M., Onderwater, E.E., Lee, C.H., et al.: Catheter-based intramyocardial injection of autologous skeletal myoblasts as a primary treatment of ischemic heart failure: clinical experience with six-month follow-up. J. Am. Coll. Cardiol. **42**(12), 2063–2069 (2003)
10. Dib, N., Diethrich, E.B., Campbell, A., Goodwin, N., Robinson, B., Gilbert, J., et al.: Endoventricular transplantation of allogenic skeletal myoblasts in a porcine model of myocardial infarction. J. Endovasc. Ther. **9**(3), 313–319 (2002)
11. Dib, N., Diethrich, E.B., Campbell, A., Goodwin, N., Robinson, B., Gilbert, J., et al.: Endoventricular transplantation of allogenic skeletal myoblasts in a porcine model of myocardial infarction. J. Endovasc. Ther. **9**(3), 313–319 (2002)
12. Fuchs, S., Kornowski, R., Weisz, G., Satler, L.F., Smits, P.C., Okubagzi, P., et al.: Safety and feasibility of transendocardial autologous bone marrow cell transplantation in patients with advanced heart disease. Am. J. Cardiol. **97**(6), 823–829 (2006)

13. Fuchs, S., Satler, L.F., Kornowski, R., Okubagzi, P., Weisz, G., Baffour, R., et al.: Catheter-based autologous bone marrow myocardial injection in no-option patients with advanced coronary artery disease: a feasibility study. J. Am. Coll. Cardiol. **41**(10), 1721–1724 (2003)
14. Dib, N., Michler, R.E., Pagani, F.D., Wright, S., Kereiakes, D.J., Lengerich, R., et al.: Safety and feasibility of autologous myoblast transplantation in patients with ischemic cardiomyopathy: four-year follow-up. Circulation **112**(12), 1748–1755 (2005)
15. Perin, E.C., Dohmann, H.F., Borojevic, R., Silva, S.A., Sousa, A.L., Mesquita, C.T., et al.: Transendocardial, autologous bone marrow cell transplantation for severe, chronic ischemic heart failure. Circulation **107**(18), 2294–2302 (2003)
16. Lunde, K., Solheim, S., Aakhus, S., Arnesen, H., Abdelnoor, M., Egeland, T., et al.: Intracoronary injection of mononuclear bone marrow cells in acute myocardial infarction. N. Engl. J. Med. **355**(12), 1199–1209 (2006)
17. Davis, M.E., Hsieh, P.C., Grodzinsky, A.J., Lee, R.T.: Custom design of the cardiac microenvironment with biomaterials. Circ. Res. **97**(1), 8–15 (2005)
18. Christman, K.L., Vardanian, A.J., Fang, Q., Sievers, R.E., Fok, H.H., Lee, R.J.: Injectable fibrin scaffold improves cell transplant survival, reduces infarct expansion, and induces neovasculature formation in ischemic myocardium. J. Am. Coll. Cardiol. **44**(3), 654–660 (2004)
19. Zhang, G., Hu, Q., Braunlin, E.A., Suggs, L.J., Zhang, J.: Enhancing efficacy of stem cell transplantation to the heart with a PEGylated fibrin biomatrix. Tissue Eng. Part A **14**(6), 1025–1036 (2008)
20. Lutolf, M.P., Hubbell, J.A.: Synthetic biomaterials as instructive extracellular microenvironments for morphogenesis in tissue engineering. Nat. Biotechnol. **23**(1), 47–55 (2005)
21. Dai, W., Wold, L.E., Dow, J.S., Kloner, R.A.: Thickening of the infarcted wall by collagen injection improves left ventricular function in rats: a novel approach to preserve cardiac function after myocardial infarction. J. Am. Coll. Cardiol. **46**(4), 714–719 (2005)
22. Huang, N.F., Yu, J., Sievers, R., Li, S., Lee, R.J.: Injectable biopolymers enhance angiogenesis after myocardial infarction. Tissue Eng. **11**(11–12), 1860–1866 (2005)
23. Thompson, C.A., Nasseri, B.A., Makower, J., Houser, S., McGarry, M., Lamson, T., et al.: Percutaneous transvenous cellular cardiomyoplasty. A novel nonsurgical approach for myocardial cell transplantation. J. Am. Coll. Cardiol. **41**(11), 1964–1971 (2003)
24. Zhang, P., Zhang, H., Wang, H., Wei, Y., Hu, S.: Artificial matrix helps neonatal cardiomyocytes restore injured myocardium in rats. Artif. Organs **30**(2), 86–93 (2006)
25. Christman, K.L., Fang, Q., Kim, A.J., Sievers, R.E., Fok, H.H., Candia, A.F., et al.: Pleiotrophin induces formation of functional neovasculature in vivo. Biochem. Biophys. Res. Commun. **332**(4), 1146–1152 (2005)
26. Christman, K.L., Fok, H.H., Sievers, R.E., Fang, Q., Lee, R.J.: Fibrin glue alone and skeletal myoblasts in a fibrin scaffold preserve cardiac function after myocardial infarction. Tissue Eng. **10**, 403–409 (2004)
27. Ryu, J.H., Kim, I.K., Cho, S.W., Cho, M.C., Hwang, K.K., Piao, H., et al.: Implantation of bone marrow mononuclear cells using injectable fibrin matrix enhances neovascularization in infarcted myocardium. Biomaterials **26**(3), 319–326 (2005)
28. Kofidis, T., Lebl, D.R., Martinez, E.C., Hoyt, G., Tanaka, M., Robbins, R.C.: Novel injectable bioartificial tissue facilitates targeted, less invasive, large-scale tissue restoration on the beating heart after myocardial injury. Circulation **112**(9 Suppl), I173–I177 (2005)
29. Davis, M.E., Motion, J.P., Narmoneva, D.A., Takahashi, T., Hakuno, D., Kamm, R.D., et al.: Injectable self-assembling peptide nanofibers create intramyocardial microenvironments for endothelial cells. Circulation **111**(4), 442–450 (2005)
30. Landa, N., Miller, L., Feinberg, M.S., Holbova, R., Shachar, M., Freeman, I., et al.: Effect of injectable alginate implant on cardiac remodeling and function after recent and old infarcts in rat. Circulation **117**(11), 1388–1396 (2008)

31. Leor, J., Tuvia, S., Guetta, V., Manczur, F., Castel, D., Willenz, U., et al.: Intracoronary injection of in situ forming alginate hydrogel reverses left ventricular remodeling after myocardial infarction in Swine. J. Am. Coll. Cardiol. **54**(11), 1014–1023 (2009)
32. Lu, W.N., Lu, S.H., Wang, H.B., Li, D.X., Duan, C.M., Liu, Z.Q., et al.: Functional improvement of infarcted heart by co-injection of embryonic stem cells with temperature-responsive chitosan hydrogel. Tissue Eng. Part A **15**(6), 1437–1447 (2009)
33. Singelyn, J.M., DeQuach, J.A., Seif-Naraghi, S.B., Littlefield, R.B., Schup-Magoffin, P.J., Christman, K.L.: Naturally derived myocardial matrix as an injectable scaffold for cardiac tissue engineering. Biomaterials **30**(29), 5409–5416 (2009)
34. Seif-Naraghi, S.B., Salvatore, M.A., Schup-Magoffin, P.J., Hu, D.P., Christman, K.L.: Design and characterization of an injectable pericardial matrix gel: a potentially autologous scaffold for cardiac tissue engineering. Tissue Eng. Part A **16**(6), 2017–2027 (2010)
35. Zimmermann, W.H., Melnychenko, I., Wasmeier, G., Didie, M., Naito, H., Nixdorff, U., et al.: Engineered heart tissue grafts improve systolic and diastolic function in infarcted rat hearts. Nat. Med. **12**(4), 452–458 (2006)
36. Sierra, D.H.: Fibrin sealant adhesive systems: a review of their chemistry, material properties and clinical applications. J. Biomater. Appl. **7**(4), 309–352 (1993)
37. Naito, M., Stirk, C.M., Smith, E.B., Thompson, W.D.: Smooth muscle cell outgrowth stimulated by fibrin degradation products. The potential role of fibrin fragment E in restenosis and atherogenesis. Thromb. Res. **98**(2), 165–174 (2000)
38. Thompson, W.D., Smith, E.B., Stirk, C.M., Marshall, F.I., Stout, A.J., Kocchar, A.: Angiogenic activity of fibrin degradation products is located in fibrin fragment E. J Pathol **168**(1), 47–53 (1992)
39. Chekanov, V., Akhtar, M., Tchekanov, G., Dangas, G., Shehzad, M.Z., Tio, F., et al.: Transplantation of autologous endothelial cells induces angiogenesis. Pacing Clin. Electrophysiol. **26**(1 Pt 2), 496–499 (2003)
40. Kleinman, H.K., McGarvey, M.L., Liotta, L.A., Robey, P.G., Tryggvason, K., Martin, G.R.: Isolation and characterization of type IV procollagen, laminin, and heparan sulfate proteoglycan from the EHS sarcoma. Biochemistry **21**(24), 6188–6193 (1982)
41. Kleinman, H.K., McGarvey, M.L., Hassell, J.R., Star, V.L., Cannon, F.B., Laurie, G.W., et al.: Basement membrane complexes with biological activity. Biochemistry **25**(2), 312–318 (1986)
42. Albini, A., Melchiori, A., Garofalo, A., Noonan, D.M., Basolo, F., Taraboletti, G., et al.: Matrigel promotes retinoblastoma cell growth in vitro and in vivo. Int. J. Cancer **52**(2), 234–240 (1992)
43. Yue, W., Brodie, A.: MCF-7 human breast carcinomas in nude mice as a model for evaluating aromatase inhibitors. J. Steroid Biochem. Mol. Biol. **44**(4–6), 671–673 (1993)
44. Laflamme, M.A., Chen, K.Y., Naumova, A.V., Muskheli, V., Fugate, J.A., Dupras, S.K., et al.: Cardiomyocytes derived from human embryonic stem cells in pro-survival factors enhance function of infarcted rat hearts. Nat. Biotechnol. **25**(9), 1015–1024 (2007)
45. Hsieh, P.C., Davis, M.E., Gannon, J., MacGillivray, C., Lee, R.T.: Controlled delivery of PDGF-BB for myocardial protection using injectable self-assembling peptide nanofibers. J. Clin. Invest. **116**(1), 237–248 (2006)
46. Narmoneva, D.A., Vukmirovic, R., Davis, M.E., Kamm, R.D., Lee, R.T.: Endothelial cells promote cardiac myocyte survival and spatial reorganization: implications for cardiac regeneration. Circulation **110**(8), 962–968 (2004)
47. Narmoneva, D.A., Oni, O., Sieminski, A.L., Zhang, S., Gertler, J.P., Kamm, R.D., et al.: Self-assembling short oligopeptides and the promotion of angiogenesis. Biomaterials **26**(23), 4837–4846 (2005)
48. Davis, M.E., Hsieh, P.C., Takahashi, T., Song, Q., Zhang, S., Kamm, R.D., et al.: Local myocardial insulin-like growth factor 1 (IGF-1) delivery with biotinylated peptide nanofibers improves cell therapy for myocardial infarction. Proc. Natl. Acad. Sci. U.S.A. **103**(21), 8155–8160 (2006)

49. Padin-Iruegas, M.E.M.D., Misao, Y.M.D., Davis, M.E.P., Segers, V.F.M.M.D.P., Esposito, G.P., Tokunou, T.M.D.P., et al.: Cardiac Progenitor Cells and Biotinylated Insulin-Like Growth Factor-1 Nanofibers Improve Endogenous and Exogenous Myocardial Regeneration After Infarction. Circulation **120**(10), 876–887 (2009)
50. Hsieh, P.C., MacGillivray, C., Gannon, J., Cruz, F.U., Lee, R.T.: Local controlled intramyocardial delivery of platelet-derived growth factor improves postinfarction ventricular function without pulmonary toxicity. Circulation **114**(7), 637–644 (2006)
51. Dubois, G., Segers, V.F., Bellamy, V., Sabbah, L., Peyrard, S., Bruneval, P., et al.: Self-assembling peptide nanofibers and skeletal myoblast transplantation in infarcted myocardium. J. Biomed. Mater. Res. B Appl. Biomater. **87**(1), 222–228 (2008)
52. Wee, S., Gombotz, W.R.: Protein release from alginate matrices. Adv. Drug Deliv. Rev. **31**(3), 267–285 (1998)
53. Lee, K.Y., Mooney, D.J.: Hydrogels for tissue engineering. Chem. Rev. **101**(7), 1869–1879 (2001)
54. Hao, X., Silva, E.A., Mansson-Broberg, A., Grinnemo, K.H., Siddiqui, A.J., Dellgren, G., et al.: Angiogenic effects of sequential release of VEGF-A165 and PDGF-BB with alginate hydrogels after myocardial infarction. Cardiovasc. Res. **75**(1), 178–185 (2007)
55. Yu, J., Gu, Y., Du, K.T., Mihardja, S., Sievers, R.E., Lee, R.J.: The effect of injected RGD modified alginate on angiogenesis and left ventricular function in a chronic rat infarct model. Biomaterials **30**(5), 751–756 (2009)
56. Khor, E., Lim, L.Y.: Implantable applications of chitin and chitosan. Biomaterials **24**(13), 2339–2349 (2003)
57. Badylak, S.F., Freytes, D.O., Gilbert, T.W.: Extracellular matrix as a biological scaffold material: Structure and function. Acta Biomater. **5**(1), 1–13 (2009)
58. Ott, H.C., Matthiesen, T.S., Goh, S.K., Black, L.D., Kren, S.M., Netoff, T.I., et al.: Perfusion-decellularized matrix: using nature's platform to engineer a bioartificial heart. Nat. Med. **14**(2), 213–221 (2008)
59. Wainright, J.M., Czajka, C.A., Patel, U.B., Freytes, D.O., Tobita, K., Gilbert, T.W., et al.: Preparation of cardiac extracellular matrix from an intact porcine heart. Tissue Eng. Part C Methods **16**(3), 525–532 (2010)
60. Gilbert, T.W., Sellaro, T.L., Badylak, S.F.: Decellularization of tissues and organs. Biomaterials **27**(19), 3675–3683 (2006)
61. Jawad, H., Ali, N.N., Lyon, A.R., Chen, Q.Z., Harding, S.E., Boccaccini, A.R.: Myocardial tissue engineering: a review. J. Tissue Eng. Regen. Med. **1**(5), 327–342 (2007)
62. Braga-Vilela, A.S., Pimentel, E.R., Marangoni, S., Toyama, M.H., de Campos Vidal, B.: Extracellular matrix of porcine pericardium: biochemistry and collagen architecture. J. Membr. Biol. **221**(1), 15–25 (2008)
63. Fuster, V.: Hurst's the Heart, 10th edn. McGraw-Hill Medical Publishing Division, New York (2001)
64. David, T.E., Feindel, C.M., Ropchan, G.V.: Reconstruction of the left ventricle with autologous pericardium. J. Thorac. Cardiovasc. Surg. **94**(5), 710–714 (1987)
65. Duran, C.M., Gometza, B., Kumar, N., Gallo, R., Martin-Duran, R.: Aortic valve replacement with freehand autologous pericardium. J. Thorac. Cardiovasc. Surg. **110**(2), 511–516 (1995)
66. Roberts, N.B., Taylor, W.H.: The preparation and purification of individual human pepsins by using diethylaminoethyl-cellulose. Biochem. J. **169**(3), 607–615 (1978)
67. Mukherjee, R., Zavadzkas, J.A., Saunders, S.M., McLean, J.E., Jeffords, L.B., Beck, C., et al.: Targeted myocardial microinjections of a biocomposite material reduces infarct expansion in pigs. Ann. Thorac. Surg. **86**(4), 1268–1276 (2008)
68. Zhang, G., Nakamura, Y., Wang, X., Hu, Q., Suggs, L.J., Zhang, J.: Controlled release of stromal cell-derived factor-1 alpha in situ increases c-kit+ cell homing to the infarcted heart. Tissue Eng. **13**(8), 2063–2071 (2007)
69. Wang, T., Wu, D.Q., Jiang, X.J., Zhang, X.Z., Li, X.Y., Zhang, J.F., et al.: Novel thermosensitive hydrogel injection inhibits post-infarct ventricle remodelling. Eur. J. Heart Fail. **11**(1), 14–19 (2009)

70. Fujimoto, K.L., Ma, Z., Nelson, D.M., Hashizume, R., Guan, J., Tobita, K., et al.: Synthesis, characterization and therapeutic efficacy of a biodegradable, thermoresponsive hydrogel designed for application in chronic infarcted myocardium. Biomaterials **30**(26), 4357–4368 (2009)
71. VandeVondele, S., Voros, J., Hubbell, J.A.: RGD-grafted poly-L-lysine-graft-(polyethylene glycol) copolymers block non-specific protein adsorption while promoting cell adhesion. Biotechnol. Bioeng. **82**(7), 784–790 (2003)
72. Dobner, S., Bezuidenhout, D., Govender, P., Zilla, P., Davies, N.: A synthetic non-degradable polyethylene glycol hydrogel retards adverse post-infarct left ventricular remodeling. J. Card. Fail. **15**(7), 629–636 (2009)
73. Jin, H., Wyss, J.M., Yang, R., Schwall, R.: The therapeutic potential of hepatocyte growth factor for myocardial infarction and heart failure. Curr. Pharm. Des. **10**(20), 2525–2533 (2004)
74. Kofidis, T., de Bruin, J.L., Hoyt, G., Lebl, D.R., Tanaka, M., Yamane, T., et al.: Injectable bioartificial myocardial tissue for large-scale intramural cell transfer and functional recovery of injured heart muscle. J. Thorac. Cardiovasc. Surg. **128**(4), 571–578 (2004)
75. Kornowski, R., Leon, M.B., Fuchs, S., Vodovotz, Y., Flynn, M.A., Gordon, D.A., et al.: Electromagnetic guidance for catheter-based transendocardial injection: a platform for intramyocardial angiogenesis therapy. Results in normal and ischemic porcine models. J. Am. Coll. Cardiol. **35**(4), 1031–1039 (2000)
76. Laham, R.J., Post, M., Rezaee, M., Donnell-Fink, L., Wykrzykowska, J.J., Lee, S.U., et al.: Transendocardial and transepicardial intramyocardial fibroblast growth factor-2 administration: myocardial and tissue distribution. Drug Metab. Dispos. **33**(8), 1101–1107 (2005)
77. Baklanov, D.V., Moodie, K.M., McCarthy, F.E., Mandrusov, E., Chiu, J., Aswonge, G., et al.: Comparison of transendocardial and retrograde coronary venous intramyocardial catheter delivery systems in healthy and infarcted pigs. Catheter. Cardiovasc. Interv. **68**(3), 416–423 (2006)
78. Chachques, J.C., Azarine, A., Mousseaux, E., El Serafi, M., Cortes-Morichetti, M., Carpentier, A.F.: MRI evaluation of local myocardial treatments: epicardial versus endocardial (Cell-Fix catheter) injections. J. Interv. Cardiol. **20**(3), 188–196 (2007)
79. Krause, K., Jaquet, K., Schneider, C., Haupt, S., Lioznov, M.V., Otte, K.M., et al.: Percutaneous intramyocardial stem cell injection in patients with acute myocardial infarction: first-in-man study. Heart **95**(14), 1145–1152 (2009)
80. Plewka, M., Krzeminska-Pakula, M., Lipiec, P., Peruga, J.Z., Jezewski, T., Kidawa, M., et al.: Effect of intracoronary injection of mononuclear bone marrow stem cells on left ventricular function in patients with acute myocardial infarction. Am. J. Cardiol. **104**(10):1336-1342
81. Laham, R.J., Rezaee, M., Post, M., Sellke, F.W., Braeckman, R.A., Hung, D., et al.: Intracoronary and intravenous administration of basic fibroblast growth factor: myocardial and tissue distribution. Drug. Metab. Dispos. **27**(7), 821–826 (1999)
82. Laham, R.J., Chronos, N.A., Pike, M., Leimbach, M.E., Udelson, J.E., Pearlman, J.D., et al.: Intracoronary basic fibroblast growth factor (FGF-2) in patients with severe ischemic heart disease: results of a phase I open-label dose escalation study. J. Am. Coll. Cardiol. **36**(7), 2132–2139 (2000)
83. Badylak, S.F.: The extracellular matrix as a biologic scaffold material. Biomaterials **28**(25), 3587–3593 (2007)
84. Uriel, S., Labay, E., Francis-Sedlak, M., Moya, M.L., Weichselbaum, R.R., Ervin, N., et al. Extraction and assembly of tissue-derived gels for cell culture and tissue engineering. Tissue Eng. Part C Methods **15**(9), 309–321 (2009)
85. Macfelda, K., Kapeller, B., Wilbacher, I., Losert, U.M.: Behavior of cardiomyocytes and skeletal muscle cells on different extracellular matrix components—relevance for cardiac tissue engineering. Artif. Organs **31**(1), 4–12 (2007)

86. Brown, L.: Cardiac extracellular matrix: a dynamic entity. Am. J. Physiol. Heart Circ. Physiol. **289**(3), H973–H974 (2005)
87. Patel, Z.S., Mikos, A.G.: Angiogenesis with biomaterial-based drug- and cell-delivery systems. J. Biomater. Sci. Polym. Ed. **15**(6), 701–726 (2004)
88. Nam, J., Huang, Y., Agarwal, S., Lannutti, J.: Improved cellular infiltration in electrospun fiber via engineered porosity. Tissue Eng. **13**(9), 2249–2257 (2007)
89. Shake, J.G., Gruber, P.J., Baumgartner, W.A., Senechal, G., Meyers, J., Redmond, J.M., et al.: Mesenchymal stem cell implantation in a swine myocardial infarct model: engraftment and functional effects. Ann. Thorac. Surg. **73**(6), 1919–1925 (2002) (Discussion 1926)
90. Strauer, B.E., Brehm, M., Zeus, T., Kostering, M., Hernandez, A., Sorg, R.V., et al.: Repair of infarcted myocardium by autologous intracoronary mononuclear bone marrow cell transplantation in humans. Circulation **106**(15), 1913–1918 (2002)

Tissue Engineering Approaches for Myocardial Bandage: Focus on Hydrogel Constructs

Marie Noëlle Giraud and Hendrik Tevaearai

Abstract Myocardial tissue engineering ambitions to regenerate, repair or replace damaged cardiac muscle by combining cellular and engineering technologies. Several issues must be addressed before this approach may one day find clinical applications for cardiac disorders such as congenital diseases or ventricular dysfunction following myocardial infarction for example. The chance of the myocardial tissue engineering approach is nevertheless real. Indeed, on the one hand, several clinical studies have recently confirmed the positive effect of stem cell therapy in patients with heart failure. On the other hand, research from several laboratories have demonstrated over the past decade that engineered muscle tissues can be created and successfully applied in models of myocardial injury. Engineering a functional myocardial graft faces with many challenges, and various approaches have been investigated. In the current chapter, we focus our review on hydrogel-based engineered tissues for myocardial application. The literature on injectable and implantable hydrogel constructs is discussed and an overview of our own experience is presented. We emphasize important aspects on development of hydrogel constructs in particular the mechanical and electrical conditioning of the construct as well as smart hydrogels.

None of the authors have conflict of interest to disclose.

M. N. Giraud (✉) and H. Tevaearai
Clinic for Cardiovascular Surgery, Inselspital Berne, Berne University Hospital and University of Berne, 3010 Berne, Switzerland
e-mail: marie-noelle.giraud-flueck@insel.ch

1 Introduction

Tissue engineering represents a group of approaches combining recent engineering technologies (material science, biotechnologies and nanotechnologies), biology and medicine and aiming to create bio-inspired artificial tissues. It is in fact the logical evolution of a modern society seeking for therapeutic options that increasingly imitate the natural healing and/or regenerative processes.

As regards to the heart, cardiovascular diseases represent a major cause of mortality and morbidity in western countries especially when they lead to heart failure (HF). It is indeed estimated that more than 17 million people died from cardiovascular disease in 2004 of which more than 7 million were attributed to coronary heart disease [1]. Several approaches are now established for the treatment of myocardial defect and dysfunctions. Traditionally, drug therapy and surgical corrections are first chosen whereas for more advanced diseases ventricular assist devices (as a bridge to transplantation or as a destination therapy) or transplantation may be the ultimate therapeutic possibility. Unfortunately, current therapeutic possibilities are limited and cardiac transplantation remains the only long-term valuable option for patients in an advanced stage of HF today. New therapeutic options are still at an experimental stage and involve cell therapy, gene therapy, xenotransplantation or the possible implantation of assist devices used here as bridge to myocardial recovery. Alternatively, an engineered contractile tissue could become a serious option for the treatment of ischemic heart if it would permit the restoration of an adequate local or global ventricular contractile function [2].

Furthermore, tissue engineering for myocardial repair will fully benefit from recent and future progresses of the majority of the traditional and new therapeutic strategies as well as from the increasing clinical interest for new areas such as nanotechnologies and material science. Indeed, while cardiac surgical procedures performed nowadays often require the use of artificial constructs such as prosthetic valves, rings, conduits and/or patches using various materials tested and commonly accepted by the surgical community, a recent trend toward bio-mimicking materials is observed. This is nicely illustrated by two clinical needs and trends. First, we observe a recent switch in cardiac valvular surgery from a high rate of mechanical valve implantation to the tendency to implant biological valves, or even to repair and preserve the native valve [3]. In this case, evolving research in tissue engineering of heart valves will also improve the shortcomings (thrombosis, infection, inability to grow for example) of current valves replacement. Second, the current practice in these young patients with a congenital malformation is traditionally to replace the failing structure by an inert, non-functional (non-contractile) and non-growing patch or prosthesis. Consequently, this part of the ventricle does not participate to the dynamic of the heart and even worse may require a later replacement as the patient grows to an adult size. Advances in engineered myocardial tissue could bring major breakthroughs for repair of the myocardial defect of infants [4].

To understand the challenge of myocardial tissue engineering, a few aspects must first be considered before developing our focus on hydrogel-based constructs.

2 Background

2.1 Bandage of the Heart

During the evolution of heart failure, the left ventricle progressively dilates and typically switches its geometry from the normal ellipsoidal to a spherical shape. According to the Laplace equation, the ventricular wall stress increases, contributing then to the further progression of the disease. One principle objective of supporting the heart by an external restraining system is thus to reshape the ventricular cavity by reducing the akinesis or dyskinesis of the infarcted area and by reducing the diameter of the failing ventricle [5]. Restraining the ventricle by physically restoring a better ventricular geometry may thus allow an improved contractile function.

The Anstadt cup may be considered as a precursor for this approach. This pneumatic activated cardiac vest was conceptualized as early as the sixties in order to provide a non-blood contact cardiac support [6]. Its elliptical form and its epicardial location made it very practical to implant via a minithoracotomy. Its principle has been confirmed in series of patients and since then several patents based on the same principle have been proposed [7].

As a second-generation restrain device, the CorCap (Acorn Cardiovascular, inc, St. Paul, MN, USA) system consist of a mesh designed to wrap the failing heart and restrain its diameter. Therefore it provides a reduction of the wall stress and a circumferential diastolic support [5]. Results of recent clinical trials support this hypothesis [8].

Following these initial approaches, the next major step in mechanical restrain was the development of the dynamic cardiomyoplasty. In this original technique, the latissimus dorsi is prepared and transposed into the thoracic cavity in order to envelop both ventricular masses. The restraining effect combined with synchronous contraction with the cardiac muscle has long been the explanation for positive clinical results [9]. This biological approach is basically a cardiac bandage that layers a normally vascularized structure directly onto the pathological tissue. In the context of myocardial tissue engineering, it can be considered as a visionary approach to the treatment of heart failure and the precursor of cellular cardiomyoplasty [10].

2.2 The Understanding of Myocardial Natural Repairs

The progressive dilatation and replacement of the injured tissue by a non-contractile scar is the adaptative response of the heart following myocardial infarction. It might be considered as a poor healing process without complete regeneration of the tissue. There are arguments to support this lack of formation of a new tissue identical in architecture, composition and function of the health myocardium. Clearly, the ventricle is perpetually subjected to a high mechanical stress therefore the infarcted area is forced into an immediate and efficient protective remodeling process, which in this case results in the formation of a solid fibrous scar that will prevent ventricular rupture. Indeed, immediately implanting a balloon pump or another ventricular assist device, in other words mechanically unloading the freshly infarcted ventricle has been shown to significantly reduce the size of the scar [11, 12]. Similarly, unloading a ventricle has also been shown to favor a reverse remodeling process eventually leading to a functional recovery [13, 14].

Furthermore, it was for long believed that the heart cannot regenerate. However, this dogma was recently challenged by the discovery of cardiac stem cells. In addition, Bergmann et al. [15] nicely demonstrated that on average 1% of the cardiac cells are renewed per year. This regenerative process stays insufficient in case of myocardial injury. Therefore, in complement to the restrain and unloading approaches, replacement of the loss cardiomyocytes by direct transplantation of various sorts of stem cells has recently been investigated and demonstrated functional benefits. The original rationale behind myocardial cell therapy was that the implantation of a large number of precursor cells within the dysfunctional ventricle may induce neovascularisation and myogenesis, eventually resulting in the regeneration of a contractile cardiac neo-tissue [16]. However, the exact mechanisms remain unclear. Nevertheless, there are evidences that the functional benefit is partly due to a paracrine or possibly an inflammatory effect. For example, humoral factors such as CXCR4 or SDF-1 have been identified to play a role in stem cell recruitment and migration toward the myocardium, and several growth factors are now known to induce neo-angiogenesis [17, 18].

Currently, it appears increasingly evident that the heart has the potential to recover a contractile function, however, this process is most probably highly dependent on a specifically favorable mechanical and chemical environment. In other words, creating an attractive environment in order to favor the regeneration of an adequate cardiac and vascular tissue appears to be the basis for cardiac tissue regeneration and represent one of the focus of tissue engineering [19]. Accordingly, three fundamental elements for the development of specific environment and engineering a contractile tissue are the selection of the matrix such as hydrogel, the cell source and the culture condition as represented in Fig. 1. The present review focus on hydrogel-based engineered muscle tissue designed for myocardium regeneration; however, few examples of solid scaffold based construct will be cited in particular to illustrate mechanical conditioning.

3 Hydrogel Constructs

Due to the several advantages they may offer, interest in hydrogels for the construction of three-dimensional cardiac patches has particularly risen over the last decade. Hydrogels are usually composed of a network of natural polymers

Fig. 1 Schematic representation of the fundamental aspects for tissue engineering

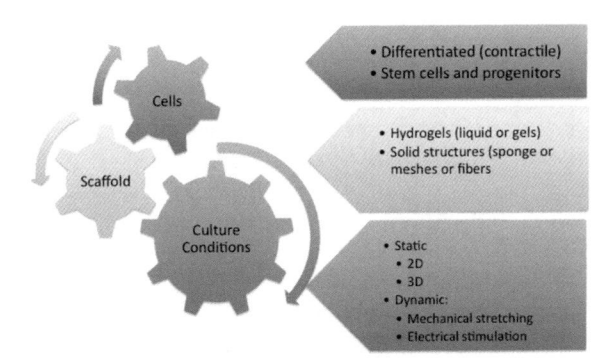

(such as collagen, fibrin, agarose, chitosan or hyaluronan) or synthetic polymers such as poly-ethylene glycol (PEG) mixed in a high content of water (Table 1). In laboratory investigations, various types of hydrogels have been described, going from simplified forms composed of only 1 or 2 myocardial ECM components, or approaches with more complex ECM compositions that include proteins, peptides, and polysaccharides (in particular glycosaminoglycan).

Hydrogels present a high degree of elasticity. Advantageously, their E-modulus (elastic characteristic) can be modulated by modifying the degree of crosslinking of the polymers (i.e. the establishment of covalent bound between polymeric chains). The adjustable elasticity is a major advantage of hydrogels. First, it allows mimicking the stiffness of various natural tissues as for example, the elasticity of the myocardium ranging between 30 and 80 kPa. Second, it permits modulation of cell infiltration and mobility within the scaffold, and consequently supports an in

Table 1 Injectable hydrogels for myocardial infarction

Injectable Hydrogel	Cell type/growth factors	References
Alginate	None	[28, 82]
Polyethylene glycol	None	[83]
Hyaluronic acid	None	[25]
Solubilized myocadial ECM	None	[30]
Solubilized pericardial ECM	None	[29]
Dex-PCL-HEMA/PNIPAAm	None	[84]
Fibrin	Bone Marrow Mononuclear Cells	[24]
	Bone marrow mechenchymal stem cells	[27]
	Bone Marrow Mononuclear Cells	[20]
	Skeletal Myoblast	[21]
Chitosan	Embryonic stem cells	[26]
Matrigel	Embryonic stem cells	[32]
Collagen/matrigel	Neonatal cardiomyocytes	[31]
Fibrin, collagen, Matrigel	None	[22]

This overview lists most of the hydrogels injected into myocardial infarct and the association with cells or biological factors vivo neo-vascularisation process via the recruitment of host cells [20–22]. Third, the elasticity of the matrix also plays a particularly important role in cell fate. As recently reported, adult stem cells such as mesenchymal stem cells (MSC) seem to demonstrate an extremely sensitive capacity to commit to specific lineage depending on the elasticity of the matrix. Soft matrices with an e-modulus that mimics brain tissues appear to be rather neurogenic whereas stiffer matrices with elastic properties similar to muscle tissues will favour a more myogenic cellular commitment, and rigid matrices that mimic collagenous bone tissues are more osteogenic [23].

Finally, the physical state of hydrogels can be modulated and two forms of hygrogel-based myocardial matrices can be considered: a viscous liquid injectable form (mostly used for cell delivery), and a viscous elastic gel form, a material used for three-dimensional constructs.

3.1 Injectable Hydrogels

The injectable form of hydrogels can be considered as an intermediate step between the direct cell injection approach and the in vitro engineered tissue. Injectable hydrogels have been developed to encapsulate clusters of transplanted cells and improve the cell engraftment [24–26]. These hydrogels are originally in a liquid phase and are thus easy to inject into a target tissue. In situ, however, they will rapidly polymerise into a gel structure in response either to temperature or pH changes. The added swelling of these polymerised hydrogels will create thus a capsule, retaining the cells at the injection site and limiting the significant loss observed following direct injection of cells alone [27]. Different types of hydrogels that have been injected in animal models of myocardial infarction are presented in Table 1.

The main advantages of using injectable hydrogels rely on several aspects. Obviously, the in situ polymerisation of the gel enables a percutaneous cell delivery into the heart [27, 28]. But maybe more interestingly, the sole injection of the hydrogel directly into an infarct area has also demonstrated the potential to positively influence the evolution of heart failure. More specifically, this acellular ECM-oriented strategy demonstrates the role of the matrix to maintain or restore the homeostasis of the myocardial ECM, and consequently the importance of its participation to the prevention of cardiac remodelling and dysfunction. For instance, several investigations have now demonstrated that injected hydrogels increase thickness of the ventricle wall, prevent fibrous tissue formation and stimulate heart function recovery. The mechanism may be related thus to the changes in scar elasticity and wall thickness and the subsequent decrease of the wall stress. But other explanations may be involved such as hydrogel-induced angiogenesis of the infarcted myocardium [29, 30].

Nevertheless, cellular hydrogels reveal superior effect on heart function as compared to either the injection of cells or the matrix alone. In particular, the groups of Zhang et al. [31] compared the injection of cardiomyocytes alone, matrigel/collagen alone, or a combination of both into a cardiac scar 3 weeks after creation of an infarct. They reported that the combination of neonatal cardiomyocytes with the artificial matrix was able to improve the heart function most effectively. As expected, cell survival was also improved in this study group. Increased restoration of heart function was confirmed in another study [32] in which a mixture of matrigel and embryonic stem cells were injected into a mouse infarct and compared to the injection of cells or matrix alone.

The principal drawback of injectable hydrogels is, however, the impossibility to control the alignment of the cells and more generally the organization of the three dimensional structure. Injection of cellular hydrogels forms bulk cellular masses at the injection sites, and result thus in the heterogeneous formation of islets of cells within the cardiac tissue.

3.2 Implantable Constructs

As opposed to their injectable counterparts, hydrogels used as a gel-based matrix already allows the creation of a 3D structure in vitro and may then be applied as a ready to implant patch directly on the target organ. This approach involves the preimplantation growth, differentiation and conditioning of the cells. The main role of hydrogels is thus to provide the cells with an extracellular matrix, which may stimulate their growth and differentiation into a tissue-like structure. In addition, they allow a high diffusion of nutrients, an homogenous distribution of the cells within the matrix and the rapid formation of an organized and possibly pulsative construct. Despite these remarkable advantages, hydrogel scaffolds have been less closely studied compared to dry and solid scaffolds such as collagen sponge or fleece.

Neonatal cardiomyocytes are the typical cells used for the engineering of implantable contractile constructs [33], but embryonic stem cells or endothelial progenitors have also been described [34, 35]. Also, myoblast cells hold promises as a readily accessible source of cells for bioartificial cardiac myocardium replacement [36].

Interestingly, pulsative tissues were first developed with skeletal myoblasts. Two different methods were initially proposed to engineer a functional 3D structure. Practically, the gel-based matrix was seeded with cells and this initial cellular mixture placed in a casting mold [33, 34, 37] in which, within a 4 to 7 days period, the cells formed a structured network and participated to the hardening of the hydrogel. Another strategy was developed by Strohman et al. [38] Dennis et al. [39] and modified by Huang et al. [40] and established the engineering of a 3D muscle from myoblast/fibroblast primary cells cultured in a thin layer of hydrogel composed of fibrin. In this technique, two artificial tendons were positioned in contact with the gel and in a co-axial way and separated from each other by about 12 mm. Upon differentiation of the myoblasts, the gel retracts and detaches from the plate before rolling itself around the tendon virtual axis into a cylindrical structure. In this set up, the engineered muscle may spontaneously contract.

Regarding the composition of the matrix, collagen/matrigel mixtures have been proposed by the group of Powel et al. [36] for the creation of bioartificial muscle (termed BAM), Eschenhagen et al. [41] applied this methods for the creation of engineered heart tissues (EHT), whereas Huang et al. [41] have used fibrin gels for the creation of a bioengineered heart muscle (BEHM). Our group used both described strategies to produce engineered skeletal muscle grafts (ESMG) [37, 42]. We examined the role of biodegradable gel-type scaffolds composed of collagen I and, matrigel in promoting myoblast growth and differentiation, and formation of 3D structures cast in a simple mold (disk shape) or anchored between 2 artificial tendons (cylindric shape).

3.2.1 Methodology for ESMG Preparation

Cell isolation Harvesting of skeletal muscle cells from adult male Lewis rats (100–150 g) is performed according to a modified method of Kosnik et al. [43]. Briefly, soleus and extensor digitorum longus muscles are surgically removed under sterile conditions, exposed to UV for 1 h and preincubated for 48 h at 37°C. Each muscle is then dissociated in 15 ml of a solution containing collagenase 100 U/ml and dispase 4U/ml for 3–4 h at 37°C. Cells are plated and incubated for 48 h. Adherent cells are rapidly harvested and frozen. The procedure gives a yield of desmin positive cells of $29 \pm 13\%$. When cells were cultured longer the content of myoblasts decreased. This may be explained by either the faster growth rate of fibroblasts compared to myoblasts and the consequent overtaking of the cell culture, or by a loss of the myoblast phenotype of the primary cells.

Matrix preparation Based on a modified protocol from Eschenhagen et al. [33] the matrix is obtained by mixing, in ice cold tubes, commercially available rat tail collagen (2 mg/ml) with Matrigel (2 mg/ml, BD Biosciences, Bedford, MA, USA) diluted with culture medium. The pH is neutralized with NaOH.

Disk shape ESMG The matrix (300 µl) is then mixed with 100 µl myoblasts (1 million/100 µl). The cell–matrix mixture is deposited in 24-well plates and incubated overnight at 37°C and 5% CO2. Thereafter, 2 ml culture medium (containing 20% FBS) is added to each dish and the culture medium is changed every second day for 7 days. The myoblasts start to differentiate into myotubes within 2 days. The matrix rapidly solidifies and forms a 3-dimensional structure, detaching from the culture plate after 4–5 days. After 7 days, randomly oriented small myotubes are clearly formed and distributed widely within the scaffold (Fig. 2). The matrix/differentiated cells form a thick gel-like structure suitable for surgical implantation (Fig. 3) [37].

Cylindric shape ESMG According to a modified procedure from Kosnik et al. [43], 35 mm culture dishes are coated with Sylgard 184 and laminin. Two 6 mm artificial tendons cut from suture silk are soaked in laminin solution, placed 12 mm

Fig. 2 Phase contrast of randomly oriented myotubes formed within the disk shape ESMG after 7 days in culture

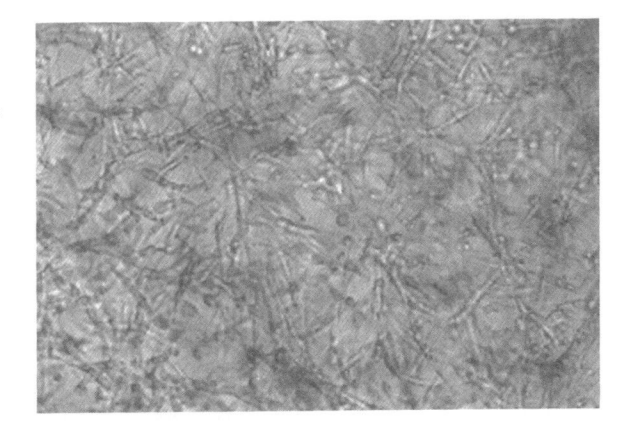

Fig. 3 Surgical implantation of disk shape ESMG. The graft was glue (using fibrin glue, Tissucool, Baxter) at the surface of a myocardial infarct of a rat model, 2 weeks post left anterior descending artery

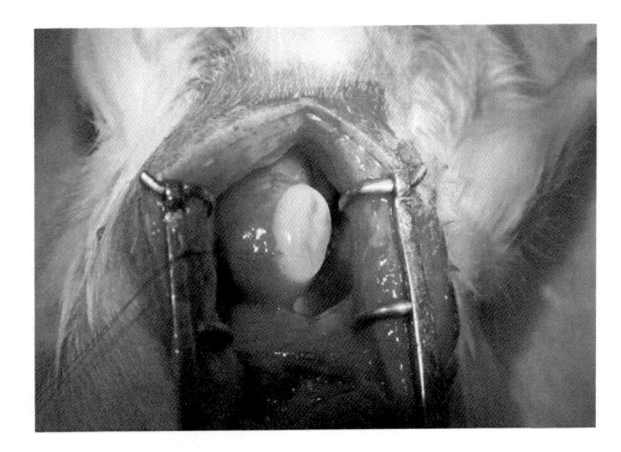

apart in the culture dish and pinned down with insect 0.1 mm pins. The culture dishes are placed under UV radiation for sterilisation for 1 h. The cell/matrix mixture (2.10^6 cells in 100 μl + 300 μl matrix) is deposited onto the slygard/laminin coating and incubated for 48 h. The medium is changed to differentiating medium (containing 5% horse serum) when cells reach confluency. The matrix starts to retract. Artificial anchors prevent tissue detachment and create passive stretch. Cylindric structures form within 15 days (Fig. 4) and start to spontaneously beat. The developed contractile forces can easily be assessed using force transducer equipped chambers and upon electrical stimulation, cylindric constructs develop a force of 272 ± 92 μN [42]. The passive stretch induced by anchorage of the construct may explain the alignment of the myotubes (Fig. 5) and the pulsative characteristic of the construct. Conversely, no spontaneous contraction has been observed in disk shaped ESMGs.

Fig. 4 Cylindric ESMG form by artificial silk anchorage of the matrix/cell mixture culture for 15 days

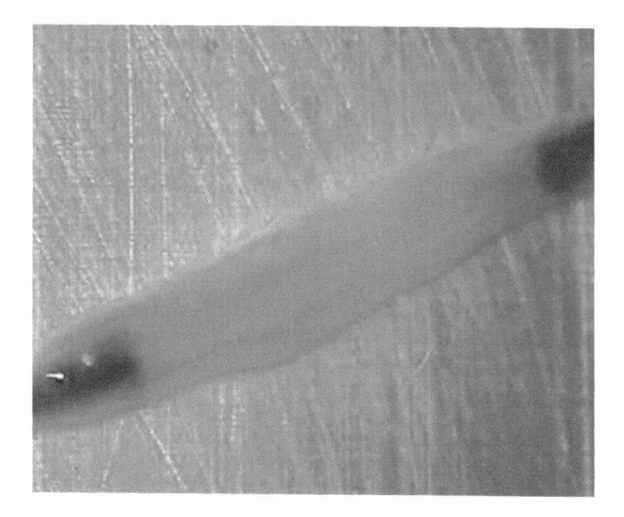

Fig. 5 Immunofluorescent staining of the desmin positive myotubes and nuclei (DAPI) of cylindric ESMG cultured for 15 days. The myotubes aligned in parallel to the passive tension created by the artifical anchors

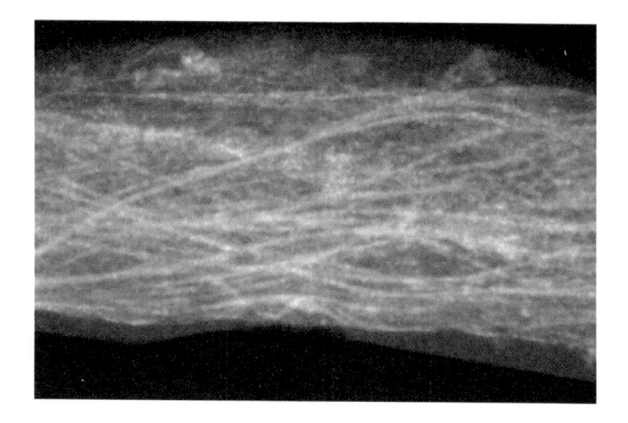

3.2.2 ESMG Implantation and In Vivo Evaluation

Standard in vivo evaluation of ESMG were performed using myocardial infarction animal models. We implanted disk-shaped ESMGs using fibrin glue at the surface of myocardial infarcted areas in a rat model and observed the heart function 4 weeks later. We have shown an improved systolic wall thickening and a significant increase in the fractional shortening as compared to control groups of infarcted animals that received either fibrin glue only or no treatment. In agreement with previous investigation, we also confirmed the neo-vascularisation of the graft. However, we found that this observed peri- and intra-infarct vascularisation was similar in both the ESMG and fibrin glue treated animals and could therefore not explain the functional improvement observed. Granulation tissue was also detected and probably inherent to the presence of matrigel [37].

For cylindric ESMGs, however, their epicardial implantation is challenging. For this reason they were not investigated in vivo despite their pulsative properties. To our knowledge, only a few studies, [44, 45] compared contracting and non-contracting construct for myocardial regeneration. The authors reported that the implantation of a construct with contractile properties is favourable for improvement of regional systolic function. In the same study, it was also demonstrated that contraction of the grafts plays a role in the recovery effect, however, the contraction properties may not account for possible scar stabilisation or paracrine effect [44].

3.3 Mechanical and Electrical Culture Conditioning

Creating an engineered tissue with contractile properties is of greatest interest. But developing a rhythmic, synchronous and sustainable contractile activity is obviously a major challenge. Several factors have been shown to influence the development of a contractile function. For instance, the cell seeding density [46], the adaptation of culture conditions by adding growth factors [47] or eliminating streptomycin [48] and the culture duration [41] are all factors that play a role in the generation of a construct with contractile properties. In addition, conditioning the construct by mechanical and/or electrical stimulation has shown to greatly improve the contractile potential [49].

3.3.1 Mechanical Conditioning

Cardiomyocytes responses to mechanical stretch have been substantially evaluated [50]. It was demonstrated that mechanical stimulation controlled cell orientation, enhanced organisation, improved intercellular connection via connexin 43 and consequently ameliorated the development of conductive properties close to native cardiac tissue [51]. Furthermore, differentiation into striated muscle can drastically be enhanced thus contributing to increase the amplitude of synchronous contractions. Indeed, it has been shown that generation of mechanical tension and forces influence the expression of contractile protein synthesis such as myosin heavy chain isoforms [52]. In addition the secretion of growth factors was also increased after cyclic stretching, this feature hold particular importance in cell therapy and its paracrine way of action [53].

Although the pathway driving response of cell cultured under loading condition is not yet fully elucidated, the mechano-chemical transduction of mechanical stress to signalling cascades and downstream nuclear and cytoplasmic processes is mediated via protein complexes such as focal adhesion sites and integrins [54]. Understanding these processes will be important for optimal tissue construct engineering.

Passive or cyclic stretch of hydrogel 3D structure has been well defined and optimized for engineered muscle such as BEHM [41], EHT [55] and BAM [36] and showed improved functional properties compared to construct lacking mechanical context. In particular, Powell et al. [36] presented real-time measurement of internal forces developed during the formation of bioartificial muscles. When followed by cyclic stretching, the constructs maintained their elasticity over the next 8 days. Practically, an optimal hydrogel-based graft should involve the development of a tissue with cohesive forces that are at least similar to the ventricular wall tensions generated during the cardiac cycle within a healthy myocardium. Additionally, the contraction forces should be similar to those generated during a normal systole. Using a mixture of collagen and Matrigel for the embedding of neonatal cardiomyocytes and culture under mechanical stimulation, Zimmermann et al. developed an EHT grafts that responded to calcium and isoprenaline stimulation and generated a maximal force of 350 µN with a specific force of 0.4 kN/m^2 [56]. Increases in construct strength and force development were obtained by the group of Birla [41]. Using a fibrin gel and neonatal cardiomyocytes, the authors developed BEHM grafts demonstrating an active force in the order of 835 µN and a specific force of 15 kN/m2.

Furthermore, combining hydrogel with solid structure scaffold, Boullik et al., submitted neonatal cardiac cells seeded on fibrin/polymeric knitted scaffolds to stretching condition [57]. The authors provide evidence that cyclic stretch induced increases in collagen and DNA. However, mechanical function of the construct such as ultimate tensile strength, failure strain and strain energy density were decreased compared to static culture conditions. Nevertheless, these properties might be directly link to the extracellular matrix and the type of scaffold. In particular, when Akhyari et al. compared the effect of stretching on grafts composed of gelatin seeded with human heart cells to unseeded scaffold and non stretched grafts, they showed an increased the tensile strength and resistance [58].

Cyclic mechanical strain for myocardial constructs has been generated via custom-made bioreactors [59–61] or commercially available ones (Flexcell or Biostretch), they regenerate uniaxial or biaxial tensile stresses (stretching), which are of particular interest for muscle construct as opposed to shear and compressive stress mostly applied to vascular constructs [62]. Commonly, stretch patterns can be modulated with varying strain magnitude as well as frequency for cyclic stretch mode or stretching and resting period for ramp stimulation. Not only hydrogel based construct benefited from mechanical stretching but also muscle construct composed of solid scaffold structure. Hydrogel based construct are generally developed in culture for several days and then fixed to the bioreactor for cyclic or ramp stimulation; in this case specific casting and anchorage processes are necessary prior to cyclic stretch. On opposite, for constructs based on solid scaffold such as Gelfoam, fixation is easier and stretching can be then applied from the beginning of the culture or following a 24 h period static time necessary for cell adhesion. In parallel, for mechanical conditioning of matrix free muscle construct, Lee and Recum [63] recently developed a new approach where cell sheet are stretched on their custom made device and subsequently intact sheet are detachment using a thermo-responsive elastic substrate.

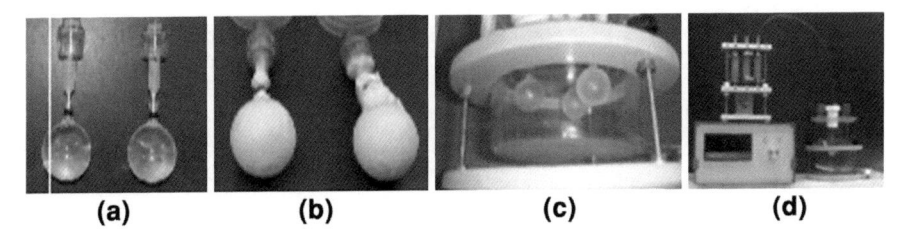

Fig. 6 Mechanical stimulation is applied by using a pumping system filled with distilled water. **a** 2 silicon bulb carriers; **b** two bulb carriers covered with electrospun PCL matrix; **c** culture chamber with three bulbs; **d** complete bioreactor setting

Techniques for mechanical stimulation of muscle construct culture are mainly performed in a planar manner [64]. However, this approach fail to mimic entirely physiological or pathological mechanical stress of the heart particularly the heart geometry as well as loading and unloading conditions. In particular, in clinical situations volume/pressure overload induce a progressive change from a normal ellipsoidal LV to a conoidal and spherical shape· accompanied by a progressive cellular and tissue remodelling showing the importance of the geometry of the left ventricle on cell response. [65, 66].

Accordingly, systems developing circumferential stress have also been designed [67, 68] The general principle is based on a flexible bulb carrier that is inflated and deflated by a pump. We developed a rapid cyclic pump for a cyclic frequency of 1–3 Hz and a slow cyclic pump to allow, respectively, cyclic and ramp stretches. A silicone bulb carrier covered with electrospun polycaprolactone (PCL) microfibers-based extracellular matrix is attached to the pump. Cells are seeded on the ECM and culture in a bioreactor. The circumferential stretch developed varies from base to apex of the bulb. Tissue structure such as thickness and orientation is dependant of the stretching conditions (Fig. 6) [68].

Furthermore, in order to reproduce in vitro all aspect of mechanical loading of the native heart, namely pulsative stretch and pressure, Giridharan et al. [69] recently presented their microfluidic cardiac cell culture model (μCCCM), however, this device has not been used so far for tissue culture but rather for monolayer cell culture.

3.3.2 Electrical Conditioning

It is well established that in vitro electrical stimulation of cardiomyocytes enhances cell maturation and coupling resulting in synchroneous contraction. Accordingly, at ultrastructural and molecular levels, gap junction formation and sarcomeric organisation are improved. Consecutively, mechanical properties and calcium transients are also enhanced. Electrical stimulation as part of the conditioning of engineered tissue is also a logical approach to improve their contractile potential. The group of Vunjak–Novakovi has intensively investigated the optimal electrical conditions to stimulate of 3D myocardial constructs composed of solid collagen sponge and neonatal cardiomyocytes [70]. They optimized a stimulation protocol aiming for the delivery of signals mimicking those observed in native heart tissues. Their results confirmed that their engineered constructs had enhanced functional performance as well as improved cell morphology and differentiation [71]. As far as skeletal myoblasts are concerned, electrical stimulation of a 3D hydrogel construct has a significant impact on myofiber diameter, composition and alignment as well as on their potential for myogenic differentiation [36, 72, 73].

In addition to contractile properties, electrical integrity of the construct hold particular importance for myocardial regeneration. Zimmermann et al. [44] provided evidence of synchronous contractions between the host myocardium and the graft and therefore confirmed the feasibility to create neo-tissues that have the potential to establish excitation coupling with the host myocardium. Synchronous beating may also be achieved with external pacing of the grafts. In this regard, the concept of Coghlan et al. [74] highlights the potential role that cardiac interstitial matrices may play in the propagation of the electrical signal within the 3 dimensional hierarchy by interconnecting ventricular myocytes layers. For example, the implantation of ions within the hydrogel would add conductive properties to the scaffold and thus allow it to directly stimulate the seeded cells. Using nanotechnologies, MacDonald et al. demonstrated that the electrical conductivity of cell-seeded collagen gels can be increased through the incorporation of carbon nanotubes [75].

3.4 Smart Hydrogels

Functionalisation of the hydrogel matrix can also be obtained by chemical modification, the addition of growth factors or incorporation of biological active compounds such as drugs. These processes aim to (i) influence the matrix itself with a change in the biomaterial properties according to external stimuli such as temperature, pH resulting in controlled and targeted drug release, (ii) increase materials biocompatibility, (iii) regulate the behaviour of the seeded cell, (iv) trigger specific response from the host organ. This approach is still in its infancy but appears very promising with multiple applications. Among them, smart hydrogels may permit investigation of the potential effect of biological molecules on cardiac repair and as well allow the control of spatial and temporal regenerative events to respond to the complexity of myocardium. In addition, smart hydrogels could stimulate in situ regenerative mechanism by creating an optimized microenvironment that will favour myogenesis, neovascularisation, and recruitment of stem cells and recovery of the heart function. Furthermore this approach could represent a wonderful platform for drug screening and delivery [76].

Various processing methods to conjugate bioactive factors to hydrogel components have been described and combinations of regenerative agents including peptides, growth factors and genes have been immobilized onto hydrogel [77].

For example, Kraehenbuehl et al. [35] reported that synthetic 3D matrix metalloproteinase (MMP)-sensitive poly ethylene glycol (PEG)-based hydrogels can direct differentiation of pluripotent cardioprogenitors in vitro. This effect can be modulated in relation to MMP sensitivity, matrix stiffness and concentration of bound peptide RGDSP. Smart hydrogels may also influence the neo-vascularisation processes since they could allow the delivery of vascular endothelial growth-factor (VEGF) and basic fibroblast growth factor (FGF), two main factors that control neoangiongenesis. Phelps et al. [78] engineered PEG-based bioartificial hydrogel matrices presenting MMP-degradable sites as growth factors release system, RGD peptide as cell-adhesion motifs, and VEGF to induce the growth of vasculature in vivo. They reported that implantation of their construct induced the growth of new vasculature into the matrix in vivo and resulted in significantly increased rate of reperfusion in a rat limb ischemic model. Interestingly, by combining hydrogel microspheres incorporating FGF and cardiomyocyte transplantation for the treatment of myocardial infarction in a rodent model, Sakakibara et al. [79] showed that FGF induced vascularisation of the scar, resulting in improved cell survival and enhanced ventricle elastance. Fractional shortening was improved compared to animals receiving no treatment but there was no significant difference between groups treated with FGF, cells alone or a combination of both.

The delivery of cardioprotective drugs via the implantation of hydrogel patches has been shown to be effective in heart regeneration. For example, Kobayashi et al. [80] used an erythropoietin (EPO)–gelatin drug delivery system (DDS) which allows local release of EPO. Continuous delivery of EPO over 14 days provided evidence for reduced MI size, improved LV remodelling and function by activation pro-survival signalling and exerting antifibrotic and angiogenic effects. In this study, the hydrogel delivery system demonstrated superior cardiac beneficial effects as compared to systemically administrated EPO and interestingly presented none of the side effect commonly observed after systemic administration.

Ultimately, a combination of hydrogel, solid scaffold, cells and biological molecules may represent the most potent construct. Recently Miyagi et al. [81] investigated this complex structure using a surgical ventricular restoration approach. Taking advantages of the respective properties of each component, the authors combined, respectively, the porous and biodegradable Gelfoam that allow cell infiltration and engraftment together with PCL in order to increase the scaffold strength. In addition, the angiogenic cytokines, stem cell factor and SDF 1, and mesenchymal stem cell were mixed in a biodegradable, temperature-sensitive hydrogel (copolymer of poly valerolactone and PEG) and injected on the scaffold during surgical replacement of the infarct scar. They provide evidence for better functional outcomes with the combined approach and in situ development of a viable tissue.

Table 2 Challenging aspects for the development of engineered myocardial tissues

Physical properties of the scaffold	Production of the construct	Function of the construct	Compatibility of the construct	Other aspects to consider
Size and porosity	Availability and quantity of cells	Contractile function	Biocompatibility	Implantation technique
Elasticity	Cell attachment and growth in vitro	Restraining function	Immunocompatibility	Regulatory aspects
Mechanical resistance	Cell differentiation	Paracrine function	Cell growth, differentiation, recruitability and migration in vivo	Ethical aspects
Biodegradability	Conditioning: mechanical (stress), physical (electrical)	Drug release function	Electrical and structural integration	Psychological aspects
Toxicity of the degradation products	Oxygen and nutrient supply		Inflammatory reaction	Economical aspects

4 Conclusion

Rapid progresses in the fields of biomaterials and in the identification of biological factors responsible for myocardial regeneration and stem cell differentiation will provide breakthroughs for replacement of the damaged myocardium. The feasibility of engineering a three-dimensional tissue for a potential myocardial application had been demonstrated, however, it also appears clear that the challenge remain enormous, especially if one aims to reproduce a fully functional myocardial tissue. It would therefore seem more reasonable to aspire to a more modest ambition and select just a few key properties that may contribute to support the heart in its functional recovery, rather than replace it. For this purpose it is important to recognize all the various aspects that challenge the engineering of a myocardial tissue as summarized in Table 2. Finally, understanding from a molecular, cellular and mechanical point of view, the mechanisms of myocardial recovery observed in animal models is paramount successful for transition toward clinical applications of myocardial tissue engineering.

Acknowledgments We thank Ms Laura Graham for excellent assistance in manuscript preparation, Ms Concetina Receputo and amd Mr Gilles Godar for excellent technical assistance. The research investigations were support by grants from the Swiss National Science Foundation (3200B0-108417 and 122334)

References

1. World Health Organisation (WHO) (2004) http://www.who.int/topics/cardiovascular_diseases/en
2. Bar, A., Haverich, A., Hilfiker, A.: Cardiac tissue engineering: "reconstructing the motor of life". Scand. J. Surg. **96**, 154–158 (2007)
3. Gammie, J.S., Sheng, S., Griffith, B.P., Peterson, E.D., Rankin, J.S., O'Brien, S.M., Brown, J.M.: Trends in mitral valve surgery in the United States: results from the Society of Thoracic Surgeons Adult Cardiac Surgery Database. Ann. Thorac. Surg. **87**, 1431–1437 (2009). Discussion 1437–1439
4. Zimmermann, W.H., Cesnjevar, R.: Cardiac tissue engineering: implications for pediatric heart surgery. Pediatr. Cardiol. **30**, 716–723 (2009)
5. Walsh, R.G.: Design and features of the Acorn CorCap cardiac support device: the concept of passive mechanical diastolic support. Heart Fail. Rev. **10**, 101–107 (2005)
6. Anstadt, G.L., Blakemore, W.S., Baue, A.E.: A new instrument for prolonged mechanical massage. Circulation **31**, S2–S43 (1965)
7. Wang, Q., Yambe, T., Shiraishi, Y., Duan, X., Nitta, S., Tabayashi, K., Umezu, M.: An artificial myocardium assist system: electrohydraulic ventricular actuation improves myocardial tissue perfusion in goats. Artif. Organs **28**, 853–857 (2004)
8. Mann, D.L., Acker, M.A., Jessup, M., Sabbah, H.N., Starling, R.C., Kubo, S.H.: Clinical evaluation of the CorCap Cardiac Support Device in patients with dilated cardiomyopathy. Ann. Thorac. Surg. **84**, 1226–1235 (2007)
9. Furnary, A.P., Chachques, J.C., Moreira, L.F., Grunkemeier, G.L., Swanson, J.S., Stolf, N., Haydar, S., Acar, C., Starr, A., Jatene, A.D., Carpentier, A.F.: Long-term outcome, survival analysis, and risk stratification of dynamic cardiomyoplasty. J. Thorac. Cardiovasc. Surg. **112**, 1640–1649 (1996). discussion 1649–1650
10. Chachques, J.C., Salanson-Lajos, C., Lajos, P., Shafy, A., Alshamry, A., Carpentier, A.: Cellular cardiomyoplasty for myocardial regeneration. Asian Cardiovasc. Thorac. Ann. **13**, 287–296 (2005)
11. Tamareille, S., Achour, H., Amirian, J., Felli, P., Bick, R.J., Poindexter, B., Geng, Y.J., Barry, W.H., Smalling, R.W.: Left ventricular unloading before reperfusion reduces endothelin-1 release and calcium overload in porcine myocardial infarction. J. Thorac. Cardiovasc. Surg. **136**, 343–351 (2008)
12. Blom, A.S., Pilla, J.J., Gorman 3rd, R.C., Gorman, J.H., Mukherjee, R., Spinale, F.G., Acker, M.A.: Infarct size reduction and attenuation of global left ventricular remodeling with the CorCap cardiac support device following acute myocardial infarction in sheep. Heart Fail. Rev. **10**, 125–139 (2005)
13. Brinks, H., Tevaearai, H., Muhlfeld, C., Bertschi, D., Gahl, B., Carrel, T., Giraud, M.N.: Contractile function is preserved in unloaded hearts despite atrophic remodeling. J. Thorac. Cardiovasc. Surg. **137**, 742–746 (2009)
14. Tevaearai, H.T., Eckhart, A.D., Walton, G.B., Keys, J.R., Wilson, K., Koch, W.J.: Myocardial gene transfer and overexpression of beta2-adrenergic receptors potentiates the functional recovery of unloaded failing hearts. Circulation **106**, 124–129 (2002)
15. Bergmann, O., Bhardwaj, R.D., Bernard, S., Zdunek, S., Barnabe-Heider, F., Walsh, S., Zupicich, J., Alkass, K., Buchholz, B.A., Druid, H., Jovinge, S., Frisen, J.: Evidence for cardiomyocyte renewal in humans. Science **324**, 98–102 (2009)
16. Passier, R., van Laake, L.W., Mummery, C.L.: Stem-cell-based therapy and lessons from the heart. Nature **453**, 322–329 (2008)
17. Zaruba, M.M., Franz, W.M.: Role of the SDF-1-CXCR4 axis in stem cell-based therapies for ischemic cardiomyopathy. Expert Opin. Biol. Ther. **10**, 321–335 (2010)
18. Banfi, A., von Degenfeld, G., Blau, H.M.: Critical role of microenvironmental factors in angiogenesis. Curr. Atheroscler. Rep. **7**, 227–234 (2005)

19. Giraud, M.N., Armbruster, C., Carrel, T., Tevaearai, H.T.: Current state of the art in myocardial tissue engineering. Tissue Eng. **13**, 1825–1836 (2007)
20. Ryu, J.H., Kim, I.K., Cho, S.W., Cho, M.C., Hwang, K.K., Piao, H., Piao, S., Lim, S.H., Hong, Y.S., Choi, C.Y., Yoo, K.J., Kim, B.S.: Implantation of bone marrow mononuclear cells using injectable fibrin matrix enhances neovascularization in infarcted myocardium. Biomaterials **26**, 319–326 (2005)
21. Christman, K.L., Fok, H.H., Sievers, R.E., Fang, Q., Lee, R.J.: Fibrin glue alone and skeletal myoblasts in a fibrin scaffold preserve cardiac function after myocardial infarction. Tissue Eng. **10**, 403–409 (2004)
22. Huang, N.F., Yu, J., Sievers, R., Li, S., Lee, R.J.: Injectable biopolymers enhance angiogenesis after myocardial infarction. Tissue Eng. **11**, 1860–1866 (2005)
23. Engler, A.J., Griffin, M.A., Sen, S., Bonnemann, C.G., Sweeney, H.L., Discher, D.E.: Myotubes differentiate optimally on substrates with tissue-like stiffness: pathological implications for soft or stiff microenvironments. J. Cell Biol. **166**, 877–887 (2004)
24. Li, X.Y., Wang, T., Jiang, X.J., Lin, T., Wu, D.Q., Zhang, X.Z., Okello, E., Xu, H.X., Yuan, M.J.: Injectable hydrogel helps bone marrow-derived mononuclear cells restore infarcted myocardium. Cardiology **115**, 194–199 (2010)
25. Yoon, S.J., Fang, Y.H., Lim, C.H., Kim, B.S., Son, H.S., Park, Y., Sun, K.: Regeneration of ischemic heart using hyaluronic acid-based injectable hydrogel. J. Biomed. Mater. Res. B Appl. Biomater **91**, 163–171 (2009)
26. Lu, W.N., Lu, S.H., Wang, H.B., Li, D.X., Duan, C.M., Liu, Z.Q., Hao, T., He, W.J., Xu, B., Fu, Q., Song, Y.C., Xie, X.H., Wang, C.Y.: Functional improvement of infarcted heart by co-injection of embryonic stem cells with temperature-responsive chitosan hydrogel. Tissue Eng. Part A **15**, 1437–1447 (2009)
27. Martens, T.P., Godier, A.F., Parks, J.J., Wan, L.Q., Koeckert, M.S., Eng, G.M., Hudson, B.I., Sherman, W., Vunjak-Novakovic, G.: Percutaneous cell delivery into the heart using hydrogels polymerizing in situ. Cell Transplant. **18**, 297–304 (2009)
28. Leor, J., Tuvia, S., Guetta, V., Manczur, F., Castel, D., Willenz, U., Petnehazy, O., Landa, N., Feinberg, M.S., Konen, E., Goitein, O., Tsur-Gang, O., Shaul, M., Klapper, L., Cohen, S.: Intracoronary injection of in situ forming alginate hydrogel reverses left ventricular remodeling after myocardial infarction in Swine. J. Am. Coll. Cardiol. **54**, 1014–1023 (2009)
29. Seif-Naraghi, S.B., Salvatore, M.A., Schup-Magoffin, P.J., Hu, D.P., Christman, K.L.: Design and characterization of an injectable pericardial matrix gel: a potentially autologous scaffold for cardiac tissue engineering. Tissue Eng. Part A **16**, 2017–2027 (2010)
30. Singelyn, J.M., DeQuach, J.A., Seif-Naraghi, S.B., Littlefield, R.B., Schup-Magoffin, P.J., Christman, K.L.: Naturally derived myocardial matrix as an injectable scaffold for cardiac tissue engineering. Biomaterials **30**, 5409–5416 (2009)
31. Zhang, P., Zhang, H., Wang, H., Wei, Y., Hu, S.: Artificial matrix helps neonatal cardiomyocytes restore injured myocardium in rats. Artif. Organs **30**, 86–93 (2006)
32. Kofidis, T., Lebl, D.R., Martinez, E.C., Hoyt, G., Tanaka, M., Robbins, R.C.: Novel injectable bioartificial tissue facilitates targeted, less invasive, large-scale tissue restoration on the beating heart after myocardial injury. Circulation **112**, I173–I177 (2005)
33. Eschenhagen, T., Didie, M., Munzel, F., Schubert, P., Schneiderbanger, K., Zimmermann, W.H.: 3D engineered heart tissue for replacement therapy. Basic Res. Cardiol. **97**(Suppl 1), I146–I152 (2002)
34. Jaconi M., Zammaretti-schaer P.: 3D-Cardiac Tissue Engineering For the Cell Therapy of Heart Failure. US (2008). http://www.freepatentsonline.com/y2008/0226726.htm
35. Kraehenbuehl, T.P., Zammaretti, P., Van der Vlies, A.J., Schoenmakers, R.G., Lutolf, M.P., Jaconi, M.E., Hubbell, J.A.: Three-dimensional extracellular matrix-directed cardioprogenitor differentiation: systematic modulation of a synthetic cell-responsive PEG-hydrogel. Biomaterials **29**, 2757–2766 (2008)
36. Powell, C.A., Smiley, B.L., Mills, J., Vandenburgh, H.H.: Mechanical stimulation improves tissue-engineered human skeletal muscle. Am. J. Physiol. Cell Physiol. **283**, C1557–C1565 (2002)

37. Giraud, M.N., Ayuni, E., Cook, S., Siepe, M., Carrel, T.P., Tevaearai, H.T.: Hydrogel-based engineered skeletal muscle grafts normalize heart function early after myocardial infarction. Artif. Organs **32**, 692–700 (2008)
38. Strohman, R.C., Bayne, E., Spector, D., Obinata, T., Micou-Eastwood, J., Maniotis, A.: Myogenesis and histogenesis of skeletal muscle on flexible membranes in vitro. In Vitro Cell. Dev. Biol. **26**, 201–208 (1990)
39. Dennis, R.G., Kosnik, P.E.: Excitability and isometric contractile properties of mammalian skeletal muscle constructs engineered in vitro. In Vitro Cell Dev. Biol. Anim. **36**, 327–335 (2000)
40. Huang, Y.C., Dennis, R.G., Larkin, L., Baar, K.: Rapid formation of functional muscle in vitro using fibrin gels. J. Appl. Physiol. **98**, 706–713 (2005)
41. Huang, Y.C., Khait, L., Birla, R.K.: Contractile three-dimensional bioengineered heart muscle for myocardial regeneration. J. Biomed. Mater. Res. A **80**, 719–731 (2007)
42. Giraud, M.N., Siepe, M., Ayuni, E., Tevaearai, H.T., Carrel, T.: A hydrogel based engineered skeletal muscle graft (ESMG) for cardiac repair. FASEB J. **19**, A1659 (2005)
43. Kosnik, P.E., Dennis, R.G.: Mesenchymal cell culture: functional mammalian skeletal muscle constructs. In: Atala, A., Lanza, R. (eds.) Methods Tissue Eng., pp. 303–305. Academic Press, San Diego (2002)
44. Zimmermann, W.H., Melnychenko, I., Wasmeier, G., Didie, M., Naito, H., Nixdorff, U., Hess, A., Budinsky, L., Brune, K., Michaelis, B., Dhein, S., Schwoerer, A., Ehmke, H., Eschenhagen, T.: Engineered heart tissue grafts improve systolic and diastolic function in infarcted rat hearts. Nat. Med. **12**, 452–458 (2006)
45. Miyagawa, S., Sawa, Y., Sakakida, S., Taketani, S., Kondoh, H., Memon, I.A., Imanishi, Y., Shimizu, T., Okano, T., Matsuda, H.: Tissue cardiomyoplasty using bioengineered contractile cardiomyocyte sheets to repair damaged myocardium: their integration with recipient myocardium. Transplantation **80**, 1586–1595 (2005)
46. Birla, R., Dhawan, V., Huang, Y.C., Lytle, I., Tiranathanagul, K., Brown, D.: Force characteristics of in vivo tissue-engineered myocardial constructs using varying cell seeding densities. Artif. Organs **32**, 684–691 (2008)
47. Huang, Y.C., Khait, L., Birla, R.K.: Modulating the functional performance of bioengineered heart muscle using growth factor stimulation. Ann. Biomed. Eng. **36**, 1372–1382 (2008)
48. Birla, R.K., Huang, Y.C., Dennis, R.G.: Effect of streptomycin on the active force of bioengineered heart muscle in response to controlled stretch. In Vitro Cell Dev. Biol. Anim. **44**, 253–260 (2008)
49. Fink, C., Ergun, S., Kralisch, D., Remmers, U., Weil, J., Eschenhagen, T.: Chronic stretch of engineered heart tissue induces hypertrophy and functional improvement. Faseb J. **14**, 669–679 (2000)
50. Terracio, L., Miller, B., Borg, T.K.: Effects of cyclic mechanical stimulation of the cellular components of the heart: in vitro. In Vitro Cell. Dev. Biol. **24**, 53–58 (1988)
51. Katare R.G., Ando M., Kakinuma, Y., Sato, T.: Engineered heart tissue: a novel tool to study the ischemic changes of the heart in vitro. PLoS One **5**, e9275 (2010). doi:10.1371/journal.pone.0009275
52. Luther, H.P., Hille, S., Haase, H., Morano, I.: Influence of mechanical activity, adrenergic stimulation, and calcium on the expression of myosin heavy chains in cultivated neonatal cardiomyocytes. J. Cell. Biochem. **64**, 458–465 (1997)
53. Ruwhof, C., van Wamel, A.E., Egas, J.M., van der Laarse, A.: Cyclic stretch induces the release of growth promoting factors from cultured neonatal cardiomyocytes and cardiac fibroblasts. Mol. Cell Biochem. **208**, 89–98 (2000)
54. Klossner, S., Durieux, A.C., Freyssenet, D., Flueck, M.: Mechano-transduction to muscle protein synthesis is modulated by FAK. Eur. J. Appl. Physiol. **106**, 389–398 (2009)
55. Zimmermann, W.H., Melnychenko, I., Eschenhagen, T.: Engineered heart tissue for regeneration of diseased hearts. Biomaterials **25**, 1639–1647 (2004)

56. Zimmermann, W.H., Schneiderbanger, K., Schubert, P., Didie, M., Munzel, F., Heubach, J.F., Kostin, S., Neuhuber, W.L., Eschenhagen, T.: Tissue engineering of a differentiated cardiac muscle construct. Circ. Res. **90**, 223–230 (2002)
57. Boublik, J., Park, H., Radisic, M., Tognana, E., Chen, F., Pei, M., Vunjak-Novakovic, G., Freed, L.E.: Mechanical properties and remodeling of hybrid cardiac constructs made from heart cells, fibrin, and biodegradable, elastomeric knitted fabric. Tissue Eng. **11**, 1122–1132 (2005)
58. Akhyari, P., Fedak, P.W., Weisel, R.D., Lee, T.Y., Verma, S., Mickle, D.A., Li, R.K.: Mechanical stretch regimen enhances the formation of bioengineered autologous cardiac muscle grafts. Circulation **106**, I137–I142 (2002)
59. Lee, A.A., Delhaas, T., Waldman, L.K., MacKenna, D.A., Villarreal, F.J., McCulloch, A.D.: An equibiaxial strain system for cultured cells. Am. J. Physiol. **271**, C1400–C1408 (1996)
60. Birla, R.K., Huang, Y.C., Dennis, R.G.: Development of a novel bioreactor for the mechanical loading of tissue-engineered heart muscle. Tissue Eng. **13**, 2239–2248 (2007)
61. Pang, Q., Zu, J.W., Siu, G.M., Li, R.K.: Design and development of a novel biostretch apparatus for tissue engineering. J. Biomech. Eng. **132**, 014503 (2010)
62. Barron, V., Lyons, E., Stenson-Cox, C., McHugh, P.E., Pandit, A.: Bioreactors for cardiovascular cell and tissue growth: a review. Ann. Biomed. Eng. **31**, 1017–1030 (2003)
63. Lee, E.L., von Recum, H.A.: Cell culture platform with mechanical conditioning and nondamaging cellular detachment. J. Biomed. Mater. Res. A **93**, 411–418 (2010)
64. Brown, T.D.: Techniques for mechanical stimulation of cells in vitro: a review. J. Biomech. **33**, 3–14 (2000)
65. Ashikaga, H., Covell, J.W., Omens, J.H.: Diastolic dysfunction in volume-overload hypertrophy is associated with abnormal shearing of myolaminar sheets. Am. J. Physiol. Heart Circ. Physiol. **288**, H2603–H2610 (2005)
66. Omens, J.H.: Stress and strain as regulators of myocardial growth. Prog. Biophys. Mol. Biol. **69**, 559–572 (1998)
67. Gonen-Wadmany, M., Gepstein, L., Seliktar, D.: Controlling the cellular organization of tissue-engineered cardiac constructs. Ann. N. Y. Acad. Sci. **1015**, 299–311 (2004)
68. Giraud, M.N., Bertschi, D., Guex, G., Näther, S., Fortunato, G., Carrel, T.P., Tevaearai, H.T.: A new stretching bioreactor for dynamic engineering of muscle tissues. In: Cikirikcioglu, M. (ed.) 45th Congress of the European Society for Surgical Research, pp. 33–36. Medimond International Proceedings, Geneva (2010)
69. Giridharan, G.A., Nguyen, M.D., Estrada, R., Parichehreh, V., Hamid, T., Ismahil, M.A., Prabhu, S.D., Sethu, P.: Microfluidic cardiac cell culture model (muCCCM). Anal. Chem. **82**, 7581–7587
70. Tandon, N., Cannizzaro, C., Chao, P.H., Maidhof, R., Marsano, A., Au, H.T., Radisic, M., Vunjak-Novakovic, G.: Electrical stimulation systems for cardiac tissue engineering. Nat. Protoc. **4**, 155–173 (2009)
71. Radisic, M., Park, H., Shing, H., Consi, T., Schoen, F.J., Langer, R., Freed, L.E., Vunjak-Novakovic, G.: Functional assembly of engineered myocardium by electrical stimulation of cardiac myocytes cultured on scaffolds. Proc. Natl. Acad. Sci. USA **101**, 18129–18134 (2004)
72. Moon du, G., Christ, G., Stitzel, J.D., Atala, A., Yoo, J.J.: Cyclic mechanical preconditioning improves engineered muscle contraction. Tissue Eng. Part A **14**, 473–482 (2008)
73. Serena, E., Flaibani, M., Carnio, S., Boldrin, L., Vitiello, L., De Coppi, P., Elvassore, N.: Electrophysiologic stimulation improves myogenic potential of muscle precursor cells grown in a 3D collagen scaffold. Neurol. Res. **30**, 207–214 (2008)
74. Coghlan, H.C., Coghlan, A.R., Buckberg, G.D., Cox, J.L.: 'The electrical spiral of the heart': its role in the helical continuum The hypothesis of the anisotropic conducting matrix. Eur. J. Cardiothorac. Surg. **29**(Suppl 1), S178–S187 (2006)
75. MacDonald, R.A., Voge, C.M., Kariolis, M., Stegemann, J.P.: Carbon nanotubes increase the electrical conductivity of fibroblast-seeded collagen hydrogels. Acta Biomater. **4**, 1583–1592 (2008)

76. Vandenburgh, H., Shansky, J., Benesch-Lee, F., Barbata, V., Reid, J., Thorrez, L., Valentini, R., Crawford, G.: Drug-screening platform based on the contractility of tissue-engineered muscle. Muscle Nerve **37**, 438–447 (2008)
77. Tessmar, J.K., Gopferich, A.M.: Matrices and scaffolds for protein delivery in tissue engineering. Adv. Drug Deliv. Rev. **59**, 274–291 (2007)
78. Phelps, E.A., Landazuri, N., Thule, P.M., Taylor, W.R., Garcia, A.J.: Regenerative Medicine Special Feature: Bioartificial matrices for therapeutic vascularization. Proc. Natl. Acad. Sci. USA **107**, 3323–3328 (2010)
79. Sakakibara, Y., Nishimura, K., Tambara, K., Yamamoto, M., Lu, F., Tabata, Y., Komeda, M.: Prevascularization with gelatin microspheres containing basic fibroblast growth factor enhances the benefits of cardiomyocyte transplantation. J. Thorac. Cardiovasc. Surg. **124**, 50–56 (2002)
80. Kobayashi, H., Minatoguchi, S., Yasuda, S., Bao, N., Kawamura, I., Iwasa, M., Yamaki, T., Sumi, S., Misao, Y., Ushikoshi, H., Nishigaki, K., Takemura, G., Fujiwara, T., Tabata, Y., Fujiwara, H.: Post-infarct treatment with an erythropoietin-gelatin hydrogel drug delivery system for cardiac repair. Cardiovasc. Res. **79**, 611–620 (2008)
81. Miyagi, Y., Zeng, F., Huang, X.P., Foltz, W.D., Wu, J., Mihic, A., Yau, T.M., Weisel, R.D., Li, R.K.: Surgical ventricular restoration with a cell- and cytokine-seeded biodegradable scaffold. Biomaterials **31**, 7684–7694 (2010)
82. Landa, N., Miller, L., Feinberg, M.S., Holbova, R., Shachar, M., Freeman, I., Cohen, S., Leor, J.: Effect of injectable alginate implant on cardiac remodeling and function after recent and old infarcts in rat. Circulation **117**, 1388–1396 (2008)
83. Dobner, S., Bezuidenhout, D., Govender, P., Zilla, P., Davies, N.: A synthetic non-degradable polyethylene glycol hydrogel retards adverse post-infarct left ventricular remodeling. J. Card. Fail. **15**, 629–636 (2009)
84. Wang, T., Wu, D.Q., Jiang, X.J., Zhang, X.Z., Li, X.Y., Zhang, J.F., Zheng, Z.B., Zhuo, R., Jiang, H., Huang, C.: Novel thermosensitive hydrogel injection inhibits post-infarct ventricle remodelling. Eur. J. Heart Fail. **11**, 14–19 (2009)

Engineering of Multifunctional Scaffolds for Myocardial Repair Through Nanofunctionalization and Microfabrication of Novel Polymeric Biomaterials

Elisabetta Rosellini, Caterina Cristallini, Niccoletta Barbani and Paolo Giusti

Abstract In this chapter the authors provide an overview of their research activity in the field of myocardial tissue engineering, focusing on the development of bioactive scaffolds able to guide cardiac tissue formation from dissociated stem cells. The chapter describes the preparation and characterization of new bioartificial polymeric systems, blends of natural polymers and a novel thermosensitive and bioresorbable copolymer. The functionalisation of selected polymers using different approaches is presented: surface modification by signalling peptides, application of bioactive molecules release systems and introduction of specific recognition sites by Molecular Imprinting technology. The processing steps to develop highly porous structures, injectable microspheres and innovative scaffolds resembling the cardiac extracellular matrix architecture are further described. Finally, results are presented in the context of the development of scaffolds with multifunctional properties for guiding stem cell plasticity towards myocardial regeneration.

1 Introduction

Myocardial infarction is the first cause of morbidity and mortality in industrialized countries [1, 2]. The ischemic event is followed by the formation of a fibrous scar, that alters the workload of the surrounding tissue and can lead to a congestive heart

---

E. Rosellini (✉), N. Barbani and P. Giusti
Department of Chemical Engineering, Industrial Chemistry and Materials Science, University of Pisa, Largo Lucio Lazzarino, 56126 Pisa, Italy
e-mail: elisabetta.rosellini@diccism.unipi.it

C. Cristallini and P. Giusti
CNR Institute for Composite and Biomedical Materials IMCB, Pisa c/o Department of Chemical Engineering, Pisa, Italy

failure [3]. Current treatments for acute myocardial infarction and subsequent heart failure include mechanical support using left ventricular assist devices and, ultimately, cardiac transplantation, but both of them present serious problems, such as the lack of organ donors and complications associated with rejection and infection. Cellular therapy has been proposed as a new approach to regenerate the injured myocardium. Cellular repair strategies include cellular cardiomioplasty and tissue engineering techniques.

The first approach focuses on repopulation of the injured myocardium by transplantation of healthy cells [4, 5]. Most studies demonstrate that this treatment can improve contractile function, but the efficacy of cell engraftment is very low.

For this reason, much effort is now conveyed to the development of tissue engineering strategies using scaffolds to successfully engraft new cells into the myocardium [6–10]. In this approach, biodegradable polymers may act as temporary scaffolds with the aim of augmenting the cell number, supporting the contractile function of the failing heart and providing a template for the new viable tissue. A critical step is therefore the engineering of suitable functionalized three-dimensional matrices, composed of natural and/or synthetic polymers, capable of guiding tissue repair.

The aim of our research activity in the field of myocardial tissue engineering is the development of bioactive scaffolds able to support and guide the cardiac tissue formation from dissociated stem cells.

The biomaterial is an important component in tissue engineering, since the scaffold temporarily replaces the extracellular matrix (ECM) for the seeded cells, until they produce their own matrix, which ultimately provides the structural integrity of the replacement tissue. Therefore, the first step of our work is the preparation and characterization of new polymeric biomaterials for application in myocardial tissue engineering. Blends of natural polymers, bioartificial polymers and newly synthesised biodegradable polymers are studied.

Besides providing a physical support, scaffolds should provide an ideal platform for cell-material communications, similarly to what occurred in the native tissue microenvironment. One strategy is to incorporate ECM molecules, involved in the regulation of stem cell processes, such as migration, adhesion, proliferation and differentiation. For this reason, the second step of our activity is the functionalization of the selected materials, in order to furnish bioactive scaffolds able to guide and control stem cell fate. Different approaches are investigated: surface modification of polymers using signalling proteins or peptides and Molecular Imprinting technology.

Cell-substrate interaction is critically influenced not only by the use of a biomimetic material, containing proteins or peptide sequences specific to cell adhesion or cell function, but also by the microtopography of the surface. Consequently, in the third phase of our research work, scaffolds are fabricated in three-dimensional structures of increasing complexity, first through freeze-drying and then with microfabrication techniques. Alternatively, injectable scaffolds deliverable through a minimally invasive procedure are studied.

The final step is the development of integrated experimental protocols for guiding stem cell plasticity. Bioactive scaffolds with specific sites for adhesion as well as for guided differentiation are prepared. The final verification of the bioactive scaffolds is carried out through in vitro cell culture tests with cardiac stem cells, in order to evaluate the effects of topology and biochemical factors on stem cell behaviour.

2 Novel Polymeric Biomaterials

Both natural and synthetic materials are currently being studied for cardiac tissue engineering. Among the natural materials are collagen type I [11–15], ECM proteins (Matrigel) [16, 17], gelatin [18], alginate [19], fibrin glue [20–22] and peptide nanofibers [23]. Synthetic materials have included polylactic acid, polyglycolic acid, polycaprolactone and their combinations [24–28], poly(glycerol sebacate) [29, 30], polyurethanes [31, 32] and poly(N-isopropylacrylamide) [33]. Despite interesting results documented by recent literature, the optimal scaffold for myocardial repair has not yet been found, because of the complex physiological properties of cardiac muscle tissue.

Over the past years, novel bioartificial systems, blends of natural polymers and synthesised biodegradable polymers were studied by our research group, for application in myocardial tissue engineering. An overview will be provided in this section.

2.1 Blends of Natural Polymers

For a long time, the ECM was considered to be a skeleton supporting the cohesion of tissues. It is now accepted that the dynamics of the molecular links, which exist between the ECM and cytoskeleton, via complexes of transmembrane proteins such as integrins and cadherins, play an important role in the activation of multiple signaling pathways [34].

In vivo stem cells are contacted by various soluble and insoluble ECM components that influence their differentiation. Therefore, scaffold microenvironment should incorporate the major components that are generally exhibited in vivo. It has been shown that the nature of ECM can greatly affect a number of phenotypic and genotypic characteristics of associated cells [35]. The ECM of each organ is distinct in its composition and hence in its interactions with the cells unique to that organ. Tissue-specific matrices influence the development lineage and may further promote organ differentiation.

The cardiac ECM is defined as a network surrounding and supporting the cells which make up the myocardium. The main components of the ECM include: structural proteins, such as collagen types I and III and elastin; adhesive proteins, such as laminin, fibronectin and collagen types IV and VI; anti-adhesive proteins, such as tenascin, thrombospondin and osteopontin; proteoglycans and enzymes, such as metalloproteinases, which regulate the organization and composition of the ECM [36].

On the basis of these considerations, we decided to prepare blends of natural polymers, mimicking the composition of the cardiac ECM. Gelatin was chosen as the protein component; alginate was used to replace the polysaccharide component. Gelatin was employed instead of collagen because while collagen expresses antigenicity in physiological condition, gelatin is known to have no such antigenicity. Moreover, gelatin is practically more convenient than collagen because a concentrated collagen solution is extremely difficult to prepare from the native collagen and furthermore gelatin is far cheaper than collagen.

Alginate/gelatin blends with different weight ratios were studied [37–39]. The two natural polymers were dissolved in bi-distilled water and then mixed, in order to obtain blends with 20:80, 30:70, 40:60, 50:50, 60:40, 70:30 and 80:20 weight ratios. Thin polymeric films were obtained by casting and subsequently treated by exposure to glutaraldehyde (GTA) vapours, for gelatin chemical cross-linking, and by immersion in $CaCl_2$ solution, for alginate ionic cross-linking. The physico-chemical characterization, performed by infrared spectroscopy (FT-IR), differential scanning calorimetry (DSC) and thermogravimetric analysis (TGA), revealed a good miscibility and the presence of interactions among the functional groups of the two biopolymers. In particular, infrared spectra showed the presence in the blends of all the typical adsorption bands of pure polymers and their displacement towards higher or lower frequencies, due to the presence of interactions among the typical functional groups of pure polymers [40]. This result, in agreement with what reported by Dong et al. [41], was also confirmed by TGA: the thermograms of the blends showed all the degradation events of pure polymers and their displacement toward higher temperatures in the blends; moreover, the total weight loss for the blends was lower than for pure components. These result suggested a better stability for the blends, with respect to thermodegradation, and confirmed the presence of interactions among the functional groups of pure components. Swelling and in vitro degradation tests were performed in different solutions, simulating body fluids: phosphate buffer solution (PBS) only, cell culture medium, PBS supplemented with collagenase enzyme. Both swelling degree and weight loss were higher in PBS and for the blends with a higher content of gelatin. These results indicated a better stability for the blends in culture medium than in PBS, probably due to the different ion content and colligative properties of the two different solutions. Moreover, since the weight losses in PBS collagenase solution were very similar to those obtained in PBS only, a mainly hydrolytic degradation process can be supposed. Cell culture tests were performed using C2C12 myoblasts cell line, as a model of a possible cell source for myocardial tissue engineering [42]. At fixed times, cell proliferation and differentiation were evaluated labeling cells with 4',6-diamidino-2-phenylindole (DAPI), for cell nuclei staining, and with rhodamine-phalloidin, for actin filament staining. The results showed a good cell proliferation for all the blends containing more than 60% of gelatin, with the alginate/gelatin 20:80 showing the best response. The same blend was also the only one that supported cell differentiation. Therefore, the alginate/gelatin 20:80 was selected as a suitable material to prepare scaffolds for myocardial tissue engineering.

2.2 Bioartificial Polymeric Systems

The second class of materials that were studied for application in myocardial tissue engineering are bioartificial polymeric systems. Polymers constitute the main class of materials used for scaffold preparation. Synthetic polymers are available with a wide variety of compositions, properties and forms and they may be readily fabricated into complexes shapes and structures. Nevertheless, their interaction with living tissue constituents remains the major problem to be solved. On the other hand, biological polymers show in general good biocompatibility, but their mechanical properties are often poor, their processability is complicated by the necessity of preserving biological properties and their production or recovery costs are very high. A new class of polymeric materials, based on blends of both synthetic and natural polymers, has been defined by our research group as "Bioartificial Polymeric Materials" and was originally conceived with the aim of developing new biomaterials combining the features of synthetic polymers with the specific tissue- and cell compatibility of biopolymers [43]. The performance of these materials relies on the relationship between the role played by the synthetic-natural polymer interactions and the cell and tissue compatibility of the resulting material. The basic idea for the development of bioartificial polymeric materials is to reduce the interactions between synthetic and living systems. A two (or more) component material is created, inside which changes at the molecular level have already occurred before the contact with the living tissue, because of the interactions between the synthetic and the biological component. Such a material, with pre-established molecular interactions, should behave better macroscopically then a fully synthetic material, with regard to the biological response of the host.

New bioartificial systems based on alginate, collagen and gelatin, as the natural component, and poly(N-isopropylacrylamide) (PNIPAAm), as the synthetic component, were studied for applications in the field of myocardial repair [37, 44]. PNIPAAm/alginate, PNIPAAm/collagen, PNIPAAm/gelatin and PNIPAAm/alginate/gelatin films, with different weight ratios, were obtained by casting and then cross-linked with calcium ions and/or GTA vapours, depending if they contained alginate and/or protein components. The morphological analysis showed a microporous surface for both the PNIPAAm/alginate and PNIPAAm/gelatin blends, with the 30:70 weight ratio. Since it was demonstrated that gradient of cell substratum rigidity can influence the direction and speed of cell motility [45], the topographical and qualitative mechanical investigation of the polymeric films was carried out through atomic force microscopy, using the force modulation microscopy technique, that permits to detect variations of the mechanical properties of a surface. PNIPAAm/alginate 30:70, PNIPAAm/gelatin 30:70 and PNIPAAm/alginate/gelatin 20:40:40 showed gradients of elasticity, that could influence cell adhesion. The physicochemical characterization, carried out by FT-IR and TGA, did not reveal phase separation among the polymers used for blend preparation and suggested the presence of interactions, mainly hydrogen bond type, between the functional groups of pure components. The biological characterization demonstrated the ability of PNIPAAm/alginate/gelatin 20:40:40 blend to sustain and promote both cell proliferation and differentiation, suggesting its possible use as scaffold material for myocardial tissue engineering.

2.3 Thermosensitive and Bioresorbable PNIPAAm-Based Copolymers

Injectable systems have received much interest in tissue engineering area because they can be delivered through a simple and relatively non-traumatic procedure, as compared with a complex surgical intervention [46, 47].

In the traditional tissue engineering approach, cells are cultured on a biomaterial scaffold in vitro and then the obtained tissue is implanted in the appropriate anatomical location. Although this approach is fascinating, it presents several limits, such as the small dimension of the engineered tissue, the poor graft survival and, above all, the need for a surgical procedure. An alternative is given by the in situ tissue engineering. In this approach, unseeded scaffolds or injectable biomaterials are implanted on the damaged tissue and they create a friendly environment for the proliferation and differentiation of implanted or migrated cells.

For a patient, injectable systems, deliverable through a minimally invasive procedure, offer the advantage of avoiding surgical procedures and complications. Additionally, injectable biomaterials that form scaffolds in situ have the advantage of being able to take the shape of a tissue defect, avoiding the need for patient specific scaffold prefabrication. Also problems of cell adhesion and bioactive molecule release are overcome as, under proper conditions, they can be easily incorporated in the solution by mixing prior to injection.

In developing an injectable system, one strategy is to employ the principle of lower critical phase separation. Polymer mixtures that demonstrate this behaviour experience a phase change from a soluble state to an insoluble state when the temperature is increased above the lower critical solution temperature (LCST) [48]. Lower critical phase separation is generally regarded as a phenomenon governed by the balance of hydrophilic and hydrophobic moieties on the polymer chain and driven by a negative entropy of mixing. In addition, the temperature dependence of certain molecular interactions, such as hydrogen bonding and hydrophobic effects, contribute to this type of phase separation. At the LCST, the hydrogen bonding between the polymer and water becomes unfavourable compared to water–water and polymer–polymer interactions, and an abrupt transition occurs as the hydrated hydrophilic macromolecule quickly dehydrates and changes to a more hydrophobic structure.

Among the different thermosensitive polymers, poly(N-isopropylacrylamide) (PNIPAAm) has received a great attention, since it exhibits a lower critical phase separation from water at an LCST close to normal body temperature, approximately 32°C [49]. It has been suggested that temperature-induced phase transition of PNIPAAm in aqueous solution is mainly driven by the thermal destruction of hydrogen bonds between water molecules and hydrophilic groups, such as -CO- and -NH- in the NIPAAm monomer and the increase of interactions between the hydrophobic segments in the polymer with the increasing of temperature. Thanks to this behaviour, it can be injected in a liquid form and creates a gel inside the body. However, the major drawback for the use of this polymer is that it is non-biodegradable. As known, the importance of the biodegradability for a biomaterial is self-evident, due to the absence of a chronic foreign body reaction, which usually occurs with the permanent presence of the non-biodegradable materials.

In order to overcome this limit, several groups have been working to make biodegradable NIPAAm copolymers. In general, the biodegradation of PNIPAAm-based thermosensitive copolymer was improved by adding hydrolytically and/or enzymatically degradable moieties [50–56]. However, these polymers are not totally biodegradable, due to the presence of non-biodegradable PNIPAAm segments. Another approach is to create copolymers with hydrolysis-dependent LCST properties. Neradovic et al. reported the synthesis of a new type of thermosensitive and bioerodible PNIPAAm-based copolymer with hydrolysable lactic acid ester side group, in which the lactate groups are released by hydrolysis, resulting in increased LCST values [57]. Similar results were obtained by Cui et al. through the synthesis of NIPAAm with dimethyl-γ-butyrolactone acrylate (DBA) [58]. 2-hydroxyethyl methacrylate-polylactide [59–61], poly(ε-caprolactone)-2-hydroxyethyl methacrylate [62, 63] and hydroxyethyl methacrylate-poly(trimethylene carbonate) [64] were also used as comonomers possessing a hydrolytic polyester side chain. In these studies, there can be undesired low molecular weight byproducts after hydrolysis. For example, decreases in pH in the local environment can be cytotoxic in the case of lactic acid [65].

The aim of our work was the synthesis of novel PNIPAAm-based copolymer, with hydrolysis-dependent LCST properties [37, 66]. N-isopropylacrylamide (NIPAAm) and 2-hydroxyethylmethacrylate-6-hydroxyhexanoate (HEMAHex) monomers were used. The first was chosen to confer the thermosensitive property, while the second to allow hydrolysis, thanks to the presence of a hydrolyzable ester group in its structure. The synthesis was performed by radical polymerization in tetrahydrofuran (THF) at 65°C for 24 h using 2-2'-azobisisobutyronitrile (AIBN) as the initiator. After polymerization, the copolymer (molecular weight = 20 KDa) was obtained by precipitation in diethyl ether, purified and then dried in a vacuum oven. To characterize the copolymerization process kinetically, the reactivity ratios were calculated by means of the Kelen–Tüdos method [67]. Since the product of the reactivity ratios was between 0 and 1, the behaviour of this comonomer system was between ideal random and alternating [68]. The characterization demonstrated that the newly synthesised poly(NIPAAm-co-HEMAHex) copolymer had temperature sensitivity, with the phase separation temperature under body temperature (exactly at 23°C), which is necessary for application as injectable scaffold for tissue engineering (Fig. 1). FT-IR and DSC results after hydrolysis tests indicated that incorporation of the HEMAHex ester groups provides the hydrolysis of the copolymer, which lead to an increase in the hydrophilicity of the copolymer and, consequently, to an increase in the LCST with time. Since the LCST increases above body temperature, the polymer becomes soluble again and can diffuse away. It was also demonstrated that the hydrolysis occurred on the peripheral ester bond of the lateral chain, with the release of 6-hydroxyhexanoic acid, whose biocompatibility has been already ascertained [69]. The mechanical characterization, performed through dynamic mechanical thermal analysis, showed that the elastic modulus increased passing from room temperature to body temperature. This increase in stiffness is suitable for tissue engineering application. At room temperature, poly(NIPAAm-co-HEMAHex) could be easily injected through a catheter or needle. As the temperature will be increased above the LCST to body temperature (37°C), the material rigidity will increase, to create a matrix that might support tissue growth. In addition, the good results obtained in the cytotoxicity and cytocompatibility tests suggested that poly(NIPAAm-co-HEMAHex) has potential for use as biomaterial for tissue engineering.

3 Advanced Functionalization Techniques

To guide the organization, growth and differentiation of cells in tissue engineered constructs, the biomaterials scaffold should be able to provide not only a physical

Fig. 1 Phase transition behaviour of poly(NIPAAm-co-HEMAHex): **a** at 20°C and **b** at 37°C

support for the cells but also the chemical and biological cues needed in forming functional tissues. In essence, the biomaterial should be able to cross-talk, on the molecular level, with the cells in a precise and controlled manner, analogous to cell–cell and cell–ECM communication and patterning during embryological development [70]. The binding domain of the ECM environment can be mimicked by a multifunctional cell-adhesive surface created by specific proteins, peptides and other biomolecules immobilized onto a material. Thus, in recent years, several methods for the bulk or surface modification of polymers with bioactive molecules have been studied [71].

To improve cellular proliferation and differentiation on synthetic materials, two different approaches were investigated by us: functionalization of polymers by immobilization or release of signalling proteins or peptides and realization of scaffolds able to control cell growth and differentiation, based on Molecular Imprinting technology. An overview of the two approaches and the obtained results will be provided in the following sub-sections.

3.1 Peptides Covalent Immobilization

In recent years, a large number of synthetic polymers with desired properties, in terms of mechanical stability and biodegradability, were made available as tissue engineering matrices. One important remaining problem is inadequate interaction between polymer and cells. Approaches to improve biomaterials include material modification by immobilization of cell recognition motives to obtain controlled interaction between cells and synthetic substrates [70]. Initially, these materials were coated with cell adhesive proteins such as fibronectin, collagen, or laminin. The use of proteins, however, bears some disadvantages in view of medical applications. First of all, proteins have to be isolated from other organisms and purified, therefore they may elicit undesirable immune responses and increase infection risks. In addition, proteins are prone to proteolytic degradation and consequently have to be refreshed continuously. Moreover, only a part of the proteins have proper orientation for cell adhesion, due to their stochastic orientation on the surface. One strategy to solve most of these problems is to use, instead of entire proteins, cell recognition motifs as small immobilized peptides. Peptides present several advantages, such as higher stability towards sterilization conditions, heat treatment and pH-variation, storage and conformational shifting as well as easier characterization and cost effectiveness. Furthermore, thanks to lower space requirement, they can be packed with higher density on surfaces. This provides a chance to compensate for possibly lower cell adhesion activity.

In multicellular organisms contacts of cells with neighboring cells and the surrounding ECM are mediated by cell adhesion receptors. Integrin family represents the most numerous and versatile group [72]. Integrins play a major role as anchoring molecules, but they are also involved in important processes like embryogenesis, cell differentiation, immune response, wound healing and hemostasis. It is well established that integrin mediated cell spreading and focal adhesion formation trigger survival and proliferation of anchorage dependent cells. In contrast, loss of attachment causes apoptosis in many cell types. Apoptosis can even be induced by the presence of immobilized ECM molecules when non-immobilized soluble ligands are added.

In order to provide a stable linking, peptides should be covalently attached to the polymer, through functional groups like hydroxyl-, amino-, or carboxyl groups. Since many polymers do not have functional groups on their surface, these need to be introduced by blending, co-polymerization, chemical or physical treatment [73–78]. Different strategies were also examined to covalently bind peptides or proteins to polymer surfaces: the most common one involves the reaction between substrate carboxylic groups and amino groups of the bioactive molecule, to generate an amide bond.

An additional point to take in consideration is the surface density of the bioactive molecules. It is well known that the number of attached cells is clearly related to peptide surface density and in particular cell attachment, as a function of peptide concentration, has a sigmoid increase. In merit to this, Massia and Hubbell demonstrated that a density of 1 fmol peptide/cm^2 is sufficient for cell spreading, while 10 fmol/cm^2 for focal contact formation, on a RGD functionalized glycophase glass surface [79].

The aim of our work was the surface modification of biodegradable synthetic polymers, via hydrolysis in sodium hydroxide solution and subsequent covalent attachment of biologically active peptide sequences [37, 80]. The materials used were commercial polycaprolactone (PCL), a tri-block poly(ester-ether-ester) copolymer and a novel biodegradable poly(ester urethane), synthesized in our lab. The tri-block polycaprolactone–polyoxyethylene–polycaprolactone copolymer, PCL–POE–PCL, was obtained by reaction of preformed poly(ethylene glycol) (PEG) with ε-caprolactone [81]. The copolymer formation occurred through a ring-opening mechanism, where the active hydrogen atoms of the preformed PEG determine a cleavage of the lactone ring. Then, the additions of monomeric lactone units occurs with the formation of two external polyester blocks. The synthesis was carried out in bulk, without the addition of any catalyst and it was thermally activated at a temperature of 185°C. The good biodegradability, biocompatibility and hemocompatibility of the copolymer were widely demonstrated [82]. The polyurethane used was a novel degradable segmented poly(ether–ester–urethane), obtained by using a L-lysine derived diisocyanate as chain extender. The synthesis, characterization and degradation studies of this polyurethane, hereafter indicated as PU, were already reported [83].

Polymeric films of the three different materials were obtained by casting from solution. The samples were immersed in sodium hydroxide solutions of appropriate concentrations and reacted for a predetermined period of time, at different temperature. The hydrolysis conditions, in terms of hydrolysis time, temperature and sodium hydroxide concentration, were optimized for the three different materials. After hydrolysis, the films were protonated with hydrochloric acid to yield polymer surfaces bearing carboxylic groups. The polymeric samples were further modified by attaching cell-recognizing peptides, through 1-(3-dimethylaminopropyl)-3-ethylcarbodimide hydrochloride (EDC)/N-hydroxysuccinimide (NHS) chemistry. The peptides sequences chosen for the functionalization were H-Gly-Arg-Gly-Asp-Ser-OH (GRGDS) from fibronectin and H-Tyr-Ile-Gly-Ser-Arg-OH (YIGSR) from laminin. Fibronectin is the main extracellular matrix adhesive glycoprotein and plays a fundamental role in the cell adhesion, migration and repair processes [70]. Laminin is a glycoprotein of the basal lamina and it has been demonstrated to increase the ability of stem cells to differentiate into beating cardiomyocytes [84].

The occurrence of the coupling reaction was demonstrated by infrared spectroscopy, as the presence on the functionalized materials of the adsorption peaks typical of the two peptides. The peptide surface density was determined measuring the corresponding residual amount in the coupling solution by chromatographic analysis and the distribution of the peptides was studied by FT-IR Chemical Imaging. The results showed a homogeneous peptide distribution, with a density several orders of magnitude above the minimum value necessary to promote cell adhesion [79]. The peptide superficial density was higher on PCL, because of the higher number of ester bonds (PCL–POE–PCL and PU contain, respectively, 70 and 80% of PCL). The introduction of the bioactive molecules had a positive effect on improving C2C12 myoblasts growth on the synthetic materials. In particular, the fibronectin GRGDS sequence showed a good ability of promoting cell proliferation, while the laminin YIGSR sequence appeared able to enhance cell differentiation, with the appearance of multinucleated myotubes in the absence of differentiation medium. This work could be considered a first step in the development of a synthetic ECM substitute, in which ligand type and density may be readily varied, in order to guide and control cardiac tissue formation.

3.2 Bioactive Molecules Release Systems

In native tissues, cells are surrounded not only by the ECM but also by soluble factors, that contributes to tissue regeneration and organogenesis. To induce tissue regeneration at the defective site, the culture microenvironment should therefore be able to deliver growth and differentiation factors for long term support of the surrounding cells, through the application of a drug delivery system [85, 86]. Typical release systems that can be applied for this task include biodegradable nano- or microparticles, that can be added to the prefabricated scaffolds or incorporated during scaffold manufacture.

The aim of our work was the preparation of alginate/gelatin microparticles to be used as delivery system in tissue engineering [37, 87]. Since it has been demonstrated that ascorbic acid promotes stem cell differentiation towards the cardiac phenotype [88], loading and releasing tests were carried out using ascorbic acid as bioactive agent. Microparticles were prepared using a single water in oil emulsion. The morphological analysis showed the formation of microspheres, with a diameter between 500 nm and 5 μm. Ascorbic acid was loaded on microspheres directly in emulsion or by adsorption method. The physicochemical analysis pointed out a more homogenous drug distribution with the adsorption method. Moreover, the morphological characterization showed that the interactions among ascorbic acid and the biopolymers did not allow the formation of microspheres in the emulsion method. The microspheres showed high encapsulation efficiency and a prolonged release, over a period of one month.

3.3 Molecular Imprinting Technology

A totally new approach for the creation of advanced synthetic support structures for cell adhesion and proliferation is represented by Molecular Imprinting (MI) technology [89–91]. This highly innovative technique permits the creation of intelligent scaffolds with recognition properties, capable of improving cell adhesion. MI technique in fact permits the production of synthetic polymers ("molecularly imprinted polymers", MIPs) capable of selectively linking themselves to a specific substance, called template. In recent years there have been various new openings for the application of MIPs and include the separation of macromolecules by chemical affinity [92, 93], the creation of artificial antibody-mimetic systems [94] and the creation of biomimetic sensors [89, 90]. Finally, it is possible to foresee very interesting development of MIPs for both the biomaterial sector in general [95, 96] and the tissue engineering sector in particular. To the best of our knowledge, our research group has been the first and, until now, the unique, to propose MI as a new nanotechnology for the creation of advanced synthetic support structures for cell adhesion and proliferation [97]. The basic idea is that intelligent matrices usable in the field of tissue engineering will be characterized by their particular capacity for molecular recognition of extracellular proteins (collagen, fibronectin, laminin, vitronectin, etc.) that favour cellular adhesion. These polymeric matrices will show a capacity for increasing the adhesion and growth characteristics of the cells on the scaffolds thanks to the presence of nanosites that are complementary and selective towards specific peptide sequences present in the ECM proteins and with whom integrin receptorial domains interact.

Considering that MI cannot be used directly on biodegradable polymers, which are a fundamental requisite for tissue engineered scaffolds, biostable polymers in the form of nanoparticles will be realized. They will be used in small quantities to modify degradable materials, so that these will retain their ability to serve as temporary scaffolds.

The use of entire proteins as template molecules in MI showed to be often unsuccessful because of the large molecular dimensions and the flexibility of the chains, which limit the polymer molecular recognition capacity and selectivity. To overcome these limitations, stable short peptide sequences, representative of an accessible fragment of a larger protein of interest, can be used, these often being located in the receptor domains or in other parts directly involved in the molecular recognition process (epitopes). Therefore, if the material can recognize a peptide that represents the exposed part of a protein structure, it will also be able to bind the entire protein [98].

The aim of our experimental work was the synthesis and characterization of MIPs capable of recognizing a peptide segment of an exposed part of a protein functional domain and their application as functionalization structures in the development of bioactive scaffolds [37, 99, 100].

The imprinted nanoparticles were prepared by polymerization of methacrylic acid (MAA) as the functional monomer in the presence of the template molecule and of an appropriate cross-linking agent, to provide a sufficient rigidity to the recognition sites in the polymer structure and an enhanced hydrophilicity. Using GRGDS or YIGSR peptides as template molecule, two formulations of MIPs were synthesized (respectively, MIP-GRGDS and MIP-YIGSR). Control particles (CP) were obtained with the same method, but in the absence of the template molecule. The full monomer and cross-linker conversion was observed after 20 h reaction time and the amount of template retained by the resin was in both cases very high (around 90%). For both formulations, polymers in the form of a macroporous monolith were prepared (Fig. 2a). The polymer monoliths were crashed and ground in a mortar and the template molecule was then removed from the polymers by solvent extraction. The amount of template removed from the particles was not complete, as a consequence of the high cross-linking degree and the rigidity of the matrix. In fact, it has been indicated that bulky templates cannot easily diffuse through a polymer network, even in the presence of a cross-linker containing a polar group that promotes extraction efficiency by increasing wettability. Moreover, this result could have been expected considering the high number of interactions among the peptides and the functional monomer. The performances

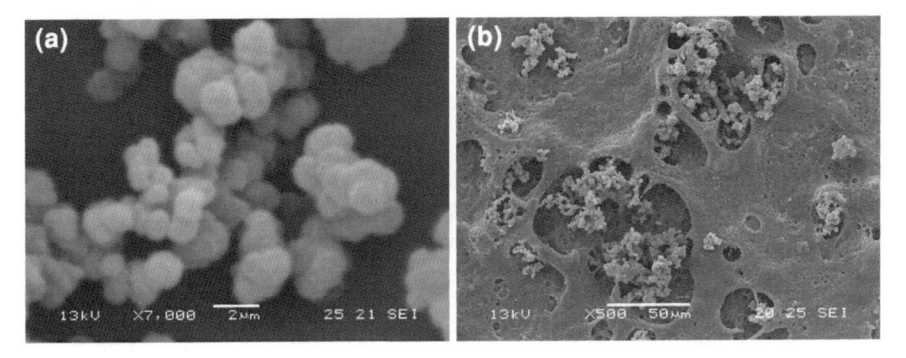

Fig. 2 SEM micrographs of MIPs with recognition properties towards fibronectin GRGDS peptide sequence (**a**) and PU film surface modified by MIPs deposition (**b**)

of the obtained polymers in selectively rebinding the template molecule were verified through recognition experiments in dynamic conditions. Both MIP–GRGDS and MIP–YIGSR showed a good performance in terms of recognition capacity and in terms of selectivity. An epitope effect was observed for MIP–GRGDS, as the extraction of the template allowed polymer to specifically recognize the fibronectin pentapeptide and also a larger fibronectin fragment, containing the GRGDS sequence. Cytotoxicity tests, performed by seeding mouse myoblasts in culture medium previously put in contact for 5 days with MIPs, showed normal vitality of C2C12 cells. The obtained results revealed possible application for MIPs as recognition materials for cell adhesive proteins; therefore, the functionalization of synthetic polymers was carried out by deposition of MIPs on their surface.

Imprinted particles were suspended in acetone and deposited on PCL, PCL–POE–PCL and PU film surfaces. The morphological analysis of bioactive scaffolds showed a quite homogenous particle distribution, even if they tend to form clusters (Fig. 2b). This result was confirmed by FT-IR Chemical Imaging. Recognition experiments were performed using polymeric films modified with CP as control. The deposition of imprinted particles did not alter their specific recognition and binding behaviour. Functionalized films showed even a higher quantitative binding than free particles, suggesting the creation of a preferred microenvironment for the rebinding process. The most remarkable result was obtained in the biological characterization: MIP-modified polymeric films significantly increased cell proliferation with respect to non functionalized materials. These results were very promising and suggested that MI can be used as an innovative functionalization technique to prepare bioactive scaffolds, with an effective capacity of improving tissue regeneration.

4 Scaffolds Fabrication

Cell-substrate interaction is critically influenced not only by the biochemical properties of the scaffolds, but also by the microtopography of the surface [101]. Scaffolds for application in the field of myocardial tissue engineering were mainly produced in the form of three-dimensional porous meshes or foams, through traditional fabrication techniques. Recently, more complex architectures that mimic cardiac tissue structure were produced by microfabrication techniques; in addition, several attempts were performed towards the development of injectable gels.

In our research activity, the materials selected for cardiac repair application were fabricated in three-dimensional structures of increasing complexity, first by freeze-drying of emulsions and then through microfabrication techniques. Moreover, microparticles based on biodegradable polymers were investigated for application as injectable scaffolds. These three different approaches will be presented below.

4.1 Highly Porous Sponges

Porous, absorbable matrices made from natural [102–105] and synthetic [19, 25, 26, 31, 106] polymers are currently being investigated as scaffolds for the purpose of cardiac tissue transplantation. Polymer scaffolds for use in cell transplantation must be highly porous with large surface/volume ratios to accommodate a large number of cells and to allow the neovascularization of the matrix. The scaffold should provide adequate sites for attachment and growth of enough cells to survive and function in vivo, yet should not limit the survival and growth of cells adjacent to the matrix surface as cells increase in number in vitro, prior to implantation and angiogenesis. Their chemistry as well as their pore characteristics, such as percentage porosity, pore size and interconnection, should allow the homogeneous distribution of cells within the matrix, their adherence and the retention of the differentiated function of attached cells. In addition to being biocompatible, scaffolds must be completely absorbable and easily eliminable from the body, when the need for an artificial support decreases. Another critical feature is their mechanical properties: besides mimicking the mechanical properties of the tissue that they are replacing, the matrices need to withstand the in vitro culturing conditions and the surgical procedure.

The aim of our work was the preparation and characterization of highly porous scaffolds obtained by freeze-drying technique [37]. Bioartificial polymeric systems and blends of natural polymers selected during the preliminary material screening were used. In particular, as bioartificial systems, PNIPAAm/alginate/gelatin with 20:40:40 weight ratio was initially investigated. In order to overcome the limit of non biodegradability of PNIPAAm, a similar bioartificial system where PNIPAAm was replaced by poly(NIPAAm-co-HEMAHex), the thermosensitive and bioresorbable copolymer newly synthesized by us, was also investigated. Concerning blends of natural polymers, considering that the final aim was to mimic the chemical composition of cardiac ECM, the selected alginate/gelatin blend with 20:80 weight ratio was compared with alginate/collagen blend, with the same composition [107]. Polymer solutions were mixed, poured in Petri dishes and freeze dried. Then they were cross-linked with GTA vapours and calcium ions and consequently immersed in a coagulation bath in acetic acid, to promote interactions among the functional groups of the different components. Samples were rinsed more times to remove traces of GTA and acetic acid, that could be dangerous for cells, and submitted to the final freeze-drying. The morphological analysis of all the prepared sponges showed a highly porous and homogeneous structure, with well interconnected pores (Fig. 3a). The presence of interactions among the functional groups of the biopolymers and a good chemical homogeneity were demonstrated by FT-IR Chemical Imaging. In vitro swelling and degradation tests demonstrated that the chemical cross-linking was necessary for a suitable stability of the sponges in aqueous environment and did not modify the biodegradability of the protein component. Stress–strain curves of all the prepared samples were recorded. It was observed that the Young's modulus of poly(NIPAAm-co-

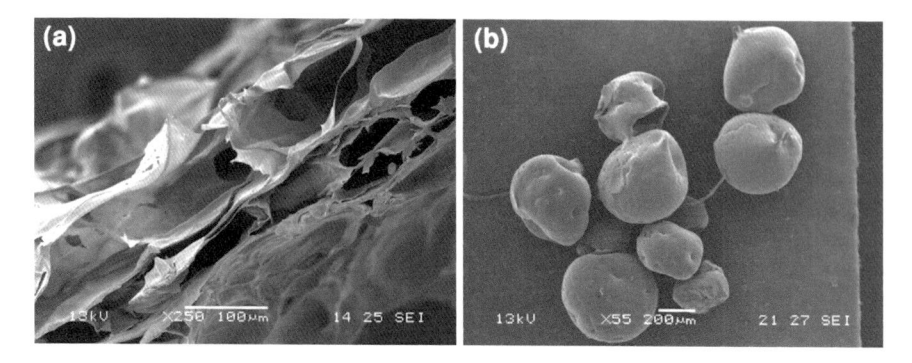

Fig. 3 SEM images of alginate/gelatin sponges obtained by freeze-drying (**a**) and alginate/gelatin microspheres obtained by water in oil emulsion (**b**)

HEMAHex)/alginate/gelatin sponges, evaluated under wet conditions, was 0.16 MPa, therefore very close to that of the human myocardium at the end of diastole [6]. In the case of PNIPAAm/alginate/gelatin samples, an interesting relationship was found among the mechanical properties of the samples and cell behaviour. Cell proliferation and differentiation tests with C2C12 myoblasts were performed on sponges with different mechanical properties, ranging from 3 to 20 MPa, depending on sample thickness. While myoblasts proliferation was not influenced by Young's modulus, samples with higher elastic modulus favored the cell differentiation process. This result was in agreement with the relationship among substrata stiffness and cell behaviour, widely described in literature [108]. The most remarkable result in the characterization of sponges based on blends of natural polymers was found in the biological characterization of alginate/gelatin sponges: in addition to a good myoblasts proliferation, the appearance of multinucleated myotubes was observed, in the absence of differentiation medium, suggesting that the material itself was able to promote cell differentiation.

4.2 Mimicking the Cardiac ECM Microarchitecture Through Microfabrication Techniques

Physical characteristics of the scaffold, defined by the microarchitecture and the surface topography, are known to extensively influence cell function and play a crucial role in tissue regeneration [109, 110].

When traditional scaffold fabrication techniques are used, macroscopic shapes are typically defined by processes such as extrusion, melt molding, and solvent casting. Material microstructure, in contrast, is often controlled by process parameters such as the choice of solvent in phase separation, doping with particulate leaching, gas foaming, woven fibers, and controlled ice crystal formation and subsequent freeze-drying to create pores; however, these scaffolds lack a well-defined organization that is found in most tissues in vivo [111–113].

Several microfabrication techniques for the development of scaffolds with controlled architecture have been described in literature, including Soft Litography [110], 3-D printing [114, 115], laser sintering [116], fused deposition modeling [117], multiphase jet solidification [118], 3-D bioplotter [119], indirect solid freeform fabrication [120] and pressure-activated microsyringe [121].

In 2003, Motlagh et al. [122] studied the cardiomyocytes behaviour on nontextured, microgrooved, micropegged or combined substrata, obtained by Soft Litography. They found that alterations in surface topography have an impact on shape, gene expression and protein localization of cardiac myocytes. More recently, Engelmayr et al. [123] developed an accordion-like honeycomb microstructure in poly(glycerol sebacate), with controllable stiffness and anisotropy. They demonstrated the formation of grafts with preferentially aligned neonatal rat heart cells and mechanical properties more closely resembling adult rat right ventricular myocardium.

The aim of our work was the preparation of three-dimensional microfabricated scaffolds, mimicking the cardiac ECM architecture, by using Soft lithography technique [37]. For the aimed purpose, samples of pig myocardium were decellularized by treatment with hypertonic solutions and trypsin/EDTA. Decellularized and non-decellularized samples were embedded in paraffin and observed under an optical microscope. From these images, geometrical data of the cardiac ECM were obtained and a model of the ECM microarchitecture, based on rectangular geometry, was designed (Fig. 4a). The material used for scaffold development was poly(NIPAAm-co-HEMAHex)/alginate/gelatin blend. Microtextured substrata were then produced by Soft Lithography, cross-linked and assembled in a three-dimensional structure. The morphological analysis showed that the geometry of the cardiac ECM model was successfully transferred to the scaffolds (Fig. 4b). Moreover, scaffolds showed a superficial microporosity, that could promote cell adhesion. Mechanical tests revealed an elastic behavior and anisotropic properties, similar to that of adult human left ventricular myocardium [47]. The biological characterization showed the ability of the microfabricated scaffolds to support myoblasts proliferation and differentiation and to promote cell alignment. As future developments, in order to obtain a fully biomimetic tissue it would be interesting to match the stiffness of the scaffold with that of the left ventricular myocardium. This could be obtained by adjusting the cross-linking process or by using an alternative polymeric system, with an elastic modulus closer to that of the human myocardium. In additon, thanks to the thermosensitive properties of poly(NIPAAm-co-HEMAHex), it would be interesting to verify the possibility to use the micropatterns as support for the cell sheet engineering approach, described by Shimizu and coworkers [33].

4.3 Injectable Systems

Different injectable systems have been studied until now for application in the field of tissue engineering, such as in situ crosslinkable systems, thermogelling systems, self-assembling peptides, micro- and nano-particles [124].

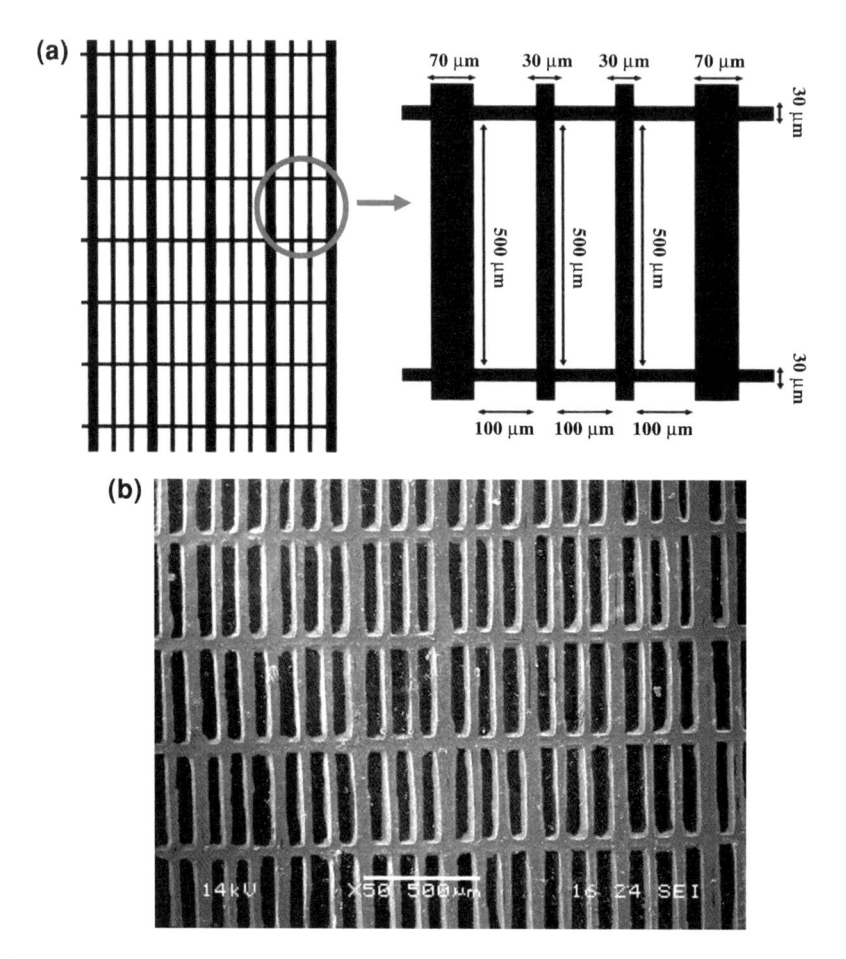

Fig. 4 Two-dimensional model of the cardiac ECM (**a**) and SEM micrograph of microfabricated scaffolds (**b**)

Despite several demonstrated advantages of biodegradable microspheres as cell culture supports [47, 125–127], such as rapid cell expansion, maintainance of a differentiated cell phenotype, possibility to be delivered through a catheter-based approach, their use in tissue engineering has only been explored to a limited extent, and nothing has been done in this direction for myocardial regeneration.

The aim of our work was the preparation and characterization of microparticles, based on alginate/gelatin blend, for application as injectable scaffold for myocardial tissue engineering [37, 87]. The microparticles were prepared by a single water in oil emulsion, using phosphatidylcholine as surfactant. The morphological analysis demonstrated the formation of microspheres (Fig. 3b). The average diameter, measured by optical microscopy, was 350 μm for dried beads and 1150 μm after swelling. The infrared analysis confirmed the presence of the typical adsorption peaks of the two biopolymers and the absence of phosphatidylcholine after washing in solution of pH = 8 and isopropyl alcohol. FT-IR Chemical Imaging investigation showed a good chemical homogeneity. The thermogravimetric analysis performed on microspheres samples at the end of the hydrolysis test showed the absence of the thermal degradation event due to the protein component, suggesting the release of gelatin during the incubation period in aqueous medium. The morphological analysis after hydrolysis showed a porous surface, which could promote cell adhesion and proliferation. Preliminary in vitro cell proliferation test, carried out in static condition, suggested the possibility to use alginate/gelatine microspheres as injectable support for cardiac tissue culture.

5 Development of Integrated Experimental Protocols for Guiding Stem Cell Plasticity

The regeneration of diseased myocardium is becoming a clinical possibility in large part due to parallel advances in modifying biomaterials and understanding stem cell behavior [128]. Stem cells are unspecialized cells that renew themselves for long periods through cell division and under proper stimuli can be induced to become cells with special and unique functions. In the human organism cell function, tissue morphogenesis, and organ development are thought to be regulated by a fine-tuned interplay of chemical, physical and topographical factors. Many of the principles guiding embryogenesis in vivo are also considered to be involved in the regulation of tissue development in vitro. Careful control of the cellular microenvironment may be as important for achieving regeneration or repair as the nature of the cells themselves [129]. One of the critical ways for controlling the local cellular microenvironment is through biomaterials that are designed to direct cellular behavior. More than being simply compatible with the host and serving a structural role, biomaterials should instruct cells through microenvironmental cues. Ultimately, one of the most challenging problems in tissue engineering and regeneration is to reproduce a precise series of spatially and temporally controlled events analogous to the remarkably coordinated events of organ development. Although single factor approaches may yield benefit, it is likely that multi-factorial strategies will have to be designed for successful cardiac repair.

The results obtained in our research work in terms of polymeric materials preparation, functionalization strategies and scaffold fabrication techniques were converged in the development of bioactive scaffolds with multifunctional properties for guiding cardiac tissue formation from dissociated stem cells [130]. Alginate/gelatin and poly(NIPAAm-co-HEMAHEx)/alginate/gelatin sponges were prepared by freeze-drying technique. The scaffolds were then functionalized to potentially promote both stem cell proliferation and differentiation, using molecularly imprinted polymers and drug release systems. For promoting stem cell proliferation, molecularly imprinted nanoparticles with recognition properties towards fibronectin GRGDS peptide sequence were deposited on the scaffold. In order to guide stem cell differentiation, molecularly imprinted nanoparticles with recognition properties towards laminin YIGSR peptide sequence or alginate/gelatin microparticles loaded with ascorbic acid were used. The morphological analysis showed a porous structure with well interconnected pores for the alginate/gelatin system while a less porous structure was observed for the poly(NIPAAm-co-HEMAHex)/alginate/gelatin scaffold. Moreover, it was found that molecularly imprinted particles and alginate/gelatin microspheres were mainly distributed in scaffold cavities (Fig. 5a). Chemical Imaging analysis confirmed the presence of all the polymeric components in the multifunctional scaffold and a sufficiently homogeneous distribution of the particles used for functionalization. The release of ascorbic acid was lower from the scaffold than from free particles, as expected since the microspheres were mainly present in scaffold cavities, limiting the outwards release of the drug. Considering together the two polymeric systems, ascorbic acid release was higher from alginate/gelatin scaffolds, probably thanks to the more porous structure and a more homogeneous particle distribution. Preliminary in vitro cell culture tests were performed using cardiac stem cells from pig myocardium. Vital cells were found both on scaffold surface and in scaffold porosity, showing that the bioactive scaffolds were able to promote cardiac stem cell adhesion and proliferation. The most remarkable result was the ability of cultured cells to migrate from the scaffold in the surrounding environment and create a cellular monolayer. The number of migrated cells was higher from alginate/gelatin scaffold, probably thanks to the higher porosity. Migrated cells were positive to α-sarcomeric actin, a marker of myocytes, showing that the multifunctional systems were also able to promote cardiac stem cell differentiation (Fig. 5b).

Even if further investigations will be necessary for quantifying, with appropriate markers, stem cell differentiation, the results obtained in this work were very promising, suggesting the possibility to use the multifunctional scaffolds as support for myocardial tissue engineering, able to guide cardiac tissue growth from cardiac stem cells.

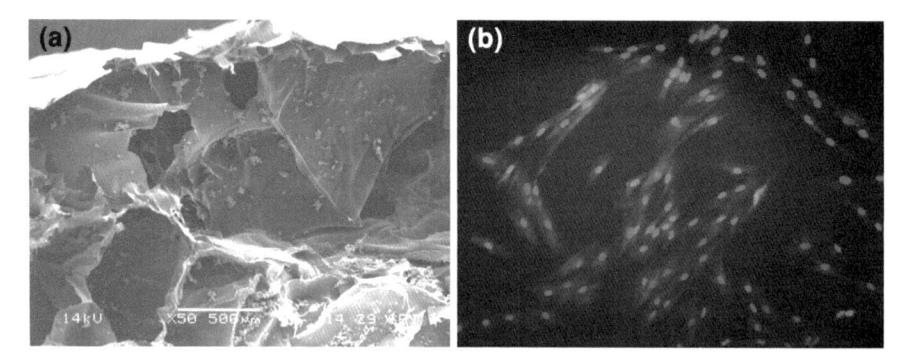

Fig. 5 SEM micrograph of alginate/gelatin sponge functionalized with MIP-GRGDS and MIP-YIGSR (a) and cardiac stem cells cultured on multifunctional scaffolds, positive to alpha-sarcomeric actin (b)

6 Conclusions

In this chapter our research activity in the field of myocardial tissue engineering was presented. The development of multifunctional polymeric scaffolds able to guide and control cardiac tissue repair was investigated, focusing our attention on three main aspects: polymer research, material functionalization through advanced techniques and scaffold fabrication.

Novel polymeric materials for both in vitro and in vivo tissue engineering applications were suggested. In particular, in addition to new blends of natural and synthetic polymers, a novel thermosensitive and bioresorbable PNIPAAm based copolymer was synthesized. In order to increase the bioactivity of the scaffolds, selected materials were functionalized using biochemical signals by surface modification of polymers and MI technology. The possibility to use this last technology as a totally new approach for bioactive scaffolds development was demonstrated for the first time by our research group. Three dimensional innovative scaffolds having complex architecture were prepared to mimic cardiac ECM by using microfabrication techniques. Injectable scaffolds in the form of biodegradable microspheres were also investigated.

Although the routine clinical treatment of impaired myocardium via tissue engineering might not become reality in a short time, our research activity over the past years has been an attempt towards this direction and it could open the door to future developments that could overcome challenges necessary for successful myocardial tissue repair.

In the future work, advanced functionalization strategies will be combined with microfabrication techniques and biomimetic materials, for developing scaffolds capable of resembling as close as possible the natural ECM.

To improve the efficiency of scaffold functionalization, advances in stem cell biology will be necessary for the identification of type and dose of bioactive molecules able to guide stem cell behaviour. Moreover, to engineer a viable cardiac patch, techniques to promote scaffold vascularization, such as the release of angiogenic growth factors, will be a necessity to allow the mass transfer of oxygen and nutrients throughout the tissue.

In vitro cell culture tests will be performed with different stem cells, i.e. cardiac and mesenchymal, in dynamic condition, using a bioreactor that combines perfusion with electrical and mechanical stimulation, resembling electrical excitation and ventricle contraction during cardiac development.

Even if in our past work most of the effort was devoted to in vitro myocardial tissue engineering, that surely is the most promising therapy for patients with impaired stem cells function and regenerative capacity, recently the attention was focused on in vivo myocardial regeneration, as an alternative approach avoiding the risk and difficulties associated with cell expansion, differentiation and biograft implantation. Therefore, it will be interesting to test the thermosensitive and bioresorbable copolymer poly(NIPAAm-co-HEMEHex), the injectable alginate/gelatine microparticles and unseeded multifunctional scaffolds as supports for in vivo myocardial tissue engineering. At this purpose, it will be necessary to additionally functionalize them with suitable molecules, known as promoting stem cell recruitment.

References

1. European Cardiovascular Disease Statistics 2008 Edition. British Heart Foundation Health Promotion Research Group, Department of Public Health, University of Oxford. http://www.heartstats.org/uploads/documents%5Cproof30NOV2007.pdf Accessed 22 March 2010
2. Statistical Fact Sheet—Populations 2009 Update. International Cardiovascular Disease Statistics. American Heart Association. http://www.americanheart.org/downloadable/heart/1236204012112INTL.pdf Accessed 22 March 2010
3. John Sutton, M.G., Sharpe, N.: Left ventricular remodeling after myocardial infarction: pathophysiology and therapy. Circulation **101**, 2981–2988 (2000)
4. Koh, G.Y., Soonpaa, M.H., Klug, M.G., et al.: Strategies for myocardial repair. J. Interv. Cardiol. **8**, 387–393 (1995)
5. Hassink, R.J., Dowell, J.D., Brutel de la Riviere, A., et al.: Stem cell therapy for ischemic heart disease. Trends Mol. Med. **9**, 436–441 (2003)
6. Chen, Q.Z., Harding, S.E., Ali, N.N., et al.: Biomaterials in cardiac tissue engineering: Ten years of research survey. Mater. Sci. Eng. **R59**, 1–37 (2008)
7. Jawad, H., Ali, N.N., Lyon, A.R., et al.: Myocardial tissue engineering: a review. J. Tissue Eng. Regen. Med. **1**, 327–342 (2007)
8. Leor, J., Amsalem, Y., Cohen, S.: Cells, scaffolds, and molecules for myocardial tissue engineering. Pharmacol. Therapeut. **105**, 151–163 (2005)
9. Christman, K.L., Lee, R.J.: Biomaterials for the treatment of myocardial infarction. J. Am. Coll. Cardiol. **48**, 907–913 (2006)
10. Zammaretti, P., Jaconi, M.: Cardiac tissue engineering: regeneration of the wounded heart. Curr. Opin. Biotech. **15**, 430–434 (2004)
11. Kofidis, T., Akhyari, P., Wachsmann, B., et al.: A novel bioartificial myocardial tissue and its prospective use in cardiac surgery. Eur. J. Cardio-Thorac. **22**, 238–243 (2002)
12. Zhong, S.P., Teo, W.E., et al.: Formation of collagen–glycosaminoglycan blended nanofibrous scaffolds and their biological properties. Biomacromolecules **6**, 2998–3004 (2005)
13. Huang, N.F., Yu, J., Sievers, R., et al.: Injectable biopolymers enhance angiogenesis after myocardial infarction. Tissue Eng. **11**, 1860–1866 (2005)
14. Thompson, C.A., Nasseri, B.A., Makower, J., et al.: Percutaneous transvenous cellular cardiomyoplasty. A novel nonsurgical approach for myocardial cell transplantation. J. Am. Coll. Cardiol. **41**, 1964–1971 (2003)
15. Dai, W., Wold, L.E., Dow, J.S., et al.: Thickening of the infarcted wall by collagen injection improves left ventricular function in rats: a novel approach to preserve cardiac function after myocardial infarction. J. Am. Coll. Cardiol. **46**, 714–719 (2005)
16. Zimmermann, W.H., Fink, C., Kralisch, D., et al.: Three-dimensional engineered heart tissue from neonatal rat cardiac myocytes. Biotechnol. Bioeng. **68**, 106–114 (2000)
17. Zhang, P., Zhang, H., Wang, H., et al.: Artificial matrix helps neonatal cardiomyocytes restore injured myocardium in rats. Artif. Organs **30**, 86–93 (2006)
18. Li, R.K., Jia, Z.Q., Weisel, R.D., et al.: Survival and function of bioengineered cardiac grafts. Circulation **100**, II63–II69 (1999)
19. Leor, J., Aboulafia-Etzion, S., Dar, A., et al.: Bioengineered cardiac grafts: a new approach to repair the infarcted myocardium? Circulation **102**, III56–II161 (2000)

20. Christman, K.L., Vardanian, A.J., Fang, Q., et al.: Injectable fibrin scaffold improves cell transplant survival, reduces infarct expansion, and induces neovasculature formation in ischemic myocardium. J. Am. Coll. Cardiol. **44**, 654–660 (2004)
21. Ryu, J.H., Kim, I.K., Cho, S.W., et al.: Implantation of bone marrow mononuclear cells using injectable fibrin matrix enhances neovascularization in infarcted myocardium. Biomaterials **26**, 319–326 (2005)
22. Christman, K.L., Fok, H.H., Sievers, R.E., et al.: Myoblasts delivered in an injectable fibrin scaffold improve cardiac function and preserve left ventricular geometry in a chronic myocardial infarction model. Circulation **108**, S246–S247 (2003)
23. Davis, M.E., Motion, J.P., Narmoneva, D.A., et al.: Injectable self-assembling peptide nanofibers create intramyocardial microenvironments for endothelial cells. Circulation **111**, 442–450 (2005)
24. Kellar, R.S., Shepherd, B.R., Larson, D.F., et al.: Cardiac patch constructed from human fibroblasts attenuates reduction in cardiac function after acute infarct. Tissue Eng. **11**, 1678–1687 (2005)
25. Pego, A.P., Poot, A.A., Grijpma, D.W., et al.: Biodegradable elastomeric scaffolds for soft tissue engineering. J. Control Release **87**, 69–79 (2003)
26. Ozawa, T., Mickle, D.A., Weisel, R.D., et al.: Optimal biomaterial for creation of autologous cardiac grafts. Circulation **106**, I176–I182 (2002)
27. Ishii, O., Shin, M., Sueda, T., et al.: In vitro tissue engineering of a cardiac graft using a degradable scaffold with an extracellular matrix-like topography. J. Thorac. Cardiovasc. Surg. **130**, 1358–1368 (2006)
28. Zong, X., Bien, H., Chung, C.Y., et al.: Electrospun fine-textured scaffolds for heart tissue constructs. Biomaterials **26**, 5330–5338 (2005)
29. Radisic, M., Deen, W., Langer, G., et al.: Mathematical model of oxygen distribution in engineered cardiac tissue with parallel channel array perfused with culture medium containing oxygen carriers. Am. J. Physiol. Heart Circ. Physiol. **288**, H1278–H1289 (2005)
30. Chen, Q.Z., Bismarck, A., Hansen, U., et al.: Characterisation of a soft elastomer poly(glycerol sebacate) designed to match the mechanical properties of myocardial tissue. Biomaterials **29**, 47–57 (2008)
31. McDevitt, T.C., Woodhouse, K.A., Hauschka, S.D., et al.: Spatially organized layers of cardiomyocytes on biodegradable polyurethane films for myocardial repair. J. Biomed. Mater. Res. **66A**, 586–595 (2003)
32. Alperin, C., Zandstra, P.W., Woodhouse, K.A.: Polyurethane films seeded with embryonic stem cell-derived cardiomyocytes for use in cardiac tissue engineering applications. Biomaterials **26**, 7377–7386 (2005)
33. Shimizu, T., Yamato, M., Kikuchi, A., et al.: Cell sheet engineering for myocardial tissue reconstruction. Biomaterials **24**, 2309–2316 (2003)
34. Libby, P., Lee, R.T.: Matrix matters. Circulation **102**, 1874–1876 (2000)
35. Simpson, D.G., Terracio, L., Terracio, M., et al.: Modulation of cardiac myocyte phenotype in vitro by the composition and orientation of the extracellular matrix. J. Cell Physiol. **161**, 89–105 (1994)
36. Ju, H., Dixon, I.M.: Extracellular matrix and cardiovascular diseases. Can. J. Cardiol. **12**, 1259–1267 (1996)
37. Rosellini, E., Bioactive scaffolds for controlling stem cell differentiation: application in myocardial tissue engineering, PhD thesis, University of Pisa-Italy, Pisa (2009)
38. Rosellini, E., Cristallini, C., Barbani, N., et al.: Preparation and characterization of alginate/gelatin blend films for cardiac tissue engineering. J. Biomed Mater Res A **91A**, 447–453 (2009)
39. Rosellini, E., Cristallini, C., Barbani, N., et al.: Alginate/gelatin blends for the preparation of biodegradable scaffolds for myocardial tissue engineering. JABB **5**, 218 (2007)
40. Daniliuc, L., De Kasel, C., David, C.: Intermolecular interactions in blends of poly(vinyl alcohol) with poly(acrylic acid)-1. FTIR and DSC studies. Eur. Polym. J. **28**, 1365–1371 (1992)

41. Dong, Z., Wang, Q., Du, Y.: Alginate/gelatin blend films and their properties for drug controlled release. J. Membr. Sci. **280**, 37–44 (2006)
42. Siminiak, T., Kurpisz, M.: Myocardial replacement therapy. Circulation **108**, 1167–1171 (2003)
43. Giusti, P., Lazzeri, L., Lelli, L.: Bioartificial polymeric materials; a new method to design biomaterials by using both biological and synthetic polymers. Trends Polym. Sci. **1**, 261–267 (1993)
44. Rosellini, E., Cristallini, C., Barbani, N., et al.: New bioartificial systems and biodegradable synthetic polymers for cardiac tissue engineering: a preliminary screening. Biomed. Eng-App. Bas C (in press)
45. Wang, H.B., Dembo, M., Hanks, S.K., et al.: Focal adhesion kinase is involved in mechanosensing during fibroblast migration. Proc. Natl. Acad. Sci. USA **98**, 11295–11300 (2001)
46. Klouda, L., Mikos, A.G.: Thermoresponsive hydrogels in biomedical applications. Eur. J. Pharm. Biopharm. **68**, 34–45 (2008)
47. Kretlow, J.D., Klouda, L., Mikos, A.G.: Injectable matrices and scaffolds for drug delivery in tissue engineering. Adv. Drug Deliver. Rev. **59**, 263–273 (2007)
48. Kamide, K.: Thermodynamics of polymer solutions: phase equilibria and critical phenomena. Elsevier, New York (1990)
49. Schild, H.G.: Poly(N-isopropylacrylamide): experiment-theory and application. Prog. Polym. Sci. **17**, 163–249 (1992)
50. Zhang, X., Wu, D., Chu, C.C.: Synthesis and characterization of partially biodegradable, temperature and pH sensitive Dex-MA/PNIPAAm hydrogels. Biomaterials **25**, 4719–4730 (2004)
51. You, Y., Hong, C., Wang, W., et al.: Preparation and characterization of thermally responsive and biodegradable block copolymer comprised of PNIPAAm and PLA by combination of ROP and RAFT methods. Macromolecules **37**, 9761–9767 (2004)
52. Huang, X., Nayak, B.R., Lowe, T.L.: Synthesis and characterization of novel thermoresponsive-co-biodegradable hydrogels composed of N-isopropylacrylamide, poly(L-lactic acid) and dextran. J. Polym. Sci. **A142**, 5054–5066 (2004)
53. Ohya, S., Nakayama, Y., Matsuda, T.: Thermoresponsive artificial extracellular matrix for tissue engineering: hyaluronic acid bioconjugated with poly(N-isopropylacrylamide) grafts. Biomacromolecules **2**, 856–863 (2001)
54. Ohya, S., Matsuda, T.: Poly(N-isopropylacrylamide) (PNIPAM)-grafted gelatin as thermoresponsive three-dimensional artificial extracellular matrix: molecular and formulation parameters vs. cell proliferation potential. J. Biomater. Sci. Polym. Ed. **16**, 809–827 (2005)
55. Kim, S., Chung, E.H., Gilbert, M., et al.: Synthetic MMP-13 degradable ECMs based on poly(N-isopropylacrylamide-co-acrylic acid) semi-interpenetrating polymer networks. I. Degradation and cell migration. J. Biomed. Mater. Res. **A75**, 73–88 (2005)
56. Li, F., Carlsson, D., Lohmann, C., et al.: Cellular and nerve regeneration within a biosynthetic extracellular matrix for corneal transplantation. Proc. Nat. Acad. Sci. USA **100**, 15346–15351 (2003)
57. Neradovic, D., Hinrichs, W.L.J., Kettenes-van den Bosch, J.J., et al.: Poly(N-isopropylacrylamide) with hydrolysable lactic acid ester side group. Macromol. Rapid Commun. **20**, 577–581 (1999)
58. Cui, Z., Lee, B.H., Vernon, B.L.: New hydrolysis-dependent thermosensitive polymer for an injectable degradable system. Biomacromolecules **8**, 1280–1286 (2007)
59. Lee, B.H., Vernon, B.: Copolymers of N-isopropylacrylamide, HEMA-lactate and acrylic acid with time-dependent lower critical solution temperature as a bioresorbable carrier. Polym. Int. **54**, 418–422 (2005)
60. Guan, J., Hong, Y., Ma, Z., et al.: Protein-reactive, thermoresponsive copolymers with high flexibility and biodegradability. Biomacromolecules **9**, 1283–1292 (2008)

61. Lee, B.H., Vernon, B.: In situ-gelling, erodible N-isopropylacrylamide copolymers. Macromol. Biosci. **5**, 629–635 (2005)
62. Wang, T., Wu, D.Q., Jiang, X.J., et al.: Novel thermosensitive hydrogel injection inhibits post-infarct ventricle remodeling. Eur. J. Heart Fail. **11**, 14–19 (2009)
63. Wu, D.Q., Qiu, F., Wang, T., et al.: Toward the development of partially biodegradable and injectable thermoresponsive hydrogels for potential biomedical applications. ACS Appl. Mater. Interf. **2**, 312–327 (2009)
64. Fujimoto, K.L., Ma, Z., Nelson, D.M., et al.: Synthesis, characterization and therapeutic efficacy of a biodegradable, thermoresponsive hydrogel designed for application in chronic infarcted myocardium. Biomaterials **30**, 4357–4368 (2009)
65. Ignatius, A.A., Claes, L.E.: In vitro biocompatibility of bioresorbable polymers: poly(l, dl-lactide) and poly(l-lactide-co-glycolide). Biomaterials **17**, 831–839 (1996)
66. Rosellini, E., Cristallini, C., Guerra, G.D., et al.: Synthesis and characterization of a novel PNIPAAm-based copolymer with hydrolysis-dependent thermosensitivity. Biomed. Mater. **5**, 035012 (2010)
67. Kelen, T., Tüdos, F.: Analysis of the linear methods for determining copolymerization reactivity ratios. I. A new improved linear graphic method. J. Macromol. Sci. A **9**, 1–27 (1975)
68. Erbil, C., Terlan, B., Akdemir, O., et al.: Monomer reactivity ratios of N-isopropylacrylamide–itaconic acid copolymers at low and high conversions. Eur. Polym. J. **45**, 1728–1737 (2009)
69. Sbarbati Del Guerra, R., Gazzetti, P., Lazzerini, G., et al.: Degradation products of poly(ester-ether-ester) block copolymers do not alter endothelial methabolism in vitro. J. Mater. Sci-Mater. M **6**, 824–828 (1995)
70. Hersel, U., Dahmen, C., Kessler, H.: RGD modified polymers: biomaterials for stimulated cell adhesion and beyond. Biomaterials **24**, 4385–4415 (2003)
71. Shin, H., Jo, S., Mikos, A.G.: Biomimetic materials for tissue engineering. Biomaterials **24**, 4353–4364 (2003)
72. Humphries, M.J.: Integrin structure. Biochem. Soc. T **28**, 311–339 (2000)
73. Rowley, J.A., Madlambayan, G., Mooney, D.J.: Alginate hydrogels as synthetic extracellular matrix materials. Biomaterials **20**, 45–53 (1999)
74. Rowley, J.A., Mooney, D.J.: Alginate type and RGD density control myoblast phenotype. J. Biomed. Mater. Res. **60**, 217–223 (2002)
75. Sun, H., Onneby, S.: Facile polyester surface functionalization via hydrolysis and cell recognizing peptide attachment. Polym. Int. **55**, 1336–1340 (2006)
76. Zhu, Y., Gao, C., Liu, X., et al.: Surface modification of polycaprolactone membrane via aminolysis and biomacromolecule immobilization for promoting cytocompatibility of human endothelial cells. Biomacromolecules **3**, 1312–1319 (2002)
77. Santiago, L.Y., Nowak, R.W., Rubin, J.P., et al.: Peptide-surface modification of poly(caprolactone) with laminin-derived sequences for adipose-derived stem cell applications. Biomaterials **27**, 2962–2969 (2006)
78. Patel, S., Thakar, R.G., Wong, J., et al.: Control of cell adhesion on poly(methyl methacrylate). Biomaterials **27**, 2890–2897 (2006)
79. Massia, S.P., Hubbell, J.A.: An RGD spacing of 440 nm is sufficient for integrin $\alpha_V\beta_3$-mediated Fibroblast Spreading and 140 nm for focal contact and stress fiber formation. J. Cell Biol. **114**, 1089–1100 (1991)
80. Rosellini, E., Cristallini, C., Barbani, N., et al.: Production of bioactive scaffolds for cardiac tissue engineering through traditional and advanced functionalization techniques. In: Burattini, R., Contro, R., Dario, P., Landini, L. (eds.) Congresso Nazionale di Bioingegneria 2008 Atti. Pàtron Editore, Bologna (2008)
81. Cerrai, P., Tricoli, M.: Block copolymers from L-lactide and poly(ethylene glycol) through a non catalyzed route. Makromol. Chem. Rapid Commun. **14**, 529–538 (1993)
82. Cerrai, P., Guerra, G.D., Lelli, L., et al.: Block copolymers of L-lactide and poly(ethylene glycol) for biomedical applications. J. Mater. Sci. Mater. Med. **5**, 308–313 (1994)

83. Ciardelli, G., Rechichi, A., Cerrai, P., et al.: Segmented polyurethanes for medical applications: synthesis, characterization and in vitro enzymatic degradation studies. Macromol. Symp. **218**, 261–271 (2004)
84. Battista, S., Guarnieri, D., Borselli, C., et al.: The effect of matrix composition of 3D constructs on embryonic stem cell differentiation. Biomaterials **26**, 6194–6207 (2005)
85. Tessmar, J.K., Göpferich, A.M.: Matrices and scaffolds for protein delivery in tissue engineering. Adv. Drug Deliver. Rev. **59**, 274–291 (2007)
86. Tabata, Y.: Significance of release technology in tissue engineering. DDT **10**, 1639–1646 (2005)
87. Rosellini, E., Barbani, N., Tinti, E., et al.: Alginate-gelatin microparticles as injectable scaffold for myocardial tissue engineering. JABB **7**, 69 (2009)
88. Takahashi, T., Lord, B., Schulze, P.C., et al.: Ascorbic acid enhances differentiation of embryonic stem cells into cardiac myocytes. Circulation **107**, 1912–1916 (2003)
89. Mosbach, K., Ramström, O.: The emerging technique of molecular imprinting and its future impact on biotechnology. Biotechnology **14**, 163–170 (1996)
90. Shea, K.J.: Molecular imprinting of synthetic network polymers: the de novo synthesis of macromolecular binding and catalytic sites. Trends Polym. Sci. **2**, 166–173 (1994)
91. Steinke, J., Sherrington, D., Dunkin, I.: Imprinting of synthetic polymers using molecular templates. Adv. Polym. Sci. **123**, 80–125 (1995)
92. Sellergren, B., Shea, K.J.: Chiral ion exchange chromatography: correlation between solute retention and a theoretical ion-exchange model using imprinted polymers. J. Chromatogr. A **654**, 17–28 (1993)
93. Hart, B.R., Shea, K.J.: Synthetic peptide receptors: molecularly imprinted polymers for the recognition of peptides using peptide-metal interactions. J. Am. Chem. Soc. **123**, 2072–2073 (2001)
94. Wulff, G.: Molecular Imprinting in cross-linked materials with the aid of molecular templates: a way towards artificial antibodies. Angew Chem. Int. Ed. Engl. **34**, 1812–1832 (1995)
95. Cristallini, C., Ciardelli, G., Giusti, P., et al.: Acrylonitrile-acrylic acid copolymer membrane imprinted with uric acid for clinical uses. Macromol. Biosci. **4**, 31–38 (2004)
96. Silvestri, D., Borrelli, C., Giusti, P., et al.: Polymeric devices containing imprinted nanospheres: a novel approach to improve recognition in water for clinical uses. Anal. Chim Acta **542**, 3–13 (2005)
97. Rechichi, A., Cristallini, C., Vitale, U., et al.: New biomedical devices with selective peptide recognition properties. Part 1: Characterization and cytotoxicity of molecularly imprinted polymers. J. Cell. Mol. Med. **11**, 1367–1376 (2007)
98. Rachkov, A., Minoura, N.: Towards molecularly imprinted polymers selective to peptides and proteins. The epitope approach. Biochim. Biophys. Acta **1544**, 255–266 (2001)
99. Rosellini, E., Barbani, N., Giusti, P., et al.: Novel bioactive scaffolds with fibronectin recognition nanosites based on molecular imprinting technology. J. Appl. Pol. Sci. **118**, 3236–3244 (2010)
100. Rosellini, E., Barbani, N., Giusti, P. et al.: Molecularly imprinted nanoparticles with recognition properties towards a laminin H-Tyr-Ile-Gly-Ser-Arg-OH sequence for tissue engineering applications. Biomed. Mater. 5, 065007 (2010)
101. Falconnet, D., Csucs, G., Grandin, H.M., et al.: Surface engineering approaches to micropattern surfaces for cell-based assays. Biomaterials **27**, 3044–3063 (2006)
102. Akhyari, P., Fedak, P.W., Weisel, R.D., et al.: Mechanical stretch regimen enhances the formation of bioengineered autologous cardiac muscle grafts. Circulation **106**, I137–I142 (2002)
103. Eschenhagen, T., Didie, M., Munzel, F., et al.: 3D engineered heart tissue for replacement therapy. Basic Res. Cardiol. **97**, I146–I152 (2002)
104. Dar, A., Shachar, M., Leor, J., et al.: Optimization of cardiac cell seeding and distribution in 3D porous alginate scaffolds. Biotechnol. Bioeng. **80**, 305–312 (2002)

105. Christman, K.L., Fok, H.H., Sievers, R.E., et al.: Fibrin glue alone and skeletal myoblasts in a fibrin scaffold preserve cardiac function after myocardial infarction. Tissue Eng. **10**, 403–409 (2004)
106. Shimizu, T., Yamato, M., Isoi, Y., et al.: Fabrication of pulsatile cardiac tissue grafts using a novel 3-dimensional cell sheet manipulation technique and temperature-responsive cell culture surfaces. Circ. Res. **90**, e40 (2002)
107. Rosellini, E., Cristallini, C., Barbani, N., et al.: Biomimetic scaffolds for bone and cardiac tissue engineering. JABB **5**(3), 205 (2007)
108. Discher, D.E., Janmey, P., Wang, Y.: Tissue cells feel and respond to the stiffness of their substrate. Science **310**, 1139–1143 (2005)
109. Tsang, V.L., Bhatia, S.N.: Three-dimensional tissue fabrication. Adv. Drug Deliver. Rev. **56**, 1635–1647 (2004)
110. Chen, C.S., Mrksich, M., Huang, S., et al.: Geometric control of cell life and death. Science **276**, 1425–1428 (1997)
111. Vozzi, G., Flaim, C., Ahluwalia, A., et al.: Fabrication of PLGA scaffolds using soft lithography and microsyringe deposition. Biomaterials **24**, 2533–2540 (2003)
112. Gallego, D., Ferrell, N., Sun, Y., et al.: Multilayer micromolding of degradable polymer tissue engineering scaffolds. Mater. Sci. Eng. C **28**, 353–358 (2008)
113. Mata, A., Kim, E.J., Boehm, C.A., et al.: A three-dimensional scaffold with precise micro-architecture and surface micro-textures. Biomaterials **30**, 4610–4617 (2009)
114. Kim, S.S., Utsunomiya, H., Koski, J.A., et al.: Survival and function of hepatocytes on a novel three-dimensional synthetic biodegradable polymer scaffold with an intrinsic network of channels. Ann. Surg. **228**, 8–13 (1998)
115. Park, A., Wu, B., Griffith, L.G.: Integration of surface modification and 3D fabrication techniques to prepare patterned poly(L-lactide) substrates allowing regionally selective cell adhesion. J. Biomater. Sci. Polym. Ed. **9**, 89–110 (1998)
116. Berry, E., Brown, J.M., Connell, M., et al.: Preliminary experience with medical applications of rapid prototyping by selective laser sintering. Med. Eng. Phys. **19**, 90–96 (1997)
117. Cao, T., Ho, K.H., Teoh, S.H.: Scaffold design and in vitro study of osteochondral coculture in a three-dimensional porous polycaprolactone scaffold fabricated by fused deposition modelling. Tissue Eng. **9**, S103–S112 (2003)
118. Klebe, R.J.: Cytoscribing: a method for micropositioning cells and the construction of two- and three-dimensional synthetic tissues. Exp. Cell. Res. **179**, 362–373 (1988)
119. Mironov, V., Boland, T., Trusk, T., et al.: Organ printing: computer-aided jet-based 3D tissue engineering. The layer-by-layer assembly of biological tissues and organs is the future of tissue engineering. Trends Biotechnol. **21**, 157–161 (2003)
120. Taboas, J.M., Maddox, R.D., Krebsbach, P.H., et al.: Indirect solid free form fabrication of local and global porous, biomimetic and composite 3D polymer-ceramic scaffolds. Biomaterials **24**, 181–194 (2003)
121. Vozzi, G., Previti, A., De Rossi, D., et al.: microsyringe based deposition of 2 and 3-D polymer scaffolds with a well defined geometry for application to tissue engineering. Tissue Eng. **8**, 1089–1098 (2002)
122. Motlagh, D., Senyo, S.E., Desai, T.A., et al.: Microtextured substrata alter gene expression, protein localization and the shape of cardiac myocytes. Biomaterials **24**, 2463–2476 (2003)
123. Engelmayr, G.C., Cheng, M., Bettinger, C.J., et al.: Accordion-like honeycombs for tissue engineering of cardiac anisotropy. Nat. Mater. **7**, 1003–1010 (2008)
124. Wang, Y., Haynor, D.R., Kim, Y.: An investigation of the importance of myocardial anisotropy in finite-element modeling of the heart: methodology and application to the estimation of defibrillation efficacy. IEEE T Bio-Med. Eng. **48**, 1377–1389 (2001)
125. Chen, R., Curran, S.J., Curran, J.M., et al.: The use of poly(L-lactide) and RGD modified microspheres as cell carriers in a flow intermittency bioreactor for tissue engineering cartilage. Biomaterials **27**, 4453–4460 (2006)

126. Malda, J., Frondoza, C.G.: Microcarriers in the engineering of cartilage and bone. TRENDS Biotechnol. **24**, 299–304 (2006)
127. Bardouille, C., Lehmann, J., Heimann, P., et al.: Growth and differentiation of permanent and secondary mouse myogenic cell lines on microcarriers. Appl. Microbiol. Biotechnol. **55**, 556–562 (2001)
128. Wu, K.H., Liu, Y.L., Zhou, B., et al.: Cellular therapy and myocardial tissue engineering: the role of adult stem and progenitor cells. Eur. J. Cardiothorac. Surg. **30**, 770–781 (2006)
129. Davis, M.E., Hsieh, P.C.H., Grodzinsky, A.J., et al.: Custom design of the cardiac microenvironment with biomaterials. Circ. Res. **97**, 8–15 (2005)
130. Rosellini, E., Cristallini, C., Barbani, N., et al.: Multifunctional scaffolds for myocardial tissue engineering. JABB **8**, 125 (2010)

Electrospun Nanocomposites and Stem Cells in Cardiac Tissue Engineering

Jorge A. Genovese, Cristiano Spadaccio, Alberto Rainer and Elvio Covino

Abstract Stem cell therapy is a leading field of research worldwide given its promising potential for recovery or replacement of tissues and organs, especially for the treatment of cardiovascular pathologies. However, despite this enormous experimental effort and the reported positive results in different models, there is no conclusive demonstration of the mechanisms involved in tissue regeneration associated to adult stem cell treatment. This represents one of the major limitations for the clinical translation of stem cell therapy. A real regenerative medicine approach should consider the importance of the extracellular matrix (ECM) and the strong biological signals that it can provide. Connective tissue atmosphere in which cells are embedded exerts a number of actions affecting cells function and supporting their proliferation and differentiation. Polymeric electrospun matrices are among the most promising ECM-mimetic biomaterials, because of their physical structure closely resembling the fibrous proteins in native ECM. Moreover, electrospun materials can be easily functionalized with bioactive molecules providing localized biochemical *stimuli* to cells seeded therein. The idea of taking advantage of both stem cells plasticity and biomaterials that actively guide and provide the correct sequence of signals to allow ongoing lineage-specific differentiation is an attractive alternative and may represent a promising answer to the treatment limitations of cardiovascular severe diseases.

J. A. Genovese · C. Spadaccio · E. Covino
Department of Cardiovascular Science, Unit of Cardiovascular Surgery,
University Campus Bio-Medico of Rome, Rome, Italy

J. A. Genovese (✉)
Cardiac Regeneration Research Laboratories, Cardiothoracic Surgery, The University of Utah, 30 N 1900 E, Suite 3C145, Salt Lake City, Utah 84132, USA
e-mail: jorge.genovese@utah.edu

A. Rainer
Center for Integrated Research, Tissue Engineering Laboratory,
University Campus Bio-Medico of Rome, Rome, Italy

1 Introduction

1.1 Current and New Perspectives in Regenerative Medicine of the Heart

Stem cell therapy is a leading field of research worldwide attracting a stunning amount of clinical applications given its promising potential for recovery or replacement of tissues and organs, especially for cardiovascular pathologies.

Despite the enormous experimental effort made in stem cell treatment reporting positive results in terms of myocardial function improvement in different pathological conditions, there is no conclusive demonstration of the mechanisms involved in tissue regeneration associated to adult stem cell treatment. Many different properties and presumable actions have been assigned to adult stem cells. Immunomodulation, paracrine activity, and capacity to differentiate or transdifferentiate have been pointed out or hypothesized, but their relevance as playing a pivotal role, accounting for the reported beneficial effect, has not been definitely clarified [1, 2]. A true in situ cardiac transdifferentiation has not been demonstrated yet. The cell viability and engrafting following cell administration have been found to be low, weakening the concept of a real tissue replacement or regeneration [3]. Moreover, issues concerning the best stem cells source, type, safety, and their harmonious and functional engraftment with an adequate electromechanical integration, remain unanswered. A number of factors related with patient's systemic conditions and the microenvironment of the injection site hamper an effective integration and differentiation of stem cells in the heart.

No insights have been provided to dampen the dramatic loss of cells occurring immediately after implantation. Trauma related to the handling and injection itself, together with the negative effects of the hostile environment in which the implant takes place, are consistently at the root of this phenomenon. No precise data are reported about the number of cells effectively viable after injection and homing to the affected site. A recent study by Assis et al. demonstrated that after injection of radiolabelled bone marrow mesenchymal stem cells in the tail vein of infracted rats, approximately 70% of the cells migrated mainly to the lungs and, in small amounts, to the heart, kidneys, spleen, and bladder [4]. Additionally, injection of living cells in a beating heart is a concern. Zhang et al. performed intramyocardial injections of stem cells in beating porcine infracted hearts and in cardioplegic arresting hearts after cardiopulmonary bypass was set up. Even after intramyocardial injection, many cells migrated to extracardiac organs, especially to the spleen, lungs and liver. However, percentage of cells retained in the heart was significantly higher in the group injected in the arresting heart than while the heart was beating [5]. These unsolved questions not only limit the possibility to define the number of cells to be injected, but also raise concerns on the actual effects of the cells, especially when an attempt is done to match this biological uncertainties with the evident improvement in cardiac function.

Undoubtedly, obstacles to translate the basic knowledge to the clinics and the necessity to perform meticulous clinical trials that allow the gathering of reliable data, are still predominant. Many reasons have conspired against the generation of productive and reliable data that could facilitate a potential clinical translation. In spite of its feasibility and attractive nature, cell therapy requires a coordinated scientific, clinical and technical effort to be deeply embedded in high quality infrastructures, necessary to carry out these techniques. Additionally, up to date, no practical guidelines or consensus of the scientific community regarding the technical aspects of the application of cell therapy in patients have been issued.

The widespread application of this promising therapeutic tool is also blocked by some issues. Patient selection, contraindications, timing of treatment, effective percentage of benefit, safety, cell type, cell dose, cell engraftment, immunogenicity are some of the issues that need to be answered. Understanding the mechanisms of normal and post-injury cardiomyocyte development and turnover will be crucial for guiding the orientation of stem cell-based therapies.

Clearly, cardiac regeneration requires a complex cascade of events that goes over the simple injection of the right type of cells in the right place. In this regard, the characterization of the factors present in the hostile microenvironment of the injured myocardium hampering the survival and functional integration of transplanted cells is also essential. Undoubtedly, along with biology of stem cells, actual effectiveness of the therapy and safety for patients requires to be further defined.

A real regenerative medicine approach should consider the importance of the extracellular matrix (ECM) and the strong biological signals that it can provide. Connective tissue atmosphere and structural microenvironment in which cells are embedded exerts a number of actions affecting cells function and supporting their proliferation and differentiation.

Several studies demonstrated that the ECM *milieu* surrounding the cells has physical and structural features in the nanometer scale. This arrangement may affect several aspects of cell behavior such as morphology, adhesion and cytoskeletal arrangement [6–8]. The so-called "tissue engineering" approach in regenerative medicine exploits these concepts with the aim to reproduce a biocompatible ECM surrogate to host cells providing a biological *stimulus* to support their survival and guide their proliferation. Thus, a great effort has been made to fabricate synthetic materials into nanometer scale structures, in attempts to simulate the matrix environment in which seeded cells can be accommodated to proliferate and differentiate towards desired lineages [9–11]. Electrospinning is one of the approaches that allow the fabrication of several natural and synthetic materials into fibrous and porous structures in the micro- and nanometer scale, just by controlling few process parameters [12, 13]. Polymeric nanofiber matrices obtained with electrospinning technique are among the most promising ECM-mimetic biomaterials because their physical structure closely resembles the fibrous proteins in native ECM. Adjusting parameters such as composition, fiber diameter, and porosity/mesh openings of the electrospun scaffold, a totally customized, controlled and optimized structure can be obtained for each individual application.

Cytocompatibility studies demonstrated that nanofiber matrices can support cell adhesion and proliferation, and maintain cell phenotypes. Additionally, post-processing surface modification with bioactive molecules can be performed, yielding to high concentration of the loaded compounds, in virtue of the relatively high scaffold surface-area-to-volume ratio [14]. Moreover, particular nanofiber systems, such as aligned nanofibers or core–shell structured nanofibers, can be obtained by electrospinning. These modification strategies aim to tailor scaffolds design to improve their functions, to acquire drug-releasing or cell orientation abilities, or to optimize their mechanical and structural properties, preventing the material from too rapid degradation when placed in in vivo settings. Therefore, considering the flexibility in material selection as well as the ability to control the scaffold properties, electrospun scaffolds have raised a considerable interest in a number of different tissue engineering applications including vascular, bone, neural, and tendon/ligament repair. Clearly, given the promising results reported in the literature, electrospinning allows us to foresee an extension of its potential applications far beyond tissue engineering. Owing to their high specific surface area, functionalized polymer nanofibers will find broad applications in the biomedical field as drug delivery carriers, biosensors, and molecular filtration membranes [15].

2 The "Tissue Engineering" Approach: The Electrospinning Technique

Nowadays, tissue engineering is considered one of the most promising fields of worldwide research, attracting a huge amount of clinical applications in virtue of its auspicious potential for recovery and replacement of organ subunits. Such interest is witnessed by over three thousand papers published on tissue engineering over the last year [16].

A plethora of experimental efforts has been devoted to develop biomaterials and manufacturing techniques, suitable for cellular biology purposes, to ameliorate and guide cell growth and tissue formation, and in an attempt to obtain a rapid clinical translation of this exciting therapeutic resource. Targeting tissue regeneration or replacement, a good proportion of the literature currently focuses on developing 3D scaffolds, using synthetic polymers, either alone or in combination with natural polymers (e.g. collagen), or inorganic fillers (e.g. hydroxyapatite) to reproduce the extracellular matrix (ECM) structure and topography. The final aim is to produce a biological surrogate as close as possible to the native histoarchitecture, and able to support and guide cell attachment and proliferation [17].

Electrospinning has been used as an effective method to fabricate biomimetic non-woven scaffolds that are comprised of a large network of interconnected fibers and pores. This high porosity allows for the efficient exchange of nutrients and metabolic waste with the surrounding environment, and provides a high surface area for local and sustained delivery of biochemical signals to the seeded cells

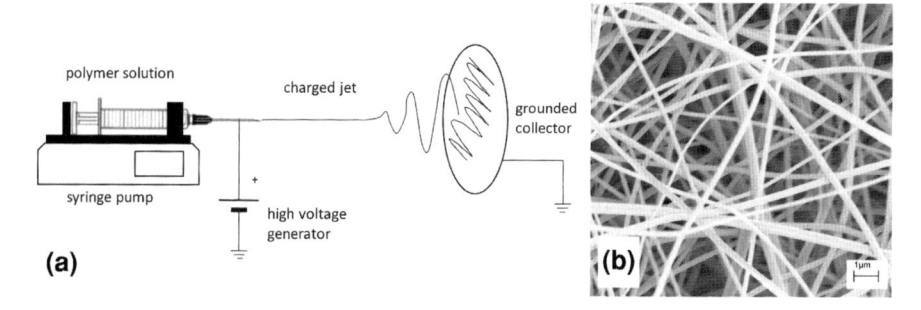

Fig. 1 a Scheme of the standard electrospinning setup; b electrospun chitosan fibers obtained with the standard setup (courtesy of Dr. L. Liverani, University Campus Bio-Medico, Rome, Italy)

[18, 19]. Electrospinning process is accomplished by applying a high electric potential to a polymer solution that is separated by some distance from an oppositely charged target to create a static electric field [20]. With the increase in the electric field, the electrostatic forces in the solution overcome the surface tension of the solution, and a thin liquid jet composed of entangled polymer chains is ejected from the polymer reservoir travelling through space toward the oppositely charged target. Instabilities within the charged jet define its orientation in space (whipping motion of the fiber) [21]. As the liquid jet travels through space, the solvent evaporates, forming a fiber that deposits as a dry, fibrous structure, and, eventually, a nonwoven mat forms on the target (Fig. 1).

Within this scenario, micro- and nanofibers, synthesized by means of electrospinning technique, have become popular candidates for tissue engineering scaffolds in recent years. Such fibers resemble the physical structure of protein fibrils in native ECM and, additionally, the fibrous structure is characterized by a high surface-area-to-volume ratio, which provides more substrates for cell attachment. Both these features are conducive for tissue culture and have fostered a number of studies that demonstrated—in in vitro culture settings—the ability of these scaffolds to mimic the ECM environment and induce attachment and proliferation of several cell types [22–25]. A variety of polymers have been electrospun, and various applications have been proposed, including nerve regeneration [26], bone tissue engineering [27], vascular engineering [28], stem cells tissue engineering [29], and skin tissue engineering [30, 31]. There is a good degree of evidence allowing us to foresee a potential advantage in the use of electrospun polymers in the clinical arena. Among these materials, poly(L-lactide) (PLLA) has been widely investigated because of its good biocompatibility [32–34] and its aptitude to be modified with inorganic materials in order to improve its biological properties for tissue engineering purposes [10, 35–38].

A number of variables can affect scaffolds production using electrospinning. Process control parameters have been classified by Doshi and Reneker in terms of solution properties, controlled variables, and ambient parameters [39]. Solution properties include the viscosity, conductivity, surface tension, polymer molecular weight, dipole moment, and dielectric constant. Controlled variables include the flow rate, electric field strength, distance between tip and collector, needle tip design, and collector composition and geometry. Ambient parameters include temperature, humidity, and air velocity.

In polymer electrospinning, two methods can be used. The polymer can be dissolved in a suitable solvent, or can be directly electrospun from a melt. In general, solution spinning results in a wider range of obtainable fiber sizes, while melt spun fibers are typically limited to micron size or larger, but the process does not require to use harsh organic solvents, which is ideal for scaled-up processes. Also, it eliminates the problem of inadequate solvent evaporation during the flight between the capillary tip and the collector, but during spinning an adequate cooling is needed to generate fibers of cylindrical morphology. As a drawback, melts need to be kept at elevated temperatures, precluding their use for drug delivery applications [14].

In addition to adjusting solution or processing parameters, the electrospinning setup can greatly influence the resulting product. In particular, this includes choices in nozzle configuration (single, single with emulsion, side-by-side, or coaxial nozzles). In the single nozzle technique, a charged polymer solution flows through a single capillary and allows spinning of polymer blends in a common solvent [28]. However, when the polymers of interest are not soluble in a common solvent, a side-by-side configuration becomes necessary. Basically, two separate polymer solutions flow through two different capillaries, which are set side-by-side. In this configuration, the solution conductivity plays a more important role in the ability to form a single fiber jet under a strong electric field (Fig. 2a).

A relatively new nozzle configuration is the coaxial configuration, which enables the simultaneous coaxial electrospinning of two different polymer solutions which flow through two different coaxial capillaries with a smaller capillary inside a larger capillary (Fig. 2b). This configuration allows a smaller fiber essentially to be encapsulated in a larger fiber, leading to what is known as a core–shell morphology, with particular interest especially in drug delivery applications as demonstrated by Zhang et al., who successfully encapsulated a fluorescein isothiocyanate-conjugated bovine serum albumin, along with poly(ethylene glycol) (PEG) in poly(ε-caprolactone) (PCL) obtaining an initial burst release associated with release from polymer/drug blends, and a longer sustained release [40, 41]. In this context, Townsend-Nicholson and Jayasinghe have demonstrated the successful encapsulation of living cells within a poly(dimethylsiloxane) (PDMS) fiber [42].

Manipulation of each of these variables leads to production of fibers with different morphology, composition, alignment, mechanical properties and degradation rates, deeply affecting the biological properties of the obtained scaffold [43]. In this extent, the degree of scaffold hydrophilicity/hydrophobicity has a profound influence on cell adhesion to synthetic polymers. To optimize the balance between hydrophilic and hydrophobic properties, and therefore cell adherence and degradation rates, one of the most effective methods is to realize

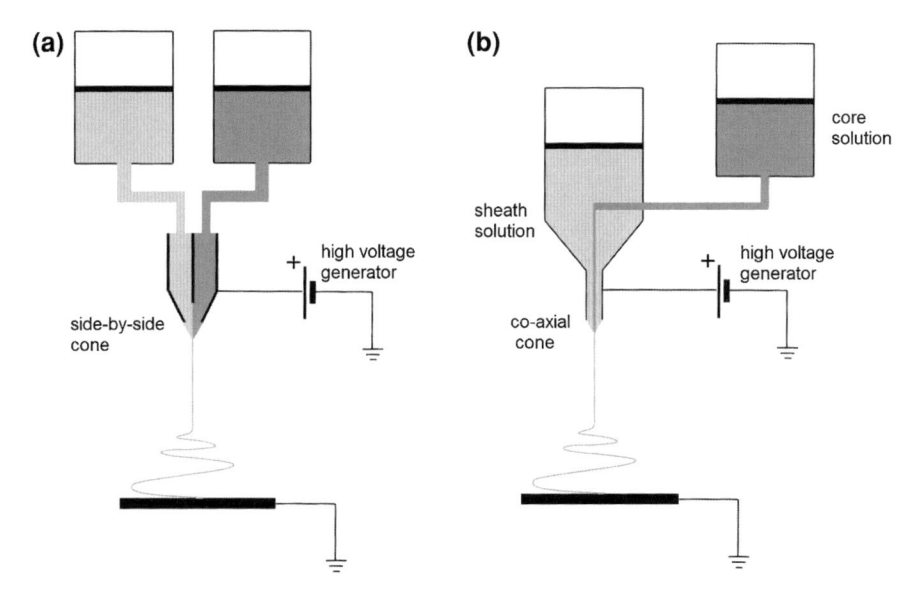

Fig. 2 Side-by-side (a) and co-axial (b) electrospinning configurations

biocomposites, blending synthetic and natural polymers [44–46]. With this aim, gelatin was blended with electrospun PCL to obtain a nanofibrous scaffold with enhanced cell adhesion capabilities and degradation properties. This kind of scaffold has been successfully used both for nerve tissue engineering and for dermal reconstitution, as confirmed by morphological and chemical analysis and cell viability assays [44, 47].

Studies on the blending of chitosan and synthetic aliphatic polyesters, such as poly-ε-caprolactone (PCL), poly(butylene succinate) (PBS), poly(lactic acid) (PLA), poly(butylene terephthalate adipate) (PBTA), and poly(butylene succinate adipate) (PBSA) have also been performed [48]. This represents the most common strategy to overcome polyesters drawbacks [49–51]. Compatibility and mechanical properties were improved by using PCL/chitosan in place of PCL [52], the degradation rate and the degree of crystallinity decreased with the chitosan content [48, 53], and PCL/chitosan scaffolds ensured a better cell viability with respect to PLC-based scaffolds [54]. Although electrospinning was first described over 70 years ago, several advances have been obtained within the past 10 years, especially as the result of the rising interest in nanoscale materials. The tissue engineering community has begun to invest on the inherent nanoscale nature of electrospun polymeric fibers as potential scaffolds to mimic native ECM. Electrospun matrices are able to support the attachment and proliferation of a wide variety of cell types; moreover, cells are able to maintain their phenotypes on these nanofiber scaffolds. Over the last several years, both natural and synthetic polymers have been electrospun for several tissue engineering applications [43, 55].

3 The Electrospinning Approach in Cardiac Tissue Engineering

3.1 Cells and Scaffolds in Cardiac Tissue Engineering: Two Faces, One Coin. The Questions

Cardiac tissue engineering represents a complex field of regenerative medicine and demands for a careful evaluation of all the components involved in heart regeneration. Tailored design of structural properties of the polymeric scaffold, conceived as a mechanical and biological support, cannot disregard the importance of the characteristics of the cells populating the material. Therefore, wide efforts need to be oriented to the optimization of the interaction between scaffold and cells, starting from the selection of the most suitable cell candidate for myocardial regeneration.

The ideal cell source to create an engineered myocardial patch should be easy to harvest, proliferative, non-immunogenic, and have the ability to differentiate into mature, functional cardiomyocyte. Several cell sources have been proposed [56]. Allogenic cells could represent a good option as they are easily recruitable, but their use entails risky immunosuppression. On the other side, autologous cells have no immunologic barriers, but are harnessed by low yield and issues of expansion [57]. Even if, ideally, the structural, contractile and electrophysiological properties of pre-differentiated cardiomyocytes might constitute the more amenable alternative, their recruitment and expansion are problematic. Additionally, these cells are extremely sensitive to ischemic insults. To date, skeletal muscle-derived progenitors, or myoblasts, and crude bone marrow mononuclear cells [58] represent the most widely used cell types for cardiac cell therapy in human patients, as they are readily available, autologous, and easily expanded in vitro. Literature and expert opinion is divided about the actual efficacy of myoblasts as their apparent inability for transdifferentiation into cardiac or endothelial cells, and the risk of arrhythmias associated with their injection [59]. On the other side, bone marrow-derived stem cells are currently gaining favour because of their high plasticity, enabling a prompt phenotypic response to cues from the target organ. In this regard, some research groups favour simple harvest and immediate reinjection of unfractionated bone marrow cells in patients with acute myocardial infarction. Reports of cardiac calcifications with bone-marrow mononuclear cells [60] and myocardial scarring with mesenchymal stem cells [61] put on hold initial enthusiasms.

The use of autologous adult stem cells is harnessed by issues concerning yield and recovery from tissues of origin, i.e. bone marrow, adipose tissue, or circulation, limiting the number of effectively available cells for treatment [62].

In addition, taking into account the clinical scenarios, the numerous comorbidities affecting patients currently in need of cardiovascular treatments, (i.e. diabetes, chronic renal insufficiency, etc.) dramatically lower cell treatment potentialities as for impaired proliferation, adhesion, and incorporation into vascular structures [63]. Moreover, cancerous transformation has been demonstrated in adult mesenchymal stem cell from aged subjects. Li et al. showed spontaneous expression of embryonic factors and p53 point mutation in aged mesenchymal stem cells in mice. Spontaneously transformed cells contributed directly to tumor formation, tumor vasculature, and tumor adipose tissue, recruited additional host bone marrow-derived cells and fused with them acting as cancer stem cells [64]. This represents a further concern, especially considering the advanced age of the patients currently in need of cardiologic treatments.

For human embryonic stem cells, risk of teratoma and cancerogenic shifts represent the major concerns [65].

Another possible recently suggested alternative could be represented by the use of a pre-differentiated phenotype, partially committed towards the tissue of interest and receiving once within the target environment the correct sequence of signals to complete its differentiation [3, 66–68].

Continuous research on the development of an efficient and reproducible method to control and direct the differentiation of stem cells to the desired cell type in vitro is still ongoing.

On the other side, the biomaterial scaffold plays a key role in most tissue engineering strategies. The scaffold is responsible for guidance of cell organization, growth, and differentiation in tissue engineered constructs, providing not only a physical support for the cells, but also the chemical and biological cues required in forming functional tissues [69]. Therefore, the biomaterial should be able to interact with cells, providing, on the molecular level, a biological crosstalk similarly to the natural interactions existing between cells and the native ECM. Additionally, this artificial ECM, while preserving all the structural properties of natural components, is required to be non-toxic and non-immunogenic, sustaining cell viability, promoting cell–matrix interaction and eliciting specific cellular responses and biomolecular recognition.

Thus, in cardiac tissue engineering, the trend has been to design bioactive materials which on one hand have the appropriate physical strength as well as the degradation kinetics of synthetic polymers, and, on the other hand, have the biological specificity of collagen, fibronectin, and laminin—the major ECM components.

Cardiac tissue naturally presents a high degree of hierarchical organization and an architecture that define a particular functional environment, characterized by a precise geometry and dense cellularity. Therefore, the primary aims in the production of a cardiac substitute are focused on the possibility to accommodate a high number of working cells maintaining a resistant structure compliant to different working loads. Interconnected porous structures appear to be the most amenable candidates to accomplish this role allowing to host a large number of cells. However, porosity needs to be tailored in order to meet the nutritional needs of the future cardiomyoyctes. A porosity of at least 50 μm is required to guarantee vascularization of the scaffold after transplantation, nutrients and oxygen diffusion and secretion removal [70]. Additionally, the possibility to provide a biological signaling similar to that of the natural ECM to promote cellular adhesion, proliferation and diffentiation, is desirable.

3.2 The Electrospinning Response

The area of tissue engineering has been driven by biomimicry-inspired design of materials to recreate the natural three-dimensional environment for better cell and tissue growth [71]. An important aspect of these efforts is to mimic the fibrillar structure of the extracellular matrix (ECM), which provides essential guidance for cell organization, survival and function. Only recently, success has been reached in developing biomaterials with sub-micron fibers and morphological similarity to the ECM, thus possibly providing an innate setting for cell assembly and growth. Electrospinning has emerged as a simple yet versatile method in manufacturing such biomaterials out of synthetic [24] or natural [72, 73] polymers. Recently, electrospun nonwoven nanofiber membranes have been demonstrated in multiple biomedical applications, including the production of scaffolds for tissue engineering, wound healing, drug delivery, and medical implants [72]. Electrospun biomaterials have a very high surface-area-to-volume ratio and high porosity useful for better cell incorporation and perfusion. Moreover, the process allows control of structure at the nano-, micro- and macro-scales for flexible tissue design. The morphological similarities to the three-dimensional ECM protein fiber network of nanofibrous electrospun materials, has been suggested to improve cellular response and biocompatibility [74]. Additionally, the electrospinning process is versatile and allows incorporation of multiple polymers and bioactive ingredients. It can also be used to enhance mechanical properties of the obtained nanofibrous materials in comparison to solid-walled equivalents [75]. A wide variety of biodegradable and biocompatible polymers have been investigated to fabricate nanofibrous membranes including synthetic polymers. Among all the biodegradable and biocompatible polymers, both synthetic and natural, poly(ethylene-co-vinyl alcohol), collagen, fibrinogen, poly(L-lactide) (PLLA), poly(glycolide) (PGA) and their copolymers, and poly(ε-caprolactone) (PCL) [18], hold a great potential for success since they are FDA approved, low-cost materials for use in humans, and they are easy to process to achieve controllable microstructure and morphology. In addition, the degradation rate and mechanical properties of polymers and copolymers can be easily tailored for different targeted applications.

The sub-micron pores and nano-topography of electrospun structures are reported as optimal in the realization of successful extracellular matrix-like scaffolds for cardiac tissue engineering [76–80] Ishii et al. have cultured primary cardiomyocytes harvested from neonatal rats onto biodegradable electrospun, nanofibrous poly(ε-caprolactone) meshes with an average fiber diameter of 250 nm, using a cell layering technique [81], with encouraging results [77]. Recently, different stem cell populations have been engrafted on commercially available PGA biodegradable patch [82–85] or collagen matrices [86], as well.

Additionally, survival and integration of transplanted cells can be improved by embedding them in matrices such as collagen [87] or Matrigel [88], or by implanting them as monolayer cell sheets [89].

Polymeric biomaterials have also been loaded with growth factors, cytokines and drugs [38, 90–93] thus obtaining drug releasing systems capable of focused and localized delivery of molecules depending on the environment requirements and the actual *milieu* in which the scaffold is placed [94]. These concepts become even more relevant in cardiac tissue engineering, where use of artificial myocardial patches is desirable to both limit geometrical and shape remodelling of the infarcted myocardium and to promote tissue restoration. The main consequence of a myocardial infarction is regarded as the proliferation of fibroblasts and the deposition of ECM, with the aim to compensate the loss of functional cardiomyocytes and prevent ventricular dilation. This adaptive mechanism results in the loss of cardiac elastomechanical properties and the decline of its dynamic compliance, with an increase in ventricular stiffness, eventually compromising diastolic function. This perpetrating circle, sustained by ECM deposition and leading to diastolic heart failure, is considered the real pathogenic mechanism at the basis of the clinical evolution of a myocardial infarction. On the other side, the loss of the systolic component, represented by the working cardiomyocytes, although tolerated within certain ranges, can further aggravate the decline in heart function. In this scenario, engineered functionalized myocardium might represent an amenable alternative as it could provide both a mean for mechanical restraint and prevention of ventricular dilation. At the same time, it could generate an environment to maintain the proliferative capacity of the cells surrounding the infarct area providing a molecular pathway to promote cell differentiation [17].

With this in mind, electrospinning represents a promising strategy for cardiac tissue engineering answering to the requirements of such a complex field of regenerative medicine. What we could call the "electrospinning response" and its relevance in the field of regenerative medicine, reflects at the level of the biomimetic inspiration of the ultrastructure of electrospun scaffolds, in the plasticity and versatility of the starting materials used, in the amenability for functionalization of the fibers with several bioactive molecules to guide tissue restoration.

3.2.1 The Influence of Structural Architecture in Electrospinning-Based Scaffold

Cardiac tissue naturally presents a high degree of hierarchical organization and architecture that define a particular functional environment characterized by precise geometry and dense cellularity. A particular macro-feature in cell organization found often in vivo—directional growth or anisotropy—can be effectively addressed using the process of electrospinning. An example was recently demonstrated by employing a spinning disk in combination with electrospinning to fabricate highly aligned scaffolds (Fig. 3) [95]. For myocardial tissue, anisotropy carries functional importance and has been the focus of extensive research to

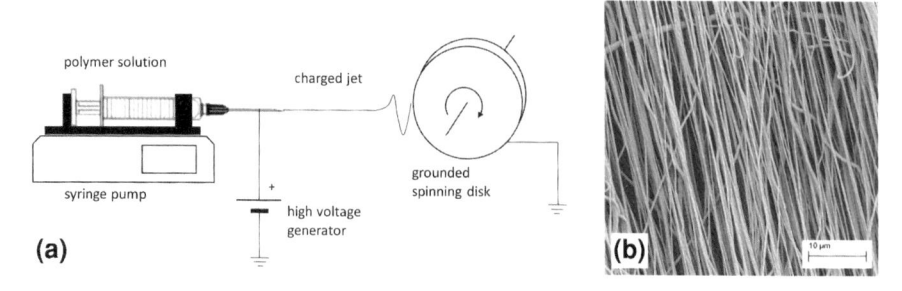

Fig. 3 a Scheme of a modified setup (spinning disk configuration) for the obtaining of aligned electrospun fibers; b electrospun aligned PCL fibers obtained with the spinning disk configuration

orient the ECM or the cell growth in a 2D setting. Electrospun oriented fibers can effectively help in solving this problem [96]. Applying a simple post-processing step, i.e. mechanical stretching of the electrospun membranes Zong et al. achieved a highly oriented scaffold texture, enforcing anisotropic cell growth. Cardiomyocytes (CMs) interacted with the provided nano- and microfibrous network and grew to follow the scaffold-prescribed direction, demonstrating that CMs are very sensitive to the composition of the electrospun PLGA-based scaffolds, with a preference to relatively hydrophobic surfaces [75]. Apparently, hydrophilic and faster degrading electrospun scaffolds allow a lower cell density. On the other hand, poly(L-lactide) scaffolds promoted better CMs adhesion and mature cytoskeleton structure with well-defined periodic units in the contractile machinery (sarcomeres). CMs reestablished cell-to-cell contacts and intercalated disks, typical of mature adult cardiac tissue, on nano- and microstructured electrospun nonwoven scaffolds. PLLA scaffolds are recognized to exert a superior behaviour on cells compared to its blends and copolymers with poly(glycolide) and poly (ethylene glycol) providing both flexibility and guidance for CM growth [75].

Fiber orientation and consequent enhancement of the mechanical properties of the scaffold is particularly important especially in the field of cardiac tissue, engineering, where elastic materials with high tensile resistance are required. Additionally, in light of a translationally approach in the clinical side, especially in the field of cardiovascular medicine in which rapid availability of repairing materials is required, electrospinning and the previous mentioned post-processing approach might present some technical difficulties, mainly related to high-voltage electric fields and low production rate limit. To address this issue, Badrossamay et al. showed a convenient method of fabrication of aligned three-dimensional nanofiber structures by utilizing high-speed, rotating polymer solution jets to extrude fibers (rotary jet-spinning). Authors showed that fiber morphology, diameter, and web porosity can be controlled by varying nozzle geometry, rotation speed, and polymer solution properties. More importantly, they were able to build anisotropic arrays of biodegradable polymer fibers and to seed the constructs with neonatal rat ventricular cardiomyocytes. The myocytes used the aligned fibers to orient their contractile cytoskeleton and to self-organize into a beating, multicellular tissue that mimics the laminar, anisotropic architecture of the heart muscle. This technique may prove advantageous for building uniaxially aligned nanofiber structures for polymers which are not amenable to fabrication by electrospinning [97].

Another further insight in understanding how scaffold architecture affects cell behavior has been added by Fromstein et al. that generated large numbers of embryonic stem cells (ESC)-derived cardiomyocytes in bioreactors and compared results of seeding on porous, three-dimensional scaffolds prepared using two different techniques: electrospinning and thermally induced phase separation (TIPS) [98]. The effect of material macro-architecture on the adhesion, viability, and morphology of the seeded cells was studied. Results suggested that material macrostructure plays a large role in modulating ESC-derived cardiomyocyte morphology. Scaffolds fabricated by electrospinning were found to promote an elongated cardiac phenotype, whereas cells cultured on TIPS scaffolds retained a rounded morphology. Cellular distribution within TIPS scaffolds was limited as most of the cells resided within the first 200 μm from the surface. On the contrary, cells were detected throughout the entire volume of electrospun scaffolds. Percentage of viable cells decreased as time progressed within TIPS scaffold, whereas electrospun scaffolds showed consistently high proportions of live to dead cells. The percentage of dead cells in the TIPS materials also increased the further the cells were from the surface. This was not a concern in electrospun scaffolds because nutrient and oxygen diffusion was sufficient to maintain the cells in a constant aerobic state. Interestingly, on both materials, the seeded cells demonstrated functionality by contracting and staining positive for sarcomeric myosin heavy chain and connexin 43. However, this study represents a further confirmation that the interactions between cells and the extracellular matrix play a significant role in cell migration, differentiation, cell cycle, and proliferation. In this case, considering the use of ESCs, is important to note that during embryogenesis, not only soluble matrix factors, such as cytokines, guide different developmental pathways, but also cues from the extracellular matrix play a vital role in deciding cell fate. It is possible that the 3D nature of scaffolds produced by electrospinning technique could result in different protein deposition patterns within the matrix, potentially exposing different binding motifs to the cells and ultimately modulating the phenotype.

Topographical features of the scaffold, in particular surface-area-to-volume ratio and pore size and interconnectivity, have a profound influence on cell proliferation and on cellular responses. The microarchitectural features of electrospun scaffolds significantly influence cell morphology, cell binding and phenotypic expression, but also control the extent and nature of nutrient diffusion and tissue ingrowth [99]. Therefore, maximizing surface-area-to-volume ratio is considered an important goal to enhance cell colonization and fluid transport, optimizing the mechanical response of the scaffold. To this extent, the scientific debate is still open. Indeed, it has been suggested that cells cannot migrate through pores smaller than 10 μm [100]. Therefore, in order for these nanofiber scaffolds to be functionally useful as tissue engineering scaffolds, cultivation of cells into a 3D framework and dynamic conditions have been attempted. Both dynamic culture conditions and in vivo studies have shown that there is cellular penetration and matrix deposition throughout nanofiber constructs although the mechanisms by which this occurs are not fully understood [73]. It has been hypothesized that in these conditions, cells are able to push against the fibers during their migration paths and thus optimize pore sizes for further cellular infiltration [101].

However, even if the use of dynamic cell seeding and cultivation devices, in particular perfusion bioreactors or spinner flasks, has been shown to enhance proliferation and distribution of stem cells into the scaffold [102], in the absence of an appropriate scaffold microarchitecture this approach could remain unsuccessful in ensuring the ingrowth of new tissue especially for cardiac purposes. Explanations accounting for this issue need to be found in the diffusion constraints into the interior of the construct inducing the development of a necrotic core. Carrier et al. [103, 104] and Radisic et al. [105–107] demonstrated the importance of the oxygen supply in producing thick and resistant new tissue on tissue engineering cardiac grafts (TECGs). They found that increasing the oxygen supply to an in vitro TECG, was necessary to improve the production of cardiac muscle [104, 108]. Increasing oxygen availability by increasing endothelialization is especially important as most TECGs and myocardial patches are directly applied to an infarcted area in in vivo experimental models to provide ventricular mechanical support function and act as a cell retaining and delivery system.

An optimal combination of pore size and shape is therefore essential to promote and guide the colonization of cells in 3D and pore dimensions have been shown to directly affect some biological events, with different tissues requiring optimal pore sizes for their regeneration [109]. Pore size represents an important requirement for cell proliferation inside the scaffold as cells preferentially invade macroporosities, while low size pores, additionally coupled with the hydrophobic nature of the majority of polymeric ester-based matrices used in tissue engineering, may hinder cells colonization. However, one of the most important limitations of cardiac tissue engineering scaffolds is that cell proliferation and extracellular matrix deposition may progressively occlude the entire porosity of the scaffold and, consequently reduce nutrient delivery to, and metabolic waste removal from, the interior of the construct [110]. Therefore, even if a large number of studies have pointed out the importance of uniform cell infiltration in 3D, the presence of a microporosity network not accessible for cells could be very important as it might ensure the transport of fluids necessary for cell biosynthesis. To this extent, Salerno et al. proposed a novel μ-bimodal approach in the realization of a PCL scaffold using a combination of gas foaming and selective polymer extraction from co-continuous blends obtaining a multiscaled microarchitecture. The results showed high cell seeding efficiency and the ability of the scaffold to promote and guide selective 3D hMSC colonization and proliferation into the macroporosity, while ensuring the presence of a separate porous network for fluid transport [111]. Given the reported results, it could be speculated that a combined approach using sequential electrospinning and gas foaming with polymer extraction could be an additional alternative to tailor scaffold porosity and address issues of nutrient diffusion.

3.2.2 The Influence of the Starting Material in Electrospinning-Based Scaffold

The electrospinning process is amenable to be used with both natural and synthetic materials or composites widening the application range of this technique.

Natural components of tissue ECM, such as fibrin and fibrinogen, can be successfully electrospun, providing a powerful device for cardiac tissue engineering given their innate ability to induce improved cellular interaction and subsequent scaffold remodeling compared to synthetic scaffolds. McManus et al. successfully electrospun fibrinogen highly porous scaffolds with fiber diameters as small as 80 nm. Fiber diameter could be adjusted directly modifying the electrospinning solution concentration, as is typically seen for other natural and synthetic polymers [112]. Use of fibrinogen as a primary scaffold component, however, has been limited by traditional processing techniques that lead to scaffolds with insufficient mechanical properties. Additionally, fibrinogen is proteolytically degraded by plasmin and other metalloproteinases found in serum, further deteriorating structural integrity of the scaffold and hampering its use in the complicated mechanical settings of cardiac tissue engineering. Thus, new techniques have been developed in order to overcome these issues.

Mechanical properties of the scaffold could be preserved in a cell culture environment acting on the regulation of fibrinogen degradation through physical crosslinking with glutaraldehyde vapour or enzymatic inhibition with aprotinin, a generic serum protease inhibitor. Scaffold degradation was successfully modulated by media supplementation with aprotinin, but glutaraldehyde fixation produced less reliable results when evaluating the mechanical properties of electrospun fibrinogen structures [112]. The same authors have further studied effects of this modification in terms of cellular interaction, demonstrating cardiac fibroblast migration throughout the matrix 2 days after cell seeding and collagen deposition 14 days after cell seeding [55]. On account of the reported concerns in cardiac tissue engineering about scaffold pore size, and the widely cited cut off of 10 μm in diameter as a lower limit for cell migration, these results clearly demonstrated cellular penetration within 2 days after cell seeding. This could be due to the unique nature of the process compared to other scaffold fabrication techniques. Since the fibers are collected as a dry, nonwoven mat, cells may be able to push individual fibers from their path during migration. Cardiac fibroblasts could freely penetrate and efficiently remodel an electrospun matrix of fibrinogen in a relatively short time period. This process could be further modulated using aprotinin, a protease inhibitor.

Interestingly, these results were obtained with a minimal initial seeding (approximately 29,000 cells/cm^2) [55]. In the perspective use of fibrinogen-based scaffolds obtained through autologous donation [113], it can be hypothesized that an electrospun fibrinogen matrix would require a much smaller host cell donation/collection.

However, with the idea to tailor mechanical properties and pore size for uniform cell invasion, composite scaffolds of more than one natural materials can also be created using electrospinning. For example, by sequentially spinning different polymer solutions, a scaffold with layers can be created with each layer being tailored for specific cell adhesion and differentiation [23]. This concept acquires an important significance when biomimicking of a complex multiscaled tissue such as the heart is the objective [14]. Boland et al. have demonstrated smooth muscle cell infiltration into a multi-layered scaffold of collagen types I and III and elastin when cultured in a rotary cell culture system [73]. Alternatively, a concurrent spinning of two or more polymer solutions can be performed, resulting in a scaffold with mixed types of fibers that mimics their in vivo ratios. Additionally, it is possible to blend polymers with different degradation rates thereby increasing the microvoid spaces for tissue in-growth. Co-spinning of PCL and gelatin solutions produced larger pores sizes obtained upon gelatin degradation, resulting in an increase in rabbit bone marrow cell migration [114, 115]. As previously described, in the context of electrospun materials for cardiac tissue engineering, pore size represents an important parameter as it strongly affects the full thickness and uniform engraftment of cells in the scaffold. Small mesh openings of electrospun scaffolds may be a limitation for tissue engineering applications as seeded cells may form a monolayer at the top of the scaffold, with little evidence of cell migration into its depths [23].

Among synthetic materials, aliphatic polyesters are largely used as temporary scaffold materials for tissue engineering and in vivo application because of their bioabsorbability. This class of polymers is completely hydrolyzable in physiological conditions, with the possibility to adjust and fine tune degradation times (ranging from days to years) depending on polymer molecular weight and composition. Poly(α-hydroxy esters) such as poly(lactide) (PLA), poly(glycolide) (PGA) and their co-polymers have been used in the field of heart tissue engineering and in other applications in which tailoring mechanical properties and bioabsorbability of the scaffold is crucial. Targeting the degradation rate of the scaffold, poly(ε-caprolactone) (PCL) is another widely used medical-grade polyester, characterized by a lower hydrophilicity leading to slower degradation kinetics than lactide/glycolide-based polymers. Copolymerization of different hydroxy acid or lactone monomers and blending of bioabsorbable polyesters with different hydrophilicity is a widely employed approach to design polymeric biomaterials with tailored properties and degradation rates [116].

To tailor mechanical properties of the polymeric scaffold, incorporation of carbon nanotubes into the electrospinning process has been performed in order to reinforce the fibers. Multi-walled carbon nanotubes were incorporated into PEO [117] and poly(acrylonitrile) (PAN) nanofibers [118]. The PAN fibers were spun using a moving collector, and, at high concentrations, the nanotubes were aligned with an increase in tensile modulus by 144% at 20 wt% nanotubes loading and tensile strength increased by 75% at 5 wt% nanotube loading. Therefore, it may be possible to tailor the mechanical properties of nanofiber scaffolds to resemble that of the target tissue.

A relatively new frontier in polymeric scaffold production regards the biocatalysis of starting monomers through lipase-catalyzed ring-opening polymerization (ROP) [119]. This biocatalytic route circumvents use of heavy metal catalysts that must otherwise be removed prior to biomaterial use and is conducted at lower temperatures than when using chemical catalysis, producing lower quantities of undesired by-products. Additionally, recent advances, based on the promiscuity of some lipases as *Candida Antarctica* lipase B (CALB), permitted to incorporate different monomers along chains so that co-polymers with defined composition and microstructure can be prepared, thus enabling material scientists to 'tailor' corresponding biomaterial [120]. Recently, Focarete et al. proposed the use of poly(ω-pentadecalactone) (PPDL)—a highly hydrophobic synthetic aliphatic polyester is obtained by biocatalysis—in electrospinning for cardiac tissue engineering applications and reported the optimization study aimed to fabricate submicrometric fibrous scaffolds. PPDL was synthesized from the macrocyclic lactone PDL through lipase-catalyzed ROP and electrospun in a submicrometric fibrous scaffold subsequently seeded with embryonic rat cardiac H9c2 cells [121]. PPDL is a crystalline polylactone (just like PCL) characterized by a long-chain polymer, with 14 methylene units per ester group, and has the advantage of being hydrolyzable, owing to the presence of ester bonds in the polymer chain, with mechanical properties comparable to those of polyethylene (PE) Differently from PCL, that melts at a lower temperature (Tm around 60° C), PPDL melts around 97° C thereby showing a broader range of applications.

The results of this study, demonstrated that the scaffolds are non-toxic towards cells and showed the ability of electrospun PPDL scaffolds to support cell adhesion and to promote cell proliferation with H9c2 cells covering the scaffold surface with a confluent cell monolayer and retaining their native, mesenchymal, spindle shaped, sheet-like morphology. These results suggest that PPDL homopolymer-based fibrous mats from electrospinning are promising slow-degrading biomaterial supports for tissue-engineering applications where long healing times are foreseen [121].

3.2.3 Biofunctionalization of Electrospun Scaffold

One of the basic ideas of TE concerns the use of an absorbable biocompatible material providing the initial temporary support to stimulate and orchestrate tissue repair, progressively disappearing and being replaced by newly developed tissue. Tailoring its structure, the scaffold would therefore constitute a leading framework for stem cell attachment, differentiation and deposition of new ECM, gradually remodeling and degrading until the final replacement with structured host tissue with all the mechanical and biological features of the native tissue. Another further step to promote stem cells differentiation within a three dimensional support and tissue regeneration concerns the association of a specific bioactive signaling to the scaffold.

The concept of fabricating a scaffold containing factors, able to induce stem cells differentiation and to exert important effects once in in vivo settings, is relatively novel [3, 17, 122]. The idea of designing an absorbable scaffold able to define an optimal microenvironment to induce tissue-oriented differentiation of stem lineages and at the same time to host cells in a structure resembling the natural ECM opens a new paradigm in TE. In this scenario, the biomaterial is intended as a differentiating system for the stem cells seeded therein, and is capable to guide and orient the differentiation process within a 3D biomimetic structure. Clearly, electrospinning is one the most suitable candidates for this purpose.

Electrospinning affords great flexibility in selecting materials for drug delivery applications. Either biodegradable or non-degradable materials can be used to control whether drug release will occur via diffusion alone or diffusion and scaffold degradation. Additionally, due to the flexibility in materials selection, a number of drugs can be delivered including: antibiotics, anticancer drugs, proteins, and DNA. Using the various electrospinning techniques, a number of different drug loading methods can also be utilized: coatings, embedding, and encapsulation (coaxial and emulsion electrospinning). These techniques can be used to give finer control over drug release kinetics [14].

However, functionalization of electrospun fibers is typically carried out either by post-processing methods to conjugate molecules to the surface of nanofibers or by incorporating the bioactive factor into the spinning solution [122, 123]. Interestingly, the previous mentioned coaxial electrospinning technique represents a potential alternative for functionalization as it allows the production of nanofibers with a core–shell with the possibility to introduce different solutions into the core. Therefore it is possible to embed drugs, growth factors, or genes into the core of biodegradable polymer nanofibers, producing polymer nanofibers able to release drugs. Adjusting the thickness of the polymer shell and the degradation speed of the polymer it is possible to finely tune the drug release kinetics. In addition, setting synthetic polymer as the core material and natural polymer such as collagen as the shell material, nanofibers with strong mechanical strength and good biocompatible surface can be produced, solving the issue of poor biocompatibility of synthetic polymer nanofibers.

Another strategy of biofunctionalization concerns the surface modification of synthetic biomaterials, in order to improve biocompatibility. Techniques such as plasma treatment or, more recently, introduction of different surface charges on electrospun PU fiber surfaces through plasma-induced surface polymerization of negatively or positively charged monomers have been tested. Sanders et al. showed that this technique might facilitate vessel ingrowth into the fibro-porous mesh biomaterials.

When degradable polymers are used, special attention is required to protect the nanofibers from rapid degradation and destruction, since strong reaction conditions such as plasma, ultraviolet, radiation, high temperature, and acidic or alkaline environments may ruin the degradable nanofibers easily. This is of extreme importance given the high specific surface area of electrospun, biodegradable nanofibers, in which degradation might occur even faster than in bulk materials. Another interesting technique involves a surface coating through "layer by layer" electrostatic interaction, that can immobilize biomolecules onto polymer surfaces under very moderate conditions. With this technique, ECM proteins such as collagen, fibronectin, and laminin have been deposited on silk fibroin nanofibers surface to promote cell adhesion [124].

Using these techniques, polymers combined with growth factors, cytokines and drugs have been developed [90, 91], generating drug releasing systems capable of focused and localized delivery of molecules according to the local environment requirements. A broad range of applications for electrospun fibers has been suggested [20, 125], ranging from drug delivery [126–130] to gene therapy [123]. A large variety of polymers has been functionalized with growth factors as drug delivery systems. More interestingly, the potential of these 3D ECM-mimicking scaffolds could be oriented in order to obtain a leading framework providing adequate signalling not only for engraftment and proliferation but also for differentiation of pre-committed or stem cells towards different phenotypes.

Recently, our group has developed a hydroxyapatite-functionalized electrospun PLLA scaffold with the aim to recapitulate the native histoarchitecture and the molecular signalling of osteochondral tissue to facilitate cell differentiation towards chondrocytes. PLLA/hydroxyapatite nanocomposites induced differentiation of hMSCs in a chondrocyte-like phenotype with generation of a proteoglycan based matrix [93]. Moreover, we produced data on scaffolds tailored for cardiovascular structures [38, 131]. Heparin functionalized electrospun PLLA scaffolds have been shown to promote endothelial differentiation of mesenchymal stem cells [122]. These data represent a proof of principle of the possibility to produce a scaffold suitable for stem cells seeding, containing the appropriate factors to induce a guided differentiation towards the desired phenotype. In these settings, differentiation would be realized within a three-dimensional ECM-like environment closely mimicking the tissue native architecture and allowing a harmonious ongoing cell growth and differentiation for tissue regeneration.

The intrinsic properties of the scaffolds could therefore provide correct sequences of signals to promote cell adhesion and matrix remodeling, creating a microenvironment able not only to assist and guide cell growth, differentiation and repopulation but also to mimic the mechanical properties of the native tissue, providing at the same time important signaling once in the in vivo settings.

A direct application of this concept has been recently reported in the field of cardiac tissue engineering [132]. Recent studies in cardiac cell therapy pointed out the lack of functional electromechanical coupling between the majority of grafted myoblasts and cardiomyocytes (CM) [133, 134]. In this context it has been recently reported the effect of Granulocyte Colonies Stimulating Factor (G-CSF) on angiogenesis and induction of the expression of Connexin 43 (Cx43) [135], a cardiac-specific gap junction protein crucial for effective electromechanical association [136, 137]. G-CSF has been shown to inhibit both apoptosis and remodeling in the failing heart following myocardial infarction through the receptor responsible for cardiac hypertrophy [138]. Moreover, it activates the Wnt and Jak2 signals in CMs up-regulating Cx43, protecting from ventricular arrhythmia induced by myocardial infarction, and ameliorating survival in a rodent model of myocardial infarction [135]. With this in mind, a PLLA electrospun scaffold functionalized with G-CSF has been developed and seeded with skeletal myoblasts. Authors demonstrated induction of a cardiac pre-commitment of skeletal myoblasts, indicated by the expression of cardiac-specific Connexin 43 and Troponin-I [132]. However, morphological and immunophenotypic changes achieved by myoblasts in this setting were not compatible with a complete differentiation to CM and cells did not acquire beating capabilities. In light of an in vivo application, a fully differentiated beating phenotype could paradoxically fail to integrate with the host because of asynchronous beating activity and potentially constitute an arrythmogenic *focus* leading to life threatening arrhythmias. These Authors reached the conclusion that inducing a partial differentiation towards a cardiac pre-committed phenotype, expressing some of the key proteins of a mature cardiomyocyte, could represent an interesting strategy to provide a better integration within the cardiac environment. The pre-differentiated cells would receive signals from the new environment thus achieving a gradual ongoing complete differentiation [3]. Given these results, clearly functionalization of ECM-mimicking electrospun scaffold with growth factors could be an interesting alternative to ameliorate cardiac tissue engineered constructs.

3.2.4 Heart Valve Tissue Engineering Using Electrospun-Based Scaffolds

The capability of the electrospinning process to tailor shape and structure of constructs to guide and optimize cell-to-matrix interaction and provide additional control on the geometry of the scaffold, has widened the fields of its application in cardiac tissue engineering. Indeed, since in this process a collector is used that has the desired shape of the scaffold, complex scaffold geometries can be easily reproduced. Additionally, it is possible to realize optimized fibers interconnectivity increasing the mechanical integrity of the scaffold. With this in mind, van Lieshout et al. developed electrospun PCL scaffolds for tissue engineering of the aortic valve and compared their properties with a Dacron knitted valvular scaffold after seeding with human myofibroblasts. Results were compatible with a good cell engraftment but poor resistance on mechanical loading [139]. Del Gaudio et al. successfully produced a PCL electrospun valvular scaffold, demonstrating, at pulse duplicator testing, synchronous opening of the leaflets and a correct coaptation in the diastolic phase, with a slight rotation of the leaflets. An *in silico* study by numerical simulation of the closed phase predicted the stress distribution within the leaflet, showing that peak levels are reached at the junctions and sustained by the structure without failure [140]. More recently, Hong and colleagues developed a hybrid valve starting from a decellularized porcine aortic heart valve further coated with basic Fibroblast Growth Factor (bFGF)/chitosan/poly-4-hydroxybutyrate using an electrospinning technique. This construct was successively reseeded with mesenchymal stem cells and cultured over a period of 14 days. Recellularization of the hybrid heart valve leaflet scaffolds was significantly improved compared to controls. Biochemical analysis revealed a significant increase of cell mass, 4-hydroxyproline and collagen, resulting in increased mechanical strength in the hybrid heart valve leaflets compared to controls [141].

4 Future Perspectives

Cardiac regeneration requires a complex cascade of events that goes over the simple injection of the right type of cells in the right place and a real regenerative medicine approach should consider the importance of the extracellular matrix (ECM) and the strong biological signals that it can provide. Connective tissue atmosphere and structural microenvironment in which cells are embedded exerts a number of actions affecting cell function and supporting their proliferation and differentiation. To this extent, electrospinning represents an amenable and versatile manufacturing technique for the design of biomaterial scaffolds with tailored architecture for cell and tissue growth. It allows the production of scaffolds that remarkably mimic the size and scale of the natural extracellular matrix. The submicron pores and the nano-topography of these structures are reported as optimal in the realization of successful extracellular matrix-like scaffolds for cardiac tissue engineering. A wide range of scaffolds can be constructed while at the same time precisely controlling fiber orientation, composition (blended fibers), and dimensions. Complex constructs can be fabricated to closely replicate the structural features and chemical composition of the native structures, including complex three-dimensional shapes. Additionally, the possibility to co-spin polymers with various additives (e.g. growth factors) gives an opportunity to tailoring the scaffold to a specific site and application.

Considering the current limitations of cell therapy, the idea of a bridge between the stem cells plasticity and biomaterials that actively guide and provide the correct sequence of signals to allow ongoing lineage-specific differentiation of these pluripotent precursor cells, is attractive and may represent a promising answer to the problems related to cardiovascular healing. Combination of stem cells with ECM-like functionalized scaffold locally releasing molecules tailored to promote in situ completion of differentiation and to improve homing, survival and function, could be an exciting approach circumventing risk of potential undesired effects of growth factor administration and improving tissue restoration.

References

1. Li, L., Zhang, S., Zhang, Y., et al.: Paracrine action mediate the antifibrotic effect of transplanted mesenchymal stem cells in a rat model of global heart failure. Mol. Biol. Rep. **36**, 725–731 (2009)
2. Dai, W., Hale, S.L., Kloner, R.A.: Role of a paracrine action of mesenchymal stem cells in the improvement of left ventricular function after coronary artery occlusion in rats. Regen. Med. **2**, 63–68 (2007)
3. Spadaccio, C., Chachques, E., Chello, M., et al.: Predifferentiated adult stem cells and matrices for cardiac cell therapy. Asian Cardiovasc. Thorac. Ann. **18**, 79–87 (2010)
4. Assis, A.C., Carvalho, J.L., Jacoby, B.A., et al.: Time-dependent migration of systemically delivered bone marrow mesenchymal stem cells to the infarcted heart. Cell Transpl. **19**(2), 219–230 (2009)

5. Zhang, H., Song, P., Tang, Y., et al.: Injection of bone marrow mesenchymal stem cells in the borderline area of infarcted myocardium: heart status and cell distribution. J. Thorac. Cardiovasc. Surg. **134**, 1234–1240 (2007)
6. Stevens, M.M., George, J.H.: Exploring and engineering the cell surface interface. Science **310**, 1135–1138 (2005)
7. Wan, Y., Wang, Y., Liu, Z., et al.: Adhesion and proliferation of OCT-1 osteoblast-like cells on micro- and nano-scale topography structured poly(L-lactide). Biomaterials **26**, 4453–4459 (2005)
8. Yim, E.K., Reano, R.M., Pang, S.W., et al.: Nanopattern-induced changes in morphology and motility of smooth muscle cells. Biomaterials **26**, 5405–5413 (2005)
9. He, W., Yong, T., Ma, Z.W., et al.: Biodegradable polymer nanofiber mesh to maintain functions of endothelial cells. Tissue Eng. **12**, 2457–2466 (2006)
10. Montjovent, M.O., Mathieu, L., Hinz, B., et al.: Biocompatibility of bioresorbable poly (L-lactic acid) composite scaffolds obtained by supercritical gas foaming with human fetal bone cells. Tissue Eng. **11**, 1640–1649 (2005)
11. Noh, H.K., Lee, S.W., Kim, J.M., et al.: Electrospinning of chitin nanofibers: degradation behavior and cellular response to normal human keratinocytes and fibroblasts. Biomaterials **27**, 3934–3944 (2006)
12. Xu, J., Liu, X., Jiang, Y., et al.: MAPK/ERK signalling mediates VEGF-induced bone marrow stem cell differentiation into endothelial cell. J. Cell. Mol. Med. **12**, 2395–2406 (2008)
13. Nair, L.S., Bhattacharyya, S., Laurencin, C.T.: Development of novel tissue engineering scaffolds via electrospinning. Expert Opin. Biol. Ther. **4**, 659–668 (2004)
14. Sill, T.J., von Recum, H.A.: Electrospinning: applications in drug delivery and tissue engineering. Biomaterials **29**, 1989–2006 (2008)
15. Ma, Z., Kotaki, M., Inai, R., et al.: Potential of nanofiber matrix as tissue-engineering scaffolds. Tissue Eng. **11**, 101–109 (2005)
16. ISI Web of Knowledge, Web of Science. http://www.isiknowledge.com. Accessed 3 July 2010
17. Spadaccio, C., Chello, M., Trombetta, M., et al.: Drug releasing systems in cardiovascular tissue engineering. J. Cell. Mol. Med. **13**, 422–439 (2009)
18. Li, W.J., Laurencin, C.T., Caterson, E.J., et al.: Electrospun nanofibrous structure: a novel scaffold for tissue engineering. J. Biomed. Mater. Res. **60**, 613–621 (2002)
19. Chew, S.Y., Wen, J., Yim, E.K., et al.: Sustained release of proteins from electrospun biodegradable fibers. Biomacromolecules **6**, 2017–2024 (2005)
20. Huang, Z.M., Zhang, Y.Z., Kotaki, M., et al.: A review on polymer nanofibers by electrospinning and their applications in nanocomposites. Compos. Sci. Technol. **63**, 2223–2253 (2003)
21. Shin, Y.M., Hohman, M.M., Brenner, M.P., et al.: Experimental characterization of electrospinning: the electrically forced jet and instabilities. Polymer **42**, 9955–9967 (2001)
22. Badami, A.S., Kreke, M.R., Thompson, M.S., et al.: Effect of fiber diameter on spreading, proliferation, and differentiation of osteoblastic cells on electrospun poly(lactic acid) substrates. Biomaterials **27**, 596–606 (2006)
23. Kidoaki, S., Kwon, I.K., Matsuda, T.: Mesoscopic spatial designs of nano- and microfiber meshes for tissue-engineering matrix and scaffold based on newly devised multilayering and mixing electrospinning techniques. Biomaterials **26**, 37–46 (2005)
24. Kim, T.G., Park, T.G.: Biomimicking extracellular matrix: cell adhesive RGD peptide modified electrospun poly(D, L-lactic-co-glycolic acid) nanofiber mesh. Tissue Eng. **12**, 221–233 (2006)
25. Mo, X.M., Xu, C.Y., Kotaki, M., et al.: Electrospun P(LLA-CL) nanofiber: a biomimetic extracellular matrix for smooth muscle cell and endothelial cell proliferation. Biomaterials **25**, 1883–1890 (2004)
26. Yang, F., Murugan, R., Wang, S., et al.: Electrospinning of nano/micro scale poly(L-lactic acid) aligned fibers and their potential in neural tissue engineering. Biomaterials **26**, 2603–2610 (2005)

27. Woo, K.M., Jun, J.H., Chen, V.J., et al.: Nano-fibrous scaffolding promotes osteoblast differentiation and biomineralization. Biomaterials **28**, 335–343 (2007)
28. Stitzel, J., Liu, J., Lee, S.J., et al.: Controlled fabrication of a biological vascular substitute. Biomaterials **27**, 1088–1094 (2006)
29. Mobasheri, A., Csaki, C., Clutterbuck, A.L., et al.: Mesenchymal stem cells in connective tissue engineering and regenerative medicine: applications in cartilage repair and osteoarthritis therapy. Histol. Histopathol. **24**, 347–366 (2009)
30. Min, B.M., Jeong, L., Nam, Y.S., et al.: Formation of silk fibroin matrices with different texture and its cellular response to normal human keratinocytes. Int. J. Biol. Macromol. **34**, 281–288 (2004)
31. Venugopal, J., Ramakrishna, S.: Biocompatible nanofiber matrices for the engineering of a dermal substitute for skin regeneration. Tissue Eng. **11**, 847–854 (2005)
32. Simon Jr., C.G., Eidelman, N., Kennedy, S.B., et al.: Combinatorial screening of cell proliferation on poly(L-lactic acid)/poly(D, L-lactic acid) blends. Biomaterials **26**, 6906–6915 (2005)
33. Su, S.H., Nguyen, K.T., Satasiya, P., et al.: Curcumin impregnation improves the mechanical properties and reduces the inflammatory response associated with poly(L-lactic acid) fiber. J. Biomater. Sci. Polym. Ed. **16**, 353–370 (2005)
34. Tsuji, H., Ogiwara, M., Saha, S.K., et al.: Enzymatic, alkaline, and autocatalytic degradation of poly(L-lactic acid): effects of biaxial orientation. Biomacromolecules **7**, 380–387 (2006)
35. Kikuchi, M., Koyama, Y., Takakuda, K., et al.: In vitro change in mechanical strength of beta-tricalcium phosphate/copolymerized poly-L-lactide composites and their application for guided bone regeneration. J. Biomed. Mater. Res. **62**, 265–272 (2002)
36. Kikuchi, M., Koyama, Y., Yamada, T., et al.: Development of guided bone regeneration membrane composed of beta-tricalcium phosphate and poly (L-lactide-co-glycolide-co-epsilon-caprolactone) composites. Biomaterials **25**, 5979–5986 (2004)
37. Oh, T., Rahman, M.M., Lim, J.H., et al.: Guided bone regeneration with beta-tricalcium phosphate and poly L-lactide-co-glycolide-co-epsilon-caprolactone membrane in partial defects of canine humerus. J. Vet. Sci. **7**, 73–77 (2006)
38. Spadaccio, C., Rainer, A., Chello, M., et al.: Drug releasing hybrid scaffold: new avenue in cardiovascular tissue engineering. Tissue Eng. Part A **14**, 691–943 (2008)
39. Doshi, J., Reneker, D.H.: Electrospinning process and applications of electrospun fibers. J. Electrostat. **35**, 151 (1995)
40. Zhao, P., Jiang, H., Pan, H., et al.: Biodegradable fibrous scaffolds composed of gelatin coated poly(epsilon-caprolactone) prepared by coaxial electrospinning. J. Biomed. Mater. Res. A **83**, 372–382 (2007)
41. Zhang, Y.Z., Wang, X., Feng, Y., et al.: Coaxial electrospinning of (fluorescein isothiocyanate-conjugated bovine serum albumin)-encapsulated poly(ε-caprolactone) nanofibers for sustained release. Biomacromolecules **7**, 1049–1057 (2006)
42. Townsend-Nicholson, A., Jayasinghe, S.N.: Cell electrospinning: a unique biotechnique for encapsulating living organisms for generating active biological microthreads/scaffolds. Biomacromolecules **7**, 3364–3369 (2006)
43. Pham, Q.P., Sharma, U., Mikos, A.G.: Electrospinning of polymeric nanofibers for tissue engineering applications: a review. Tissue Eng. **12**, 1197–1211 (2006)
44. Chong, E.J., Phan, T.T., Lim, I.J., et al.: Evaluation of electrospun PCL/gelatin nanofibrous scaffold for wound healing and layered dermal reconstitution. Acta. Biomater. **3**, 321–330 (2007)
45. Ciardelli, G., Chiono, V., Vozzi, G., et al.: Blends of poly-(ε-caprolactone) and polysaccharides in tissue engineering applications. Biomacromolecules **6**, 1961–1976 (2005)
46. Schnell, E., Klinkhammer, K., Balzer, S., et al.: Guidance of glial cell migration and axonal growth on electrospun nanofibers of poly-ε-caprolactone and a collagen/poly-ε-caprolactone blend. Biomaterials **28**, 3012–3025 (2007)

47. Ghasemi-Mobarakeh, L., Prabhakaran, M.P., Morshed, M., et al.: Electrospun poly (ε-caprolactone)/gelatin nanofibrous scaffolds for nerve tissue engineering. Biomaterials **29**, 4532–4539 (2008)
48. Correlo, V.M., Boesel, L.F., Bhattacharya, M., et al.: Properties of melt processed chitosan and aliphatic polyester blends. Mater. Sci. Eng. A **403**, 57–68 (2005)
49. Kweon, H., Yoo, M.K., Park, I.K., et al.: A novel degradable polycaprolactone networks for tissue engineering. Biomaterials **24**, 801–808 (2003)
50. Zhu, Y., Gao, C., Liu, X., et al.: Surface modification of polycaprolactone membrane via aminolysis and biomacromolecule immobilization for promoting cytocompatibility of human endothelial cells. Biomacromolecules **3**, 1312–1319 (2002)
51. Olabarrieta, I., Forsstrom, D., Gedde, U., et al.: Transport properties of chitosan and whey blended with poly(ε-caprolactone) assessed by standard permeability measurements and microcalorimetry. Polymer **42**, 4401 (2001)
52. Wan, Y., Wu, H., Cao, X., et al.: Compressive mechanical properties and biodegradability of porous poly(caprolactone)/chitosan scaffolds. Polym. Degrad. Stab. **93**, 1736–1741 (2008)
53. Wu, C.S.: A comparison of the structure, thermal properties, and biodegradability of polycaprolactone/chitosan and acrylic acid grafted polycaprolactone/chitosan. Polymer **46**, 147–155 (2005)
54. Sarasam, A., Madihally, S.V.: Characterization of chitosan-polycaprolactone blends for tissue engineering applications. Biomaterials **26**, 5500–5508 (2005)
55. McManus, M.C., Boland, E.D., Simpson, D.G., et al.: Electrospun fibrinogen: feasibility as a tissue engineering scaffold in a rat cell culture model. J. Biomed. Mater. Res. A **81**, 299–309 (2007)
56. Leor, J., Amsalem, Y., Cohen, S.: Cells, scaffolds, and molecules for myocardial tissue engineering. Pharmacol. Ther. **105**, 151–163 (2005)
57. Leor, J., Landa, N., Cohen, S.: Renovation of the injured heart with myocardial tissue engineering. Expert Rev. Cardiovasc. Ther. **4**, 239–252 (2006)
58. Lee, M.S., Makkar, R.R.: Stem-cell transplantation in myocardial infarction: a status report. Ann. Intern. Med. **140**, 729–737 (2004)
59. Smits, P.C., van Geuns, R.J., Poldermans, D., et al.: Catheter-based intramyocardial injection of autologous skeletal myoblasts as a primary treatment of ischemic heart failure: clinical experience with six-month follow-up. J. Am. Coll. Cardiol. **42**, 2063–2069 (2003)
60. Yoon, Y.S., Park, J.S., Tkebuchava, T., et al.: Unexpected severe calcification after transplantation of bone marrow cells in acute myocardial infarction. Circulation **109**, 3154–3157 (2004)
61. Vulliet, P.R., Greeley, M., Halloran, S.M., et al.: Intra-coronary arterial injection of mesenchymal stromal cells and microinfarction in dogs. Lancet **363**, 783–784 (2004)
62. Scheubel, R.J., Zorn, H., Silber, R.E., et al.: Age-dependent depression in circulating endothelial progenitor cells in patients undergoing coronary artery bypass grafting. J. Am. Coll. Cardiol. **42**, 2073–2080 (2003)
63. Dimmeler, S., Vasa-Nicotera, M.: Aging of progenitor cells: limitation for regenerative capacity? J. Am. Coll. Cardiol. **42**, 2081–2082 (2003)
64. Li, H., Fan, X., Kovi, R.C., et al.: Spontaneous expression of embryonic factors and p53 point mutations in aged mesenchymal stem cells: a model of age-related tumorigenesis in mice. Cancer Res. **67**, 10889–10898 (2007)
65. Thomson, J.A., Itskovitz-Eldor, J., Shapiro, S.S., et al.: Embryonic stem cell lines derived from human blastocysts. Science **282**, 1145–1147 (1998)
66. Genovese, J.A., Spadaccio, C., Chachques, E., et al.: Cardiac pre-differentiation of human mesenchymal stem cells by electrostimulation. Front Biosci. **14**, 2996–3002 (2009)
67. Genovese, J.A., Spadaccio, C., Langer, J., et al.: Electrostimulation induces cardiomyocyte predifferentiation of fibroblasts. Biochem. Biophys. Res. Commun. **370**, 450–455 (2008)
68. Genovese, J.A., Spadaccio, C., Rivello, H.G., et al.: Electrostimulated bone marrow human mesenchymal stem cells produce follistatin. Cytotherapy **11**, 448–456 (2009)

69. Langer, R., Tirrell, D.A.: Designing materials for biology and medicine. Nature **428**, 487–492 (2004)
70. Yang, S., Leong, K.F., Du, Z., et al.: The design of scaffolds for use in tissue engineering. Part I. Traditional factors. Tissue Eng. **7**, 679–689 (2001)
71. Hubbell, J.A.: Tissue and cell engineering. Curr. Opin. Biotechnol. **15**, 381–382 (2004)
72. Boland, E.D., Coleman, B.D., Barnes, C.P., et al.: Electrospinning polydioxanone for biomedical applications. Acta. Biomater. **1**, 115–123 (2005)
73. Boland, E.D., Matthews, J.A., Pawlowski, K.J., et al.: Electrospinning collagen and elastin: preliminary vascular tissue engineering. Front. Biosci. **9**, 1422–1432 (2004)
74. Liao, I.C., Liu, J.B., Bursac, N., et al.: Effect of electromechanical stimulation on the maturation of myotubes on aligned electrospun fibers. Cell. Mol. Bioeng. **1**, 133–145 (2008)
75. Zong, X., Bien, H., Chung, C.Y., et al.: Electrospun fine-textured scaffolds for heart tissue constructs. Biomaterials **26**, 5330–5338 (2005)
76. Courtney, T., Sacks, M.S., Stankus, J., et al.: Design and analysis of tissue engineering scaffolds that mimic soft tissue mechanical anisotropy. Biomaterials **27**, 3631–3638 (2006)
77. Ishii, O., Shin, M., Sueda, T., et al.: In vitro tissue engineering of a cardiac graft using a degradable scaffold with an extracellular matrix-like topography. J. Thorac. Cardiovasc. Surg. **130**, 1358–1363 (2005)
78. Shin, M., Ishii, O., Sueda, T., et al.: Contractile cardiac grafts using a novel nanofibrous mesh. Biomaterials **25**, 3717–3723 (2004)
79. Stankus, J.J., Guan, J., Wagner, W.R.: Fabrication of biodegradable elastomeric scaffolds with sub-micron morphologies. J. Biomed. Mater. Res. A **70**, 603–614 (2004)
80. Zhong, S., Teo, W.E., Zhu, X., et al.: Formation of collagen-glycosaminoglycan blended nanofibrous scaffolds and their biological properties. Biomacromolecules **6**, 2998–3004 (2005)
81. Shimizu, T., Yamato, M., Isoi, Y., et al.: Fabrication of pulsatile cardiac tissue grafts using a novel 3-dimensional cell sheet manipulation technique and temperature-responsive cell culture surfaces. Circ. Res. **90**, e40 (2002)
82. Ke, Q., Yang, Y., Rana, J.S., et al.: Embryonic stem cells cultured in biodegradable scaffold repair infarcted myocardium in mice. Sheng Li Xue Bao **57**, 673–681 (2005)
83. Zimmermann, W.H., Eschenhagen, T.: Embryonic stem cells for cardiac muscle engineering. Trends Cardiovasc. Med. **17**, 134–140 (2007)
84. Zimmermann, W.H., Melnychenko, I., Eschenhagen, T.: Engineered heart tissue for regeneration of diseased hearts. Biomaterials **25**, 1639–1647 (2004)
85. Zimmermann, W.H., Schneiderbanger, K., Schubert, P., et al.: Tissue engineering of a differentiated cardiac muscle construct. Circ. Res. **90**, 223–230 (2002)
86. Cortes-Morichetti, M., Frati, G., Schussler, O., et al.: Association between a cell-seeded collagen matrix and cellular cardiomyoplasty for myocardial support and regeneration. Tissue Eng. **13**(11), 2681–2687 (2007)
87. Menasche, P.: Skeletal myoblasts as a therapeutic agent. Prog. Cardiovasc. Dis. **50**, 7–17 (2007)
88. Laflamme, M.A., Chen, K.Y., Naumova, A.V., et al.: Cardiomyocytes derived from human embryonic stem cells in pro-survival factors enhance function of infarcted rat hearts. Nat. Biotechnol. **25**, 1015–1024 (2007)
89. Miyahara, Y., Nagaya, N., Kataoka, M., et al.: Monolayered mesenchymal stem cells repair scarred myocardium after myocardial infarction. Nat. Med. **12**, 459–465 (2006)
90. Zhang, Y., Cheng, X., Wang, J., et al.: Novel chitosan/collagen scaffold containing transforming growth factor-beta1 DNA for periodontal tissue engineering. Biochem. Biophys. Res. Commun. **344**, 362–369 (2006)
91. Luong-Van, E., Grondahl, L., Chua, K.N., et al.: Controlled release of heparin from poly (ε-caprolactone) electrospun fibers. Biomaterials **27**, 2042–2050 (2006)
92. Zhang, G., Suggs, L.J.: Matrices and scaffolds for drug delivery in vascular tissue engineering. Adv. Drug Deliv. Rev. **59**, 360–373 (2007)

93. Spadaccio, C., Rainer, A., Trombetta, M., et al.: Poly-L-lactic acid/hydroxyapatite electrospun nanocomposites induce chondrogenic differentiation of human MSC. Ann. Biomed. Eng. **37**, 1376–1389 (2009)
94. Zilla, P., Deutsch, M., Meinhart, J., et al.: Clinical in vitro endothelialization of femoropopliteal bypass grafts: an actuarial follow-up over three years. J. Vasc. Surg. **19**, 540–548 (1994)
95. Khil, M.S., Bhattarai, S.R., Kim, H.Y., et al.: Novel fabricated matrix via electrospinning for tissue engineering. J. Biomed. Mater. Res. B Appl. Biomater. **72**, 117–124 (2005)
96. Kim, G.H.: Electrospun PCL nanofibers with anisotropic mechanical properties as a biomedical scaffold. Biomed. Mater. **3**, 25010 (2008)
97. Badrossamay, M.R., McIlwee, H.A., Goss, J.A., et al.: Nanofiber assembly by rotary jet-spinning. Nano. Lett. **10**, 2257–2261 (2010)
98. Fromstein, J.D., Zandstra, P.W., Alperin, C., et al.: Seeding bioreactor-produced embryonic stem cell-derived cardiomyocytes on different porous, degradable, polyurethane scaffolds reveals the effect of scaffold architecture on cell morphology. Tissue Eng. Part A **14**, 369–378 (2008)
99. Hollister, S.J.: Porous scaffold design for tissue engineering. Nat. Mater. **4**, 518–524 (2005)
100. Boland, E.D., Telemeco, T.A., Simpson, D.G., et al.: Utilizing acid pretreatment and electrospinning to improve biocompatibility of poly(glycolic acid) for tissue engineering. J. Biomed. Mater. Res. B Appl. Biomater. **71**, 144–152 (2004)
101. Bhattarai, S.R., Bhattarai, N., Yi, H.K., et al.: Novel biodegradable electrospun membrane: scaffold for tissue engineering. Biomaterials **25**, 2595–2602 (2004)
102. Glowacki, J., Mizuno, S., Greenberger, J.S.: Perfusion enhances functions of bone marrow stromal cells in three-dimensional culture. Cell Transpl. **7**, 319–326 (1998)
103. Carrier, R.L., Rupnick, M., Langer, R., et al.: Perfusion improves tissue architecture of engineered cardiac muscle. Tissue Eng. **8**, 175–188 (2002)
104. Carrier, R.L., Rupnick, M., Langer, R., et al.: Effects of oxygen on engineered cardiac muscle. Biotechnol. Bioeng. **78**, 617–625 (2002)
105. Radisic, M., Malda, J., Epping, E., et al.: Oxygen gradients correlate with cell density and cell viability in engineered cardiac tissue. Biotechnol. Bioeng. **93**, 332–343 (2006)
106. Radisic, M., Park, H., Chen, F., et al.: Biomimetic approach to cardiac tissue engineering: oxygen carriers and channeled scaffolds. Tissue Eng. **12**, 2077–2091 (2006)
107. Radisic, M., Park, H., Gerecht, S., et al.: Biomimetic approach to cardiac tissue engineering. Philos. Trans. R Soc. Lond. B Biol. Sci. **362**, 1357–1368 (2007)
108. Zimmermann, W.H., Fink, C., Kralisch, D., et al.: Three-dimensional engineered heart tissue from neonatal rat cardiac myocytes. Biotechnol. Bioeng. **68**, 106–114 (2000)
109. Zeltinger, J., Sherwood, J.K., Graham, D.A., et al.: Effect of pore size and void fraction on cellular adhesion, proliferation, and matrix deposition. Tissue Eng. **7**, 557–572 (2001)
110. Karande, T.S., Ong, J.L., Agrawal, C.M.: Diffusion in musculoskeletal tissue engineering scaffolds: design issues related to porosity, permeability, architecture, and nutrient mixing. Ann. Biomed. Eng. **32**, 1728–1743 (2004)
111. Salerno, A., Guarnieri, D., Iannone, M., et al.: Engineered μ-bimodal poly(ε-caprolactone) porous scaffold for enhanced hMSC colonization and proliferation. Acta. Biomater. **5**, 1082–1093 (2009)
112. McManus, M.C., Boland, E.D., Koo, H.P., et al.: Mechanical properties of electrospun fibrinogen structures. Acta. Biomater. **2**, 19–28 (2006)
113. Haisch, A., Loch, A., David, J., et al.: Preparation of a pure autologous biodegradable fibrin matrix for tissue engineering. Med. Biol. Eng. Comput. **38**, 686–689 (2000)
114. Sahoo, S., Ouyang, H., Goh, J.C., et al.: Characterization of a novel polymeric scaffold for potential application in tendon/ligament tissue engineering. Tissue Eng. **12**, 91–99 (2006)
115. Zhang, Y., Ouyang, H., Lim, C.T., et al.: Electrospinning of gelatin fibers and gelatin/PCL composite fibrous scaffolds. J. Biomed. Mater. Res. B Appl. Biomater. **72**, 156–165 (2005)
116. Middleton, J.C., Tipton, A.J.: Synthetic biodegradable polymers as orthopedic devices. Biomaterials **21**, 2335–2346 (2000)

117. Salalha, W., Dror, Y., Khalfin, R.L., et al.: Single-walled carbon nanotubes embedded in oriented polymeric nanofibers by electrospinning. Langmuir **20**, 9852–9855 (2004)
118. Ge, J.J., Hou, H., Li, Q., et al.: Assembly of well-aligned multiwalled carbon nanotubes in confined polyacrylonitrile environments: electrospun composite nanofiber sheets. J. Am. Chem. Soc. **126**, 15754–15761 (2004)
119. Yamamoto, Y., Kaihara, S., Toshima, K., et al.: High-molecular-weight polycarbonates synthesized by enzymatic ROP of a cyclic carbonate as a green process. Macromol. Biosci. **9**, 968–978 (2009)
120. Ceccorulli, G., Scandola, M., Kumar, A., et al.: Cocrystallization of random copolymers of omega-pentadecalactone and epsilon-caprolactone synthesized by lipase catalysis. Biomacromolecules **6**, 902–907 (2005)
121. Focarete, M.L., Gualandi, C., Scandola, M., et al.: Electrospun scaffolds of a polyhydroxyalkanoate consisting of ω-hydroxylpentadecanoate repeat units: fabrication and in vitro biocompatibility studies. J. Biomater. Sci. Polym. Ed. **21**, 1283–1296 (2010)
122. Spadaccio, C., Rainer, A., Centola, M., et al.: Heparin-releasing scaffold for stem cells: a differentiating device for vascular aims. Regen. Med. **5**, 645–657 (2010)
123. Luu, Y.K., Kim, K., Hsiao, B.S., et al.: Development of a nanostructured DNA delivery scaffold via electrospinning of PLGA and PLA-PEG block copolymers. J. Control Release **89**, 341–353 (2003)
124. Min, B.M., Lee, G., Kim, S.H., et al.: Electrospinning of silk fibroin nanofibers and its effect on the adhesion and spreading of normal human keratinocytes and fibroblasts in vitro. Biomaterials **25**, 1289–1297 (2004)
125. Zhang, Y.Z., Venugopal, J., Huang, Z.M., et al.: Characterization of the surface biocompatibility of the electrospun PCL-collagen nanofibers using fibroblasts. Biomacromolecules **6**, 2583–2589 (2005)
126. Katti, D.S., Robinson, K.W., Ko, F.K., et al.: Bioresorbable nanofiber-based systems for wound healing and drug delivery: optimization of fabrication parameters. J. Biomed. Mater. Res. B Appl. Biomater. **70**, 286–296 (2004)
127. Kim, K., Luu, Y.K., Chang, C., et al.: Incorporation and controlled release of a hydrophilic antibiotic using poly(lactide-co-glycolide)-based electrospun nanofibrous scaffolds. J. Control Release **98**, 47–56 (2004)
128. Zeng, J., Xu, X., Chen, X., et al.: Biodegradable electrospun fibers for drug delivery. J. Control Release **92**, 227–231 (2003)
129. Kenawy, E.-R., Bowlin, G.L., Mansfield, K., et al.: Release of tetracycline hydrochloride from electrospun poly(ethylene-co-vinylacetate), poly(lactic acid), and a blend. J. Control Release **81**, 57–64 (2002)
130. Verreck, G., Chun, I., Rosenblatt, J., et al.: Incorporation of drugs in an amorphous state into electrospun nanofibers composed of a water-insoluble, nonbiodegradable polymer. J. Control Release **92**, 349–360 (2003)
131. Centola, M., Rainer, A., Spadaccio, C., et al.: Combining electrospinning and fused deposition modeling for the fabrication of a hybrid vascular graft. Biofabrication **2**, 014102 (2010)
132. Spadaccio, C., Rainer, A., Trombetta, M. et al.: A G-CSF functionalized scaffold for stem cells seeding: a differentiating device for cardiac purposes. J. Cell. Mol. Med. doi: 10.1111/j.1582-4934.2010.01100.x (2010)
133. Rubart, M., Soonpaa, M.H., Nakajima, H., et al.: Spontaneous and evoked intracellular calcium transients in donor-derived myocytes following intracardiac myoblast transplantation. J. Clin. Invest. **114**, 775–783 (2004)
134. Leobon, B., Garcin, I., Menasche, P., et al.: Myoblasts transplanted into rat infarcted myocardium are functionally isolated from their host. Proc. Natl. Acad. Sci. U S A **100**, 7808–7811 (2003)
135. Kuwabara, M., Kakinuma, Y., Katare, R.G., et al.: Granulocyte colony-stimulating factor activates Wnt signal to sustain gap junction function through recruitment of beta-catenin and cadherin. FEBS Lett. **581**, 4821–4830 (2007)

136. Lerner, D.L., Yamada, K.A., Schuessler, R.B., et al.: Accelerated onset and increased incidence of ventricular arrhythmias induced by ischemia in Cx43-deficient mice. Circulation **101**, 547–552 (2000)
137. van Rijen, H.V., Eckardt, D., Degen, J., et al.: Slow conduction and enhanced anisotropy increase the propensity for ventricular tachyarrhythmias in adult mice with induced deletion of connexin43. Circulation **109**, 1048–1055 (2004)
138. Harada, M., Qin, Y., Takano, H., et al.: G-CSF prevents cardiac remodeling after myocardial infarction by activating the Jak-Stat pathway in cardiomyocytes. Nat. Med. **11**, 305–311 (2005)
139. van Lieshout, M.I., Vaz, C.M., Rutten, M.C., et al.: Electrospinning versus knitting: two scaffolds for tissue engineering of the aortic valve. J. Biomater. Sci. Polym. Ed. **17**, 77–89 (2006)
140. Del Gaudio, C., Bianco, A., Grigioni, M.: Electrospun bioresorbable trileaflet heart valve prosthesis for tissue engineering: in vitro functional assessment of a pulmonary cardiac valve design. Ann. Ist Super Sanita **44**, 178–186 (2008)
141. Hong, H., Dong, N., Shi, J., et al.: Fabrication of a novel hybrid heart valve leaflet for tissue engineering: an in vitro study. Artif Organs **33**, 554–558 (2009)

Heart Valve Tissue Engineering

Adrian H. Chester, Magdi H. Yacoub and Patricia M. Taylor

Abstract Extensive research into the structural and functional properties of heart valves shows that they perform sophisticated functions which depend on their unique properties at the tissue, cellular and molecular levels. Furthermore, there is accumulating evidence that these complex functional properties translate into clinically relevant end points to all patients undergoing valve replacement surgery, particularly in children and young adults. Tissue engineering a heart valve offers the potential of replicating these functions and benefits via the production of a 'living' valve that resembles the shape and function of the native valve. This chapter will review the biology of heart valves and discuss strategies and advances that have been made towards the goal of producing a living tissue engineered heart valve.

1 Introduction

Until relatively recently heart valves were considered to be inert flaps of tissue that opened and closed in response to the passage of blood through and out of the heart. This simplistic view of heart valves vastly understated their complex function and importance to cardiac function and coronary flow, which only became fully evident when surgeons began to replace diseased aortic valves with purely mechanical devices or tissue valves that did not contain living cells. While many of these different valve substitutes perform well over a 10–15 year period [1], none

---

A. H. Chester (✉), M. H. Yacoub and P. M. Taylor
Heart Science Centre, National Heart and Lung Institute, Imperial College London, Harefield UB9 6JH, UK
e-mail: a.chester@imperial.ac.uk

Fig. 1 Graph showing the risks of reoperation (*bars*) and bleeding (*lines*) after aortic valve replacement with bioprostheses (BP) and mechanical prosthesis (MP) (from [139])

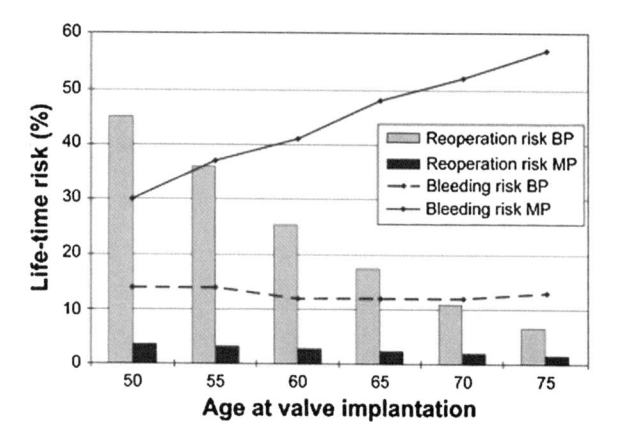

of them replicates the sophisticated function of the native valve and their clinical benefit is limited by concomitant anti-coagulation therapy or tissue degeneration (Fig. 1). Studies carried out over the last couple of decades have identified the reliance of valve cell function in maintaining the integrity and function of heart valves. Thus, within the field of tissue engineering there is now great interest in attempting to tissue engineer a living and fully functional heart valve.

The rationale for this comes not only from the experience with existing valve substitutes, but also from the Ross Procedure. This operation for aortic valve replacement involves repositioning the patient's own pulmonary valve into the aortic position, and placing a homograft valve into the pulmonary position. The diseased aortic valve is thus replaced with a living valve containing the patient's own cells. The outcome for this operation has been shown to be superior to any other type of valve replacement operation. Indeed, the life expectancy of patients who have had a Ross operation is identical to the general population [2]. This "proof of principle" highlights the need for clearly understanding the role of the cells that reside within and on the valve in order to be able to reproduce their function with cells used to tissue engineer a heart valve.

This chapter will briefly review the function of the aortic valve and the characteristics and role of valve interstitial and endothelial cells. It will then cover the principles and advances that have been made in the choice of suitable cell sources, scaffolds and mechanical conditioning strategies used in tissue engineering heart valves. Lastly, the chapter will discuss some of the future challenges that will be faced with heart valve tissue engineering as valves begin to undergo animal and clinical testing.

2 Aortic Valve Function

The aortic root functions in a co-ordinated fashion, with movements of specific structures during the cardiac cycle to ensure laminar flow of blood out of the left ventricle [3]. Its complex function is achieved via co-ordinated movements of its constituent parts [4]. The cellular components express a specific pattern of cell markers, exhibit contractile responses, have the ability to secrete extra-cellular matrix (ECM) proteins, express enzymes involved in re-modelling the surrounding matrix as well as express a wide range of molecules involved in communication between cells and between the cell and their matrix [5–12].

The optimal function of the aortic root is fundamental to the maintenance of left ventricular performance, coronary perfusion and a dynamic circulation. To achieve this role it may be expected that the aortic root should be capable of responding to changes in the performance of the myocardium as well as the vasculature. Regulation of the size and shape of the aortic root are key factors for its successful function and it is now known that the aortic root acts as a dynamic structure with communication and cross-talk between its component parts. This is demonstrated by the finding that the valve anticipates the exit of blood from the left ventricle by the fact that it starts to open prior to any detectable forward flow of blood or when there is only a very small pressure gradient across the valve [13, 14]. The initial opening movement is believed to be assisted by movements of specific parts of the root. In support of this theory it is known that changes in the shape of the root, from the sinotubular junction down to the annulus, occur during normal function of the valve [4, 13, 15–18]. The role of valve interstitial and endothelial cells in maintaining optimal function, durability and the mechanical properties of the valve are only beginning to be understood.

3 Valve Cells and Matrix

3.1 Valve Interstitial Cells

Valve interstitial cells are a heterogeneous and dynamic population of specific cell types that are phenotypically different from dermal fibroblasts [19–25]. It is likely that a family of fibroblast-like cells exists that varies its phenotype as an adaptive response to its microenvironment, as dictated by the ECM, mechanical force and soluble factors [26]. The mechanical environment confers specific functions on the cells within the valve that set these cells apart from other fibroblast-type cells. Both synthetic and contractile cell phenotypes can be identified (Fig. 2). These cells synthesise matrix components, such as collagen, elastin, proteoglycans, and glycoproteins; growth factors, cytokines and chemokines; as well as matrix remodelling enzymes, the matrix metalloproteinases (MMPs) and their tissue inhibitors (TIMPs) [27, 28].

Communication between the cells as well as between the cells and the ECM is an important part of their function with respect to the regulation of valve function. Cell surface proteins such as cadherins (N-cadherin), desmosomal junctions (desmoglein), and to a lesser extent gap junction proteins (connexin-26, 43 and 45)

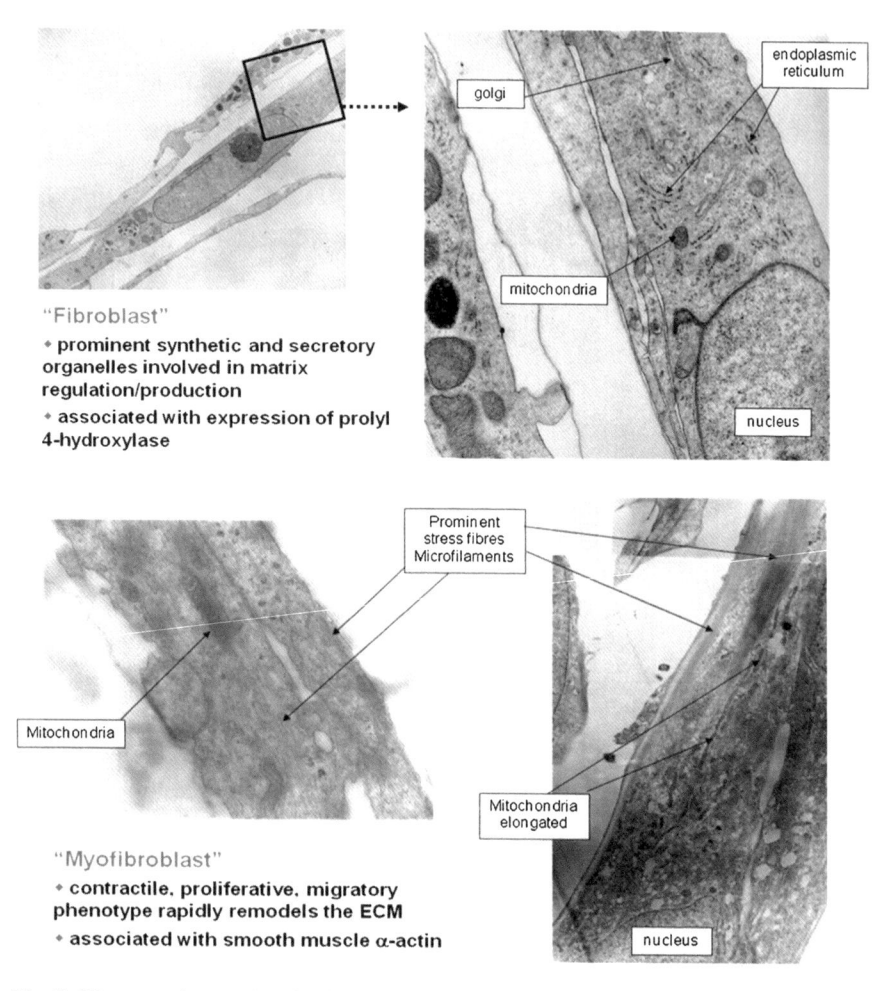

Fig. 2 Electron micrographs of different valve interstitial cell phenotypes

are all expressed by valve interstitial cells and are thought to form an integrated communication network [29]. In other tissues, this been shown to control architectural organization, cell proliferation, signalling, differentiation and death [30]. Cells in intact valves and those maintained in culture were reported to express a range of integrin molecules, including $\alpha 1$, $\alpha 2$, $\alpha 3$, $\alpha 4$, $\alpha 5$, $\beta 1$ and, to a lesser extent, $\alpha 6$ and αV. Neither $\beta 3$ nor $\beta 4$, integrins were expressed [9]. Integrins have the ability to act as sensors and transducers of mechanical signals in a bidirectional manner, through interaction with specific receptors on the ECM. These networks allow the cells to sense and respond to the mechanical forces imposed on the valve by the circulating blood, which varies throughout the cardiac cycle.

The presence of smooth muscle α-actin positive cells within the cusp has been demonstrated by immunohistochemical and molecular methods [5, 24, 31].

The functional contractile responses of valve interstitial cells in response to a wide range of vasoactive agents has been demonstrated, including 5-hydroxytryptamine (5-HT), endothelin-1, histamine, thromboxane A2, and catecholamines [6]. The association between nerves and contractile elements in the valve suggest that neurotransmitters, hormones, and pharmacological agents should have the capacity to affect the tension of cusp tissue. The cusp tissue has been shown to have a neuronal supply that includes sympathetic, parasympathetic and sensory neurotransmitters [32]. In vitro stimulation of these nerves leads to both contractile and dilator responses of the sinotubular junction, sinuses, annular and cusp tissue. In addition, intracellular calcium changes as well as induction of collagen synthesis and mitogenesis were observed in cultured valve interstitial cells in response to vasoactive agents, demonstrating that other functional responses were promoted [23, 33].

3.2 Valve Cusp Endothelial Cells

Endothelial cells function as a regulatory interface between the blood and the underlying tissue by being able to respond to their mechanical and humoral environment. This is achieved by the expression of cellular adhesion molecules and surface receptors, as well as the synthesis and expression of a range of biologically active molecules. Under physiological conditions, vascular endothelial cells limit vascular inflammation and smooth muscle cell proliferation, and maintain vascular tone and the fluidity of the blood, preventing thrombosis [34]. The endothelial cells that cover the valve cusps are exposed to a haemodynamic and mechanical environment that is unique in the vascular system. This has led to the suggestion that valve endothelial cells are unlike other endothelial cells in the vascular system and their function is regulated by an as yet undefined series of mechano-sensitive signal transduction pathways, that regulate their response to various mechanical or inflammatory stimuli [35]. For example, unlike vascular endothelial cells, which are aligned in parallel with the direction of blood flow, aortic valve endothelial cells are aligned circumferentially, perpendicular to the direction of flow[36] an effect that involves Rho-kinase and calpain, but not PI-3 kinase and is retained by cultured cells [37].

The complex haemodynamic forces imposed on the aortic valve in vivo are different on the aortic and ventricular surfaces of the valve [17, 18, 36]. The ventricular surface of the valve is exposed to pulsatile high shear laminar flow, whereas the aortic or outflow surface is exposed to disturbed, low shear flow during diastole, due to the vortices formed by the sinuses of Valsalva.

It has been shown that the expression of genes varies in cells isolated from either side of the valve. Out of 584 genes, 285 genes had higher expression and 299 had lower expression on the aortic side of the valve relative to the ventricular side [38]. Analysis of these genes revealed that the aortic surface of the valve was associated with greater expression of genes implicated in valve calcification.

The interaction between these different flow patterns on valve endothelial cells from either side of the valve are therefore of great interest. Understanding how aortic and ventricular valve endothelial cells react to laminar and disturbed flow will help elucidate the higher vulnerability of aortic-side valve endothelial cells to inflammation and calcification.

3.3 Valve Extracellular Matrix

The ECM plays a crucial role in valve function, dictating the physical and mechanical properties of the valve, maintaining the spatial arrangement of the cells that reside within it and mediating the complex crosstalk that exists between the cells, the matrix and external forces. The principle ECM components of the native valve are collagen (type I (74%), type III (24%), type V (2%)), elastic fibres and proteoglycans. Each component confers unique physical and mechanical properties, crucial for valve function. The aortic valve ECM is spatially arranged to give the leaflet three distinct regions: the fibrosa (primarily load-bearing collagen) on the aortic side of the cusp; the ventricularis (primarily elastic fibres) on the ventricular side; and the spongiosa (glycosaminoglycan-containing proteoglycans) in between [39]. Collagen fibrils confer mechanical and tensile strength, elastic fibres extensibility and the highly negatively charged glycosaminoglycans provide resistance against compressive forces, allow shearing between the layers and in addition are a source of bound growth factors and cytokines [40]. The distinct layers of the valve develop in the post-natal period, inferring that the increase in aortic pressure after birth may contribute to the thickening and further development of the aortic valve [41]. The ECM and cellular structure of the aortic valve continue to develop after birth and throughout life [41–44].

In addition to the three layered structure, complex meso- and micro-structural arrangements have been described in aortic valves that may play an important role in valve function [45, 46]. Valve cusps like other load-bearing soft tissues exhibit non-linear stress–strain relationships, viscoelasticity and importantly mechanical anisotropy, with different micro- and gross-mechanical responses in the circumferential and radial directions. These attributes have a major impact on tissue biomechanics and presumably subsequent cellular responses. We and others have shown that the mechanical properties of the cusp differ greatly in the circumferential and radial directions [47]. In order to achieve optimal function, any tissue engineered valve will need to display the same structural and mechanical properties as the native valve.

4 Principles of Tissue Engineering a Heart Valve

Tissue engineering is a multi-disciplinary field which relies on interaction between biologists, engineers, material scientists and clinicians. A tissue engineered heart valve should be a structure that resembles the size and shape of the native valve; it should be capable of growing with the recipient and be able to respond to mechanical and biological cues in such a way that permits it to function in an identical fashion to that of the native valve. The valve should be able to adapt to the changing physiological conditions experienced throughout life. Ideally the valve should be non-immunogenic, non-inflammatory, non-thrombogenic and non-obstructive. In addition, the valve should be durable and provide good haemodynamics.

This defines the gold standard for a tissue engineered valve, however, in reality it is unlikely that all these criteria can be fulfilled. It is the job of the tissue engineers to elucidate which properties of the native valve can be comprised in order to replicate the function of the native valve as near as possible. This is a challenge that faces the re-creation of a wide range of biological tissues and organs including the myocardium, blood vessels, skin, bone, bladder, cartilage and teeth [48–54].

There are two primary components to a tissue engineered valve, namely cells and a scaffold material (Fig. 3). These are then combined in a third step that mechanically conditions the valve in a bioreactor, in order to give it the mechanical properties required for to it to function optimally. Unlike some other tissues for which tissue engineering may be applicable, a heart valve will have to be sufficiently strong to withstand the haemodynamic forces as soon as it is implanted. Failure to do so will result in rapid failure of the new valve and most likely death of the patient. Approaches with different combinations of cell sources and scaffold materials are being investigated. However, the major challenge that

Fig. 3 Heart valve tissue engineering strategy

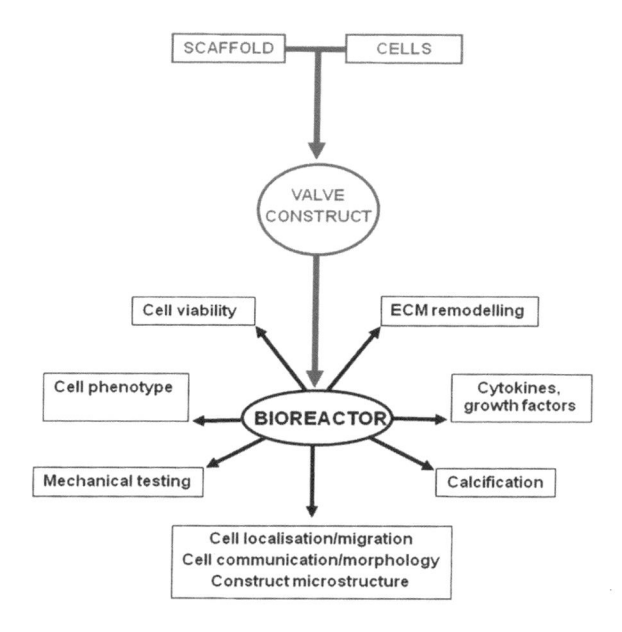

faces each of these strategies is to be able to produce a valve that is able to replicate the complex function and durability of the native valve.

4.1 Cell Sources

A number of different cells have the potential to act as a source for a tissue engineered heart valve. These include fibroblasts and vascular smooth muscle cells, amniotic derived progenitor cells and stem cells derived from bone marrow [55–65]. Ideally, cells that are considered for use should be readily available and become 'valve-like' in response to appropriate environmental, mechanical and haemodynamic cues.

There are broadly two resources of cells for tissue engineering projects. Firstly, autologous cells taken from the patient for whom the valve is to be made. This has the advantage of circumventing immunological reactions with the valve. However, such a clear advantage must be considered against a number of other technical issues. There is the issue of the severity of the procedure used to harvest the cells, the time from the harvest of the cells to producing an implantable valve and the fact that the cells that are being used come from an already sick patient. The alternative is the use of allogenic cells from donors. This strategy would allow the production of an 'off the shelf' product making it a more attractive commercial proposition. However the immunological issues relating to the use of allogenic cells and the abundance of suitable donors are a potentially severe limitation. Since mesenchymal stem cells have been reported to down-regulate immune responses their use as a cell source may be a prerequisite [66–68].

Stem cells of various types have been given a great deal of interest by researchers and clinicians working in the field of tissue engineering and regenerative medicine. Mesenchymal stem cells (MSCs) represent a source of cells that is easily available via isolation from peripheral blood, bone marrow and adipose tissue [69–72]. These cells are able to undergo self-renewal, and differentiate into other cell types with a different tissue origin, phenotype and function [73]. Initial studies using bone marrow derived MSCs have shown them to share some of the biological properties and phenotypic markers with those of valve interstitial cells (Fig. 4) [10]. These include the ability to secrete collagen in response to mechanical stretch [60].

Bioabsorbable polymers seeded with MSCs and grown in vitro in a pulsatile-flow-bioreactor demonstrated, secretionally active cell elements [57]. Collagen types I, III, smooth muscle α-actin, and vimentin were detected in the tissue engineered leaflets. Importantly the mechanical properties of the leaflets were comparable to those of native tissue. Human fetal mesenchymal progenitors grown from prenatal chorionic villus specimens have also been assessed for their suitability for tissue engineering heart valves. After culture and seeding onto synthetic biodegradable leaflet scaffolds these were conditioned in a bioreactor. Histologically, the tissue engineered valves resembled the native neonatal valve. However, native

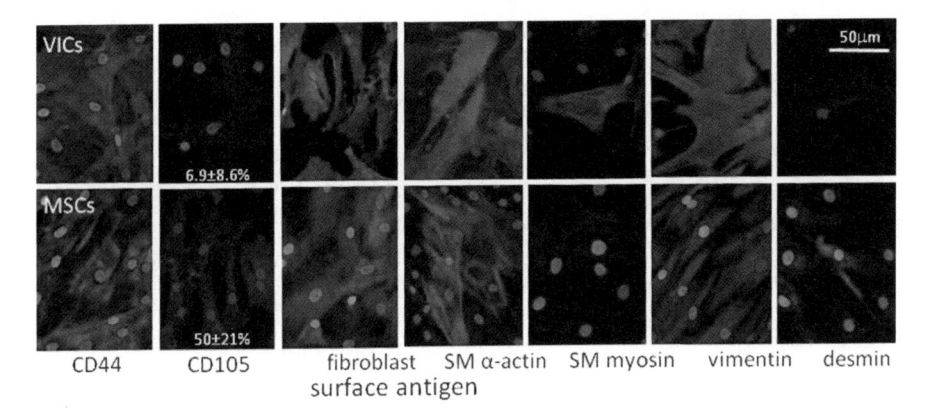

Fig. 4 Phenotypic similarity between valve interstitial cells and bone marrow derived mesenchymal cells (from [10])

neonatal pulmonary leaflets showed more homogenous distribution of collagen fibers and glycosaminoglycan content than the tissue engineered structures [62].

Arterial and venous vascular myofibroblasts also have the potential for use in tissue engineered valves. Cells from saphenous veins released more collagen than aortic cells, but both cell types had similar proliferation rates and morphological appearance when grown in biodegradable polyurethane scaffolds [74]. Implantation of a biodegradable polymer scaffold, seeded with arterial smooth muscle cells, into the pulmonary position of sheep resulted in progressive cellular and ECM formation over a 24 week period (the study endpoint) [75].

The experience with different cell types illustrates the ability of some cells to be able to lay down new matrix. The primary job of cells seeded onto a scaffold material will be to remodel the existing scaffold and lay down a matrix that resembles that of the native valve. The ability of the cells to secrete desired amounts of collagens, elastin and proteoglycans in response to the mechanical environment will govern heavily the best choice of cells. However, the longer term response of the cells (proliferation, apoptosis and differentiation) also needs to be considered.

4.2 Choice of Scaffold Material

The scaffold or matrix used for tissue engineering applications may be naturally derived or synthetic but must be compatible with the cells that are to be seeded and grown onto it to form a 'tissue construct'. Successful scaffold materials will be amenable to modification, have a controlled rate of degradation and possess properties that will promote cellular population and maintain matrix integrity. Importantly, the scaffold and its degradation products should lack cytotoxicity and not elicit an immune or inflammatory response. Three main strategies have been employed to develop a suitable scaffold for engineering a heart valve: decellularised native valves; synthetic biodegradable polymers; and biological biodegradable polymers. The biodegradable scaffold serves as a support for the cells providing the necessary mechanical integrity and preferably signalling molecules. Ideally the degradation rate of biodegradable scaffolds should be comparable with the synthesis of new ECM by the cells. The decellularised valve matrix does not require to be replaced, but will require to be remodelled and repaired. The advantages and disadvantages of each strategy are discussed below.

4.2.1 Decellularised Valves

Human (homografts) and porcine valves (bioprosthetic xenografts) have better haemodynamics than mechanical valves and have been used successfully as valve replacements for many years. Xenografts are widely used because of their availability, however they must first be treated with glutaraldehyde to reduce immunogenicity and generate sterility. Glutaraldehyde treatment increases tissue stiffness due to increased cross-linking, protects the tissue from proteolytic degradation and has been associated with calcification, particularly in younger recipients [76]. Glutaraldehyde treated valves are also susceptible to infection. Homografts, which are treated with antibiotics and may be cryopreserved are subject to calcification. The major disadvantage of homografts and xenografts is that they contain few if any (homografts) viable cells and therefore lack the potential for regeneration and growth. Glutaraldehyde treated tissues do not become repopulated in vivo, which may be the consequence of increased crosslinking.

Thus strategies are being investigated to decellularise a valve in such a way to provide a matrix that will promote efficient cell repopulation [77]. It has been reported that decellularised matrices retain the unique spatial organisation and composition of the original ECM and present biologically functional molecules or "morphogens", providing an extremely favourable environment for tissue regeneration by cells [78]. These decellularised valves may be implanted and allowed to repopulate in vivo or repopulated with endothelial and interstitial cells prior to implantation. Due to the limited availability of human valves, efforts have been directed towards the use of porcine valves. An efficient method is therefore required for the removal of cells, sterilisation of the tissue and alteration of its immunogenicity that does not destroy the composition, structure and arrangement of the ECM. It is generally believed that most of the antigenicity is cell associated, and antigens such as alpha-gal must be removed to avoid an immune or inflammatory reaction. The cells must be completely removed as cell remnants have been associated with impaired repair and remodelling as well as calcification [79, 80]. Decellularisation can be achieved using hypotonic/hypertonic solutions, or anionic detergents such as sodium dodecyl sulphate or sodium deoxycholate, or non-ionic detergents such as Triton X-100 or Tween-20 to disrupt the membranes, and sometimes a combination is used [77]. Enzymes are then used to digest nucleic acids and cell membrane remnants. A balance exists between optimal cell removal and minimal ECM damage. The valves must also be sterilised to avoid the potential transmission of microbiological hazards using an alternative method to glutaraldehyde such as irradiation, ethanol or peracetic-ethanol treatment and antibiotics.

Studies have shown that unlike tissue engineered valves derived from synthetic polymers, decellularised valves have biomechanical properties similar to those of the native valve [81]. This will be crucial for the immediate and optimal function of the valve on implantation. A lack of appropriate biomechanical integrity would quickly lead to structural deterioration in the high shear, high pressure environment of the aortic valve, potentially resulting in patient death. However, concerns exist over the phenotype of the cells that may repopulate decellularised valves in vivo and this may occur even if the decellularised valves have been repopulated by suitable cell types prior to implantation. As discussed above, the valve interstitial and endothelial cells possess unique and important characteristics. If the decellularised matrix attracted fibroblasts, this might result in the development of a fibrotic valve, whereas an infiltration of leukocytes would lead to an inflammatory event and possible damage to the matrix components. Neither would lead to the longevity of a fully functional valve. Clinical use of decellularised xenogenic valves previously led to tragic results due to a severe inflammatory reaction thought to be initiated by the xenogenic collagen matrix [82]. Clearly more investigations were required before decellularised valves could be re-considered as a feasible alternative to currently available valve replacements. Research focussed on better methods of decellularisation, which would maintain the integrity of the ECM and promising results have been reported [77, 79, 83, 84]. Decellularised xenogenic valves have been reported to perform well both in vitro and in animal models, when placed in the pulmonary or aortic position [83, 85, 86]. Clinically, decellularised allografts and xenografts have been successfully used to replace the pulmonary valve during the Ross procedure: decellularised allografts were reported to show no advantage over conventionally prepared allografts, but showed a reduced immunogenic response when compared to cryopreserved valves [83, 87, 88].

4.2.2 Synthetic Polymers

Another approach is to fabricate a valve from a synthetic polymer [63, 64]. Polymer scaffolds can be fabricated using a variety of techniques such as salt-leaching, rapid prototyping, phase separation, electrospinning and preparation of woven and non-woven meshes. Many different polymers have been tested as possible valve scaffolds: aliphatic polyesters (polyglactin, polyglycolic acid, polylactic acid), whose limitations are thickness, stiffness and non-pliability; and polyhydroxyalkanoates (polyhydroxyoctanoate, poly-4-hydroxybutyrate), which possess good thermoplastic properties, are easily mouldable into the desired shape,

Fig. 5 Tissue engineered heart valves made from synthetic polymers and seeded with ovine myofibroblasts and endothelial cells (from [55, 56])

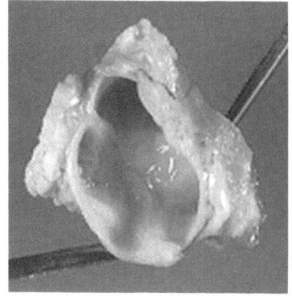

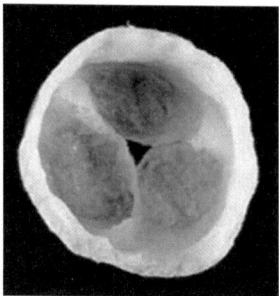

but have a slow degradation rate. Thus composite polymers of aliphatic polyesters and polyhydroxyalkanoates have generally been favoured and investigated. The advantage of using synthetic polymers is that they can be easily modified to have a wide range of mechanical and chemical properties, including rate of degradation, which must be comparable with the rate of tissue/ECM formation. Conversely, their disadvantage is their lack of biocompatibility and biological signalling molecules. However it may be feasible to overcome this limitation, as it has been demonstrated that pre-coating polyglycolic acid scaffolds with human ECM proteins improved scaffold population and increased attachment of human aortic myofibroblasts [89].

The polymer scaffold should be porous and possess an interconnected network of pores to promote cell migration and growth, as well as allow efficient supply of nutrients and removal of waste products. A variety of cell types and sources have been used to seed polymer scaffolds as previously discussed [57, 61–64, 90]. Although promising results have been obtained from in vivo animal studies, tissue engineered valves made from synthetic biodegradable polymer scaffolds have been unable to withstand aortic pressures and fail to reproduce many of the sophisticated functions of the native aortic valve (Fig. 5) [55, 56, 75, 91, 92]. In addition studies have reported problems over scaffold shrinkage, fragmentation, thickening, poor extensibility, failure to co-apt and inability to achieve a confluent endothelial layer [55–57, 59, 91–96].

4.2.3 Biological Polymers

It is becoming apparent that important signalling and adhesion molecules should be inherent or embedded within the scaffold to direct cell phenotype and function during tissue development [97]. There are several biological polymers that may be suitable for tissue engineering heart valves such as collagen, hyaluronan and fibrin. Of these, type I collagen is probably one of the most appropriate, since it is the major ECM protein of the native valve and provides most of its mechanical and tensile strength. In addition, collagens have a low antigenicity, being only weakly immunogenic largely due to their homology across species, and are biodegradable due to their proteinaceous nature. Our studies have demonstrated that valve interstitial cells and MSCs proliferated within type I collagen scaffolds under static conditions and that collagen enhanced the capacity of valve interstitial cells to express their native phenotype [24, 98, 99]. Others reported the migration of smooth muscle cells into collagen scaffolds, the development of a tissue-like morphology with highly organised newly synthesized ECM proteins and the production of collagen and proteoglycans in collagen scaffolds seeded with valve myofibroblasts [100–102].

These studies were all carried out under static conditions, but clearly demonstrate the ability of collagen to promote cell responses. We later reported MSCs seeded in a collagen scaffold produced collagen in response to mechanical force and that this was enhanced by the addition of fibronectin to the scaffold [103]. Interaction between specific recognition sequences displayed on collagen triple helices and other ECM proteins with cell membrane receptors such as integrins is responsible for triggering a variety of specific cellular functions. The recently emerging and expanding field of bionanotechnology promises great potential for the development of scaffolds with material properties, which could be designed and modified at the molecular level incorporating recognition sequences to promote the desired cellular function. Koide et al. elegantly demonstrated how optimised scaffolds may be produced from peptide-based supramolecules specifically designed to mimic the structure and function of collagen [104–107].

Hyaluronan is known to play an important role in matrix structure, lubrication, cell movement and differentiation, as well as being an essential component in cardiac morphogenesis and a predominant glycosaminoglycan of the native valve leaflet [108]. Hydrogels can be made by cross-linking hyaluronan and its characteristics make it an attractive scaffold material for tissue engineering heart valves. It has been reported that hyaluronan hydrogels increased ECM production by valve interstitial cells and induced synthesis of elastin [109, 110]. Composite type I collagen–chondroitin sulphate hydrogels have also been investigated and were reported to promote coverage by endothelial cells and increase ECM production [111].

Recently it was reported that fibrin-based tissue engineered valves seeded with human dermal fibroblasts subjected to cyclic stretching with incrementally increasing strain amplitude resulted in the production of valves that could withstand cyclic pulmonary pressures. Furthermore the leaflets displayed tensile stiffness and stiffness anisotropy similar to leaflets from sheep pulmonary valves [112]. However, since it has also been reported that fibrin may promote calcification, it may not be a suitable matrix for tissue engineering heart valves [113].

4.3 In Vitro Conditioning

Since the tissue engineered valve must possess the strength and integrity to function as the native valve on implantation, seeded synthetic and biological polymer scaffolds will need to be conditioned in vitro using a dynamic conditioning protocol to allow the development of a mature extracellular matrix characteristic of the native valve. A number of different bioreactor designs have been described for tissue engineered heart valves [55, 56, 114–118]. It has been demonstrated that subjecting tissue engineered valve constructs to mechanical conditioning results in the development of anisotropy and the formation of a tri-layered matrix similar to that of the native valve [61, 63, 64]. Importantly, it was reported that exposing tissue engineered valves to a conditioning protocol that mimics only the diastolic phase of the cardiac cycle by applying a dynamic pressure difference over a closed tissue engineered valve, promoted the development of tissue formation and non-linear tissue-like mechanical properties [117]. In addition it has been demonstrated that collagen remodelling is straining mode dependent, with dynamic or intermittent straining enhancing the crosslinking of collagen and therefore improving the quality and structural integrity of the tissue [119, 120].

It is well known that mechanical forces such as shear stress, stretch or strain and pressure can induce a variety of cellular responses, including the production of ECM [121–124].The specific 3-dimensional microstructure of the valve is influenced and directed by haemodynamic force, the replication of which will play a crucial role in the success of tissue engineered valves [55, 56, 125, 126]. However, the response of the cells to external stimuli critically depends on their phenotype and most likely their lineage, which highlights the fundamental importance of choosing the appropriate cell type for tissue engineering. To be commercially viable in vitro conditioning of seeded polymer scaffolds will need to be optimised, so that a tissue engineered valve with mature ECM possessing the mechanical properties of the native valve is produced within a reasonable length of time. This is particularly true for autologous rather than "off-the shelf" valves. Whether this will be feasible remains to be proven.

5 Future Objectives

5.1 Animal Testing

Several issues arise when trying to determine an appropriate animal model to test the tissue engineered valves prior to clinical use. Sheep are known to have a high rate of calcification, so may not be a useful model for testing tissue engineered valves, but would be a useful model for studying calcification of tissue engineered valves and strategies to prevent it. Although tissue engineered polymer valves cannot as yet withstand systemic pressure, when optimised they will need to be tested in the aortic position. This again would preclude the use of sheep, as their anatomy would make implantation extremely difficult and probably lead to high morbidity and mortality. Thus it is more likely that a porcine model would be chosen. Clearly, and has been demonstrated by the disastrous results of the clinical use of the decellularised Synergraft [82], the response of an animal to a tissue engineered valve will not necessarily reflect that of a human. In addition, testing a tissue engineered valve composed of human cells in a porcine or ovine model would not be an accurate reflection of the response in a human and is unlikely to provide reliable information with regards to its fate when implanted in a human. Thus the ideal method of testing pre-clinically remains elusive.

5.2 Accelerated Durability Testing

The tissue engineered valve will be viable and rely upon the cells to replace and repair the ECM. Thus accelerated durability testing would not be appropriate. It is more likely that the valves would be tested under physiological conditions in a bioreactor for an extended length of time prior to clinical use. A balance will need to be made between the in vitro and in vivo conditioning of the valve. Under in vivo conditions a more physiological conditioning can occur under the mechanical and flow environment in which the valve resides [127]. However, this will not be achieved until a minimum level of mechanical strength and durability has been achieved in vitro.

5.3 Valve Calcification

Pathways that mediate calcification in the aortic valves are being widely investigated. These studies have relevance to the disease process in native valves as well as any potential changes that may occur in tissue engineered heart valves as and when they are used clinically. It now believed that the effects of cytokines and growth factors such as TGF-β, bone morphogenetic proteins (BMPs), Wnts and extracellular nucleotides can stimulate specific signalling pathways that cause valve interstitial cells to differentiate into osteoblast-like cells [128–130]. Valve calcification can be characterised as an inflammatory condition by the expression of cell adhesion molecules by valve endothelium and the presence of inflammatory cells within the cusp [131]. Early lesions are characterised by inflammatory infiltrates composed of non-foam cells, foam cell macrophages and some T-cells [132]. As the disease progresses, the spotty or finely distributed calcification spreads to form radially orientated lesions that are associated with the regions of high mechanical stress [17, 18]. Eventually, the valve cells adopt an osteoblast phenotype, with increased expression of osteopontin, bone sialoprotein, osteocalcin and osteoblast-specific transcription factors [133, 134].

While some of the cell types that are being assessed for use in tissue engineered valves have been shown phenotypically and functionally to resemble native valve interstitial cells, they also possess varying degrees of multipotency. Cells in tissue engineered heart valves will be expected to function in a unfamiliar and unique mechanical environment. While these conditions may be used ex vivo to condition neo-valve cusp tissue, this change from the cells' natural mechanical environment may also stimulate deleterious responses, leading to changes that would impair their ability to function as a heart valve cell. For example when saphenous veins are placed in the arterial circulation as bypass grafts, they develop an accelerated fibro-proliferative disease that resembles atherosclerosis [135]. Importantly, under some mechanical conditions MSCs are known to be able to differentiate into specific cell types that are a feature of calcified valves, namely osteoblasts and chrondrocytes [136, 137]. Thus consideration must be given to the potential for tissue engineered valve to calcify. Development of pharmacological and molecular strategies that will regulate calcification in native valves need to be also applied to tissue engineered valve constructs in order to protect tissue engineered heart valves from degenerative calcific disease.

5.4 Growth Potential

One of the major potential applications for a tissue engineered valve is to use it in children or young adults. This is due to the concept that because the implanted valve is a living structure it should be able to grow with the patient, negating the requirement for future replacement as the patients develop. However, none of the in vivo studies with tissue engineered valves have demonstrated the ability of the valve to grow. This is most likely due to the relatively short time scale of most of these studies. Attention will need to be paid to how growth will be assessed. It will be important to be able to differentiate between dilatation of the valve and true growth.

5.5 Preparation of the Valve for Clinical Use

With the advent of percutaneous technologies for valve implantation, attention needs to be paid as to whether these methods are applicable for use with tissue engineered valves. In addition the types of sterilisation and packaging used for the valve need careful assessment. Since the aim is to provide patients with a living valve, care needs to be taken in handling these valves so as to preserve cell viability, function and structure. For example it is known that some strong antibiotic solutions used for homograft valves are toxic to cells [138].

6 Summary

The concept of tissue engineering a living, fully functional heart valve is a unique challenge in the field of tissue engineering. This concept presents problems relating to control of cell function, compatibility of scaffold material and importantly the need to produce a structure that is strong enough at the time of implantation to withstand the haemodynamic forces imposed upon it by the systemic circulation. However, the clinical benefits that would be achieved in producing a living valve that will replicate (or reproduce) the structure and function of the native valve are significant.

References

1. van Geldorp, M.W., Jamieson, W.R., Kappetein, A.P., Puvimanasinghe, J.P., Eijkemans, M.J., Grunkemeier, G.L., Takkenberg, J.J., Bogers, A.J.: Usefulness of microsimulation to translate valve performance into patient outcome: patient prognosis after aortic valve replacement with the Carpentier-Edwards supra-annular valve. J. Thorac. Cardiovasc. Surg. **134**, 702–709 (2007)
2. El-Hamamsy, I., Eryigit, Z., Stevens, L.M., Sarang, Z., George, R., Clark, L., Melina, G., Takkenberg, J.J., Yacoub, M.H.: Long-term outcomes after autograft versus homograft aortic root replacement in adults with aortic valve disease: a randomised controlled trial. Lancet. **376**, 524–531 (2010)
3. Yacoub, M.H., Kilner, P.J., Birks, E.J., Misfeld, M.: The aortic outflow and root: a tale of dynamism and crosstalk. Ann. Thorac. Surg. **68**, S37–S43 (1999)
4. Dagum, P., Green, G.R., Nistal, F.J., Daughters, G.T., Timek, T.A., Foppiano, L.E., Bolger, A.F., Ingels, N.B., Jr., Miller, D.C.: Deformational dynamics of the aortic root: modes and physiologic determinants. Circulation **100**, II54–II62 (1999)
5. Brand, N.J., Roy, A., Hoare, G., Chester, A., Yacoub, M.H.: Cultured interstitial cells from human heart valves express both specific skeletal muscle and non-muscle markers. Int. J. Biochem. Cell. Biol. **38**, 30–42 (2006)
6. Chester, A.H., Misfeld, M., Yacoub, M.H.: Receptor-mediated contraction of aortic valve leaflets. J. Heart Valve Dis. **9**, 250–254 (2000)
7. Dreger, S.A., Taylor, P.M., Allen, S.P., Yacoub, M.H.: Profile and localization of matrix metalloproteinases (MMPs) and their tissue inhibitors (TIMPs) in human heart valves. J. Heart Valve Dis. **11**, 875–880 (2002)
8. Latif, N., Sarathchandra, P., Taylor, P.M., Antoniw, J., Yacoub, M.H.: Localization and pattern of expression of extracellular matrix components in human heart valves. J. Heart Valve Dis. **14**, 218–227 (2005a)
9. Latif, N., Sarathchandra, P., Taylor, P.M., Antoniw, J., Yacoub, M.H.: Molecules mediating cell-ECM and cell-cell communication in human heart valves. Cell Biochem. Biophys. **43**, 275–287 (2005b)
10. Latif, N., Sarathchandra, P., Thomas, P.S., Antoniw, J., Batten, P., Chester, A.H., Taylor, P.M., Yacoub, M.H.: Characterization of structural and signaling molecules by human valve interstitial cells and comparison to human mesenchymal stem cells. J. Heart Valve Dis. **16**, 56–66 (2007)
11. Yacoub, M.H., Cohn, L.H.: Novel approaches to cardiac valve repair: from structure to function: Part I. Circulation **109**, 942–950 (2004)
12. Yacoub, M.H., Cohn, L.H.: Novel approaches to cardiac valve repair: from structure to function: Part II. Circulation **109**, 1064–1072 (2004)
13. Higashidate, M., Tamiya, K., Beppu, T., Imai, Y.: Regulation of the aortic valve opening. In vivo dynamic measurement of aortic valve orifice area. J. Thorac. Cardiovasc. Surg. **110**, 496–503 (1995)
14. Thubrikar, M., Bosher, L.P., Nolan, S.P.: The mechanism of opening of the aortic valve. J. Thorac. Cardiovasc. Surg. **77**, 863–870 (1979)

15. Thubrikar, M., Nolan, S.P., Bosher, L.P., Deck, J.D.: The cyclic changes and structure of the base of the aortic valve. Am. Heart J. **99**, 217–224 (1980)
16. Thubrikar, M., Piepgrass, W.C., Shaner, T.W., Nolan, S.P.: The design of the normal aortic valve. Am. J. Physiol. **241**, H795–H801 (1981)
17. Thubrikar, M.J., Aouad, J., Nolan, S.P.: Comparison of the in vivo and in vitro mechanical properties of aortic valve leaflets. J. Thorac. Cardiovasc. Surg. **92**, 29–36 (1986)
18. Thubrikar, M.J., Aouad, J., Nolan, S.P.: Patterns of calcific deposits in operatively excised stenotic or purely regurgitant aortic valves and their relation to mechanical stress. Am. J. Cardiol. **58**, 304–308 (1986)
19. Della, R.F., Sartore, S., Guidolin, D., Bertiplaglia, B., Gerosa, G., Casarotto, D., Pauletto, P.: Cell composition of the human pulmonary valve: a comparative study with the aortic valve–the VESALIO Project. Vitalitate Exornatum Succedaneum Aorticum labore Ingegnoso Obtinebitur. Ann. Thorac. Surg. **70**, 1594–1600 (2000)
20. Lester, W., Rosenthal, A., Granton, B., Gotlieb, A.I.: Porcine mitral valve interstitial cells in culture. Lab. Invest. **59**, 710–719 (1988)
21. Messier, R.H., Jr., Bass, B.L., Aly, H.M., Jones, J.L., Domkowski, P.W., Wallace, R.B., Hopkins, R.A.: Dual structural and functional phenotypes of the porcine aortic valve interstitial population: characteristics of the leaflet myofibroblast. J. Surg. Res. **57**, 1–21 (1994)
22. Mulholland, D.L., Gotlieb, A.I.: Cell biology of valvular interstitial cells. Can. J. Cardiol. **12**, 231–236 (1996)
23. Taylor, P.M., Allen, S.P., Yacoub, M.H.: Phenotypic and functional characterization of interstitial cells from human heart valves, pericardium and skin. J. Heart Valve Dis. **9**, 150–158 (2000)
24. Taylor, P.M., Allen, S.P., Dreger, S.A., Yacoub, M.H.: Human cardiac valve interstitial cells in collagen sponge: a biological three-dimensional matrix for tissue engineering. J. Heart Valve Dis. **11**, 298–306 (2002)
25. Taylor, P.M., Batten, P., Brand, N.J., Thomas, P.S., Yacoub, M.H.: The cardiac valve interstitial cell. Int. J. Biochem. Cell. Biol. **35**, 113–118 (2003)
26. Komuro, T.: Re-evaluation of fibroblasts and fibroblast-like cells. Anat. Embryol. (Berl) **182**, 103–112 (1990)
27. Sappino, A.P., Schurch, W., Gabbiani, G.: Differentiation repertoire of fibroblastic cells: expression of cytoskeletal proteins as marker of phenotypic modulations. Lab. Invest. **63**, 144–161 (1990)
28. Smith, R.S., Smith, T.J., Blieden, T.M., Phipps, R.P.: Fibroblasts as sentinel cells. Synthesis of chemokines and regulation of inflammation. Am. J. Pathol. **151**, 317–322 (1997)
29. Latif, N., Sarathchandra, P., Taylor, P.M., Antoniw, J., Brand, N., Yacoub, M.H.: Characterization of molecules mediating cell-cell communication in human cardiac valve interstitial cells. Cell Biochem. Biophys. **45**, 255–264 (2006)
30. Matter, K., Balda, M.S.: Signalling to and from tight junctions. Nat. Rev. Mol. Cell Biol. **4**, 225–236 (2003)
31. Roy, A., Brand, N.J., Yacoub, M.H.: Molecular characterization of interstitial cells isolated from human heart valves. J. Heart Valve Dis. **9**, 459–464 (2000)
32. Marron, K., Yacoub, M.H., Polak, J.M., Sheppard, M.N., Fagan, D., Whitehead, B.F., de Leval, M.R., Anderson, R.H., Wharton, J.: Innervation of human atrioventricular and arterial valves. Circulation **94**, 368–375 (1996)
33. Hafizi, S., Taylor, P.M., Chester, A.H., Allen, S.P., Yacoub, M.H.: Mitogenic and secretory responses of human valve interstitial cells to vasoactive agents. J. Heart Valve Dis. **9**, 454–458 (2000)
34. Quyyumi, A.A.: Endothelial function in health and disease: new insights into the genesis of cardiovascular disease. Am. J. Med. **105**, 32S–39S (1998)
35. Butcher, J.T., Simmons, C.A., Warnock, J.N.: Mechanobiology of the aortic heart valve. J. Heart Valve Dis. **17**, 62–73 (2008)
36. Deck, J.D.: Endothelial cell orientation on aortic valve leaflets. Cardiovasc Res **20**, 760–767 (1986)

37. Butcher, J.T., Penrod, A.M., Garcia, A.J., Nerem, R.M.: Unique morphology and focal adhesion development of valvular endothelial cells in static and fluid flow environments. Arterioscler. Thromb. Vasc. Biol. **24**, 1429–1434 (2004)
38. Simmons, C.A., Grant, G.R., Manduchi, E., Davies, P.F.: Spatial heterogeneity of endothelial phenotypes correlates with side-specific vulnerability to calcification in normal porcine aortic valves. Circ. Res. **96**, 792–799 (2005)
39. Bairati, A., DeBiasi, S.: Presence of a smooth muscle system in aortic valve leaflets. Anat. Embryol. (Berl) **161**, 329–340 (1981)
40. Mulloy, B., Rider, C.C.: Cytokines and proteoglycans: an introductory overview. Biochem. Soc. Trans. 34, 409–413 (2006)
41. Colvee, E., Hurle, J.M.: Maturation of the extracellular material of the semilunar heart values in the mouse. A histochemical analysis of collagen and mucopolysaccharides. Anat. Embryol. (Berl) **162**, 343–352 (1981)
42. Aikawa, E., Whittaker, P., Farber, M., Mendelson, K., Padera, R.F., Aikawa, M., Schoen, F.J.: Human semilunar cardiac valve remodeling by activated cells from fetus to adult: implications for postnatal adaptation, pathology, and tissue engineering. Circulation 113, 1344–1352 (2006)
43. McDonald, P.C., Wilson, J.E., McNeill, S., Gao, M., Spinelli, J.J., Rosenberg, F., Wiebe, H., McManus, B.M.: The challenge of defining normality for human mitral and aortic valves: geometrical and compositional analysis. Cardiovasc. Pathol. **11**, 193–209 (2002)
44. Sell, S., Scully, R.E.: Aging changes in the aortic and mitral valves. Histologic and histochemical studies, with observations on the pathogenesis of calcific aortic stenosis and calcification of the mitral annulus. Am. J. Pathol. **46**, 345–365 (1965)
45. Doehring, T.C., Kahelin, M., Vesely, I.: Mesostructures of the aortic valve. J. Heart Valve Dis. **14**, 679–686 (2005)
46. Sacks, M.S., Smith, D.B., Hiester, E.D.: The aortic valve microstructure: effects of transvalvular pressure. J. Biomed. Mater. Res. **41**, 131–141 (1998)
47. Kershaw, J.D., Misfeld, M., Sievers, H.H., Yacoub, M.H., Chester, A.H.: Specific regional and directional contractile responses of aortic cusp tissue. J. Heart Valve Dis. **13**, 798–803 (2004)
48. Earthman, J.C., Sheets, C.G., Paquette, J.M., Kaminishi, R.M., Nordland, W.P., Keim, R.G., Wu, J.C.: Tissue engineering in dentistry. Clin. Plast. Surg. **30**, 621–639 (2003)
49. Flanagan, T.C., Pandit, A.: Living artificial heart valve alternatives: a review. Eur. Cell Mater. **6**, 28–45 (2003)
50. Kaufman, M.R., Tobias, G.W.: Engineering cartilage growth and development. Clin. Plast. Surg. **30**, 539–546 (2003)
51. Leor, J., Amsalem, Y., Cohen, S.: Cells, scaffolds, and molecules for myocardial tissue engineering. Pharmacol. Ther. **105**, 151–163 (2005)
52. Mistry, A.S., Mikos, A.G.: Tissue engineering strategies for bone regeneration. Adv. Biochem. Eng. Biotechnol. **94**, 1–22 (2005)
53. Nugent, H.M., Edelman, E.R.: Tissue engineering therapy for cardiovascular disease. Circ. Res. **92**, 1068–1078 (2003)
54. Pomahac, B., Svensjo, T., Yao, F., Brown, H., Eriksson, E.: Tissue engineering of skin. Crit. Rev. Oral Biol. Med. **9**, 333–344 (1998)
55. Hoerstrup, S.P., Sodian, R., Daebritz, S., Wang, J., Bacha, E.A., Martin, D.P., Moran, A.M., Guleserian, K.J., Sperling, J.S., Kaushal, S., Vacanti, J.P., Schoen, F.J., Mayer, J.E., Jr.: Functional living trileaflet heart valves grown in vitro. Circulation **102**, III44–III49 (2000)
56. Hoerstrup, S.P., Sodian, R., Sperling, J.S., Vacanti, J.P., Mayer, J.E., Jr.: New pulsatile bioreactor for in vitro formation of tissue engineered heart valves. Tissue Eng. **6**, 75–79 (2000)
57. Hoerstrup, S.P., Kadner, A., Melnitchouk, S., Trojan, A., Eid, K., Tracy, J., Sodian, R., Visjager, J.F., Kolb, S.A., Grunenfelder, J., Zund, G., Turina, M.I.: Tissue engineering of functional trileaflet heart valves from human marrow stromal cells. Circulation **106**, I143–I150 (2002)

58. Jockenhoevel, S., Chalabi, K., Sachweh, J.S., Groesdonk, H.V., Demircan, L., Grossmann, M., Zund, G., Messmer, B.J.: Tissue engineering: complete autologous valve conduit–a new moulding technique. Thorac. Cardiovasc. Surg. **49**, 287–290 (2001)
59. Kadner, A., Hoerstrup, S.P., Tracy, J., Breymann, C., Maurus, C.F., Melnitchouk, S., Kadner, G., Zund, G., Turina, M.: Human umbilical cord cells: a new cell source for cardiovascular tissue engineering. Ann. Thorac. Surg. **74**, S1422–S1428 (2002)
60. Ku, C.H., Johnson, P.H., Batten, P., Sarathchandra, P., Chambers, R.C., Taylor, P.M., Yacoub, M.H., Chester, A.H.: Collagen synthesis by mesenchymal stem cells and aortic valve interstitial cells in response to mechanical stretch. Cardiovasc. Res. **71**, 548–556 (2006)
61. Mol, A., Rutten, M.C., Driessen, N.J., Bouten, C.V., Zund, G., Baaijens, F.P., Hoerstrup, S.P.: Autologous human tissue-engineered heart valves: prospects for systemic application. Circulation **114**, I152–I158 (2006)
62. Schmidt, D., Mol, A., Breymann, C., Achermann, J., Odermatt, B., Gossi, M., Neuenschwander, S., Pretre, R., Genoni, M., Zund, G., Hoerstrup, S.P.: Living autologous heart valves engineered from human prenatally harvested progenitors. Circulation **114**, I125–I131 (2006)
63. Schmidt, D., Achermann, J., Odermatt, B., Breymann, C., Mol, A., Genoni, M., Zund, G., Hoerstrup, S.P.: Prenatally fabricated autologous human living heart valves based on amniotic fluid derived progenitor cells as single cell source. Circulation **116**, I64–I70 (2007)
64. Schmidt, D., Stock, U.A., Hoerstrup, S.P.: Tissue engineering of heart valves using decellularized xenogeneic or polymeric starter matrices. Philos. Trans. R. Soc. Lond. B Biol. Sci. **362**, 1505–1512 (2007)
65. Shinoka, T., Shum-Tim, D., Ma, P.X., Tanel, R.E., Langer, R., Vacanti, J.P., Mayer, J.E., Jr. Tissue-engineered heart valve leaflets: does cell origin affect outcome? Circulation **96**, II-7 (1997)
66. Batten, P., Sarathchandra, P., Antoniw, J.W., Tay, S.S., Lowdell, M.W., Taylor, P.M., Yacoub, M.H.: Human mesenchymal stem cells induce T cell anergy and downregulate T cell allo-responses via the TH2 pathway: relevance to tissue engineering human heart valves. Tissue Eng. **12**, 2263–2273 (2006)
67. Bradley, J.A., Bolton, E.M., Pedersen, R.A.: Stem cell medicine encounters the immune system. Nat. Rev. Immunol. **2**, 859–871 (2002)
68. Frank, M.H., Sayegh, M.H.: Immunomodulatory functions of mesenchymal stem cells. Lancet **363**, 1411–1412 (2004)
69. Kassem, M.: Mesenchymal stem cells: biological characteristics and potential clinical applications. Cloning Stem Cells **6**, 369–374 (2004)
70. Lee, M.W., Yang, M.S., Park, J.S., Kim, H.C., Kim, Y.J., Choi, J.: Isolation of mesenchymal stem cells from cryopreserved human umbilical cord blood. Int. J. Hematol. **81**, 126–130 (2005)
71. Sotiropoulou, P.A., Perez, S.A., Salagianni, M., Baxevanis, C.N., Papamichail, M.: Characterization of the optimal culture conditions for clinical scale production of human mesenchymal stem cells. Stem Cells **24**, 462–471 (2006)
72. Zuk, P.A., Zhu, M., Mizuno, H., Huang, J., Futrell, J.W., Katz, A.J., Benhaim, P., Lorenz, H.P., Hedrick, M.H.: Multilineage cells from human adipose tissue: implications for cell-based therapies. Tissue Eng. **7**, 211–228 (2001)
73. Verfaillie, C.M.: Adult stem cells: assessing the case for pluripotency. Trends Cell Biol. **12**, 502–508 (2002)
74. Schnell, A.M., Hoerstrup, S.P., Zund, G., Kolb, S., Sodian, R., Visjager, J.F., Grunenfelder, J., Suter, A., Turina, M.: Optimal cell source for cardiovascular tissue engineering: venous vs. aortic human myofibroblasts. Thorac. Cardiovasc. Surg. **49**, 221–225 (2001)
75. Stock, U.A., Nagashima, M., Khalil, P.N., Nollert, G.D., Herden, T., Sperling, J.S., Moran, A., Lien, J., Martin, D.P., Schoen, F.J., Vacanti, J.P., Mayer, J.E., Jr.: Tissue-engineered valved conduits in the pulmonary circulation. J. Thorac. Cardiovasc. Surg. **119**, 732–740 (2000)

76. Simionescu, D.T.: Prevention of calcification in bioprosthetic heart valves: challenges and perspectives. Expert Opin. Biol. Ther. **4**, 1971–1985 (2004)
77. Knight, R.L., Wilcox, H.E., Korossis, S.A., Fisher, J., Ingham, E.: The use of acellular matrices for the tissue engineering of cardiac valves. Proc. Inst. Mech. Eng. H **222**, 129–143 (2008)
78. Badylak, S.F.: Xenogeneic extracellular matrix as a scaffold for tissue reconstruction. Transpl. Immunol. **12**, 367–377 (2004)
79. Booth, C., Korossis, S.A., Wilcox, H.E., Watterson, K.G., Kearney, J.N., Fisher, J., Ingham, E.: Tissue engineering of cardiac valve prostheses I: development and histological characterization of an acellular porcine scaffold. J. Heart Valve Dis. **11**, 457–462 (2002)
80. Shinoka, T., Shum-Tim, D., Ma, P.X., Tanel, R.E., Isogai, N., Langer, R., Vacanti, J.P., Mayer, J.E., Jr.: Creation of viable pulmonary artery autografts through tissue engineering. J. Thorac. Cardiovasc. Surg. **115**, 536–545 (1998)
81. Korossis, S.A., Booth, C., Wilcox, H.E., Watterson, K.G., Kearney, J.N., Fisher, J., Ingham, E.: Tissue engineering of cardiac valve prostheses II: biomechanical characterization of decellularized porcine aortic heart valves. J. Heart Valve Dis. **11**, 463–471 (2002)
82. Simon, P., Kasimir, M.T., Seebacher, G., Weigel, G., Ullrich, R., Salzer-Muhar, U., Rieder, E., Wolner, E.: Early failure of the tissue engineered porcine heart valve SYNERGRAFT in pediatric patients. Eur. J. Cardiothorac. Surg. **23**, 1002–1006 (2003)
83. Erdbrugger, W., Konertz, W., Dohmen, P.M., Posner, S., Ellerbrok, H., Brodde, O.E., Robenek, H., Modersohn, D., Pruss, A., Holinski, S., Stein-Konertz, M., Pauli, G.: Decellularized xenogenic heart valves reveal remodeling and growth potential in vivo. Tissue Eng. **12**, 2059–2068 (2006)
84. Kasimir, M.T., Rieder, E., Seebacher, G., Silberhumer, G., Wolner, E., Weigel, G., Simon, P.: Comparison of different decellularization procedures of porcine heart valves. Int. J. Artif. Organs **26**, 421–427 (2003)
85. Baraki, H., Tudorache, I., Braun, M., Hoffler, K., Gorler, A., Lichtenberg, A., Bara, C., Calistru, A., Brandes, G., Hewicker-Trautwein, M., Hilfiker, A., Haverich, A., Cebotari, S.: Orthotopic replacement of the aortic valve with decellularized allograft in a sheep model. Biomaterials **30**, 6240–6246 (2009)
86. Dohmen, P.M., Costa, F., Lopes, S.V., Yoshi, S., Souza, F.P., Vilani, R., Costa, M.B., Konertz, W.: Results of a decellularized porcine heart valve implanted into the juvenile sheep model. Heart Surg. Forum **8**, E100–E104 (2005)
87. Bechtel, J.F., Stierle, U., Sievers, H.H.: Fifty-two months' mean follow up of decellularized SynerGraft-treated pulmonary valve allografts. J. Heart Valve Dis. 17: 98–104 (2008)
88. Konertz, W., Dohmen, P.M., Liu, J., Beholz, S., Dushe, S., Posner, S., Lembcke, A., Erdbrugger, W.: Hemodynamic characteristics of the Matrix P decellularized xenograft for pulmonary valve replacement during the Ross operation. J. Heart Valve Dis. **14**, 78–81 (2005)
89. Ye, Q., Zund, G., Jockenhoevel, S., Schoeberlein, A., Hoerstrup, S.P., Grunenfelder, J., Benedikt, P., Turina, M.: Scaffold precoating with human autologous extracellular matrix for improved cell attachment in cardiovascular tissue engineering. ASAIO J **46**, 730–733 (2000)
90. Rabkin, E., Hoerstrup, S.P., Aikawa, M., Mayer, J.E., Schoen, F.J.: Evolution of cell phenotype and extracellular matrix in tissue-engineered heart valves during in vitro maturation and in vivo remodelling. J. Heart Valve Dis. **11**, 308–314 (2002)
91. Sodian, R., Hoerstrup, S.P., Sperling, J.S., Daebritz, S., Martin, D.P., Moran, A.M., Kim, B.S., Schoen, F.J., Vacanti, J.P., Mayer, J.E., Jr.: Early in vivo experience with tissue-engineered trileaflet heart valves. Circulation **102**, III22–III29 (2000)
92. Sutherland, F.W., Perry, T.E., Yu, Y., Sherwood, M.C., Rabkin, E., Masuda, Y., Garcia, G.A., McLellan, D.L., Engelmayr, G.C., Jr., Sacks, M.S., Schoen, F.J., Mayer, J.E., Jr.: From stem cells to viable autologous semilunar heart valve. Circulation **111**, 2783–2791 (2005)
93. Kim, W.G., Cho, S.K., Kang, M.C., Lee, T.Y., Park, J.K.: Tissue-engineered heart valve leaflets: an animal study. Int. J. Artif. Organs **24**, 642–648 (2001)

94. Shinoka, T.: Tissue engineered heart valves: autologous cell seeding on biodegradable polymer scaffold. Artif. Organs **26**, 402–406 (2002)
95. Shinoka, T., Breuer, C.K., Tanel, R.E., Zund, G., Miura, T., Ma, P.X., Langer, R., Vacanti, J.P., Mayer, J.E., Jr.: Tissue engineering heart valves: valve leaflet replacement study in a lamb model. Ann. Thorac. Surg. **60**, S513–S516 (1995)
96. Shinoka, T., Ma, P.X., Shum-Tim, D., Breuer, C.K., Cusick, R.A., Zund, G., Langer, R., Vacanti, J.P., Mayer, J.E., Jr.: Tissue-engineered heart valves. Autologous valve leaflet replacement study in a lamb model. Circulation **94**, II164–II168 (1996)
97. Taylor, P.M.: Biological matrices and bionanotechnology. Philos. Trans. R. Soc. Lond. B Biol. Sci. **362**, 1313–1320 (2007)
98. Taylor, P.M., Sachlos, E., Dreger, S.A., Chester, A.H., Czernuszka, J.T., Yacoub, M.H.: Interaction of human valve interstitial cells with collagen matrices manufactured using rapid prototyping. Biomaterials **27**, 2733–2737 (2006)
99. Terrovitis, J.V., Bulte, J.W., Sarvananthan, S., Crowe, L.A., Sarathchandra, P., Batten, P., Sachlos, E., Chester, A.H., Czernuszka, J.T., Firmin, D.N., Taylor, P.M., Yacoub, M.H.: Magnetic resonance imaging of ferumoxide-labeled mesenchymal stem cells seeded on collagen scaffolds-relevance to tissue engineering. Tissue Eng. **12**, 2765–2775 (2006)
100. Rothenburger, M., Vischer, P., Volker, W., Glasmacher, B., Berendes, E., Scheld, H.H., Deiwick, M.: In vitro modelling of tissue using isolated vascular cells on a synthetic collagen matrix as a substitute for heart valves. Thorac. Cardiovasc. Surg. **49**, 204–209 (2001)
101. Rothenburger, M., Volker, W., Vischer, J.P., Berendes, E., Glasmacher, B., Scheld, H.H., Deiwick, M.: Tissue engineering of heart valves: formation of a three-dimensional tissue using porcine heart valve cells. ASAIO J. **48**, 586–591 (2002)
102. Rothenburger, M., Volker, W., Vischer, P., Glasmacher, B., Scheld, H.H., Deiwick, M.: Ultrastructure of proteoglycans in tissue-engineered cardiovascular structures. Tissue Eng. **8**, 1049–1056 (2002)
103. Dreger, S.A., Chester, A.H., Bowles, C.T., Yacoub, M.H., Taylor, P.M.: Modification of biological scaffolds with specific extracellular matrix proteins enhances collagen synthesis by mesenchymal stem cells. Tissue Eng. **13**, 1729 (2007)
104. Koide, T.: Triple helical collagen-like peptides: engineering and applications in matrix biology. Connect. Tissue Res. **46**, 131–141 (2005)
105. Koide, T.: Designed triple-helical peptides as tools for collagen biochemistry and matrix engineering. Philos. Trans. R. Soc. Lond. B Biol. Sci. **362**, 1281–1291 (2007)
106. Koide, T., Homma, D.L., Asada, S., Kitagawa, K.: Self-complementary peptides for the formation of collagen-like triple helical supramolecules. Bioorg. Med. Chem. Lett. **15**, 5230–5233 (2005)
107. Yamazaki, C.M., Asada, S., Kitagawa, K., Koide, T.: Artificial collagen gels via self-assembly of de novo designed peptides. Biopolymers **90**, 816–823 (2008)
108. Turley, E.A., Noble, P.W., Bourguignon, L.Y.: Signaling properties of hyaluronan receptors. J. Biol. Chem. **277**, 4589–4592 (2002)
109. Masters, K.S., Shah, D.N., Walker, G., Leinwand, L.A., Anseth, K.S.: Designing scaffolds for valvular interstitial cells: cell adhesion and function on naturally derived materials. J. Biomed. Mater. Res. A **71**, 172–180 (2004)
110. Ramamurthi, A., Vesely, I.: Evaluation of the matrix-synthesis potential of crosslinked hyaluronan gels for tissue engineering of aortic heart valves. Biomaterials **26**, 999–1010 (2005)
111. Flanagan, T.C., Wilkins, B., Black, A., Jockenhoevel, S., Smith, T.J., Pandit, A.S.: A collagen-glycosaminoglycan co-culture model for heart valve tissue engineering applications. Biomaterials **27**, 2233–2246 (2006)
112. Syedain, Z.H., Tranquillo, R.T.: Controlled cyclic stretch bioreactor for tissue-engineered heart valves. Biomaterials **30**, 4078–4084 (2009)
113. Gu, X., Masters, K.S.: Regulation of valvular interstitial cell calcification by adhesive peptide sequences. J. Biomed. Mater. Res. A **93**, 1620–1630 (2010)

114. Breuer, C.K., Mettler, B.A., Anthony, T., Sales, V.L., Schoen, F.J., Mayer, J.E.: Application of tissue-engineering principles toward the development of a semilunar heart valve substitute. Tissue Eng. **10**, 1725–1736 (2004)
115. Dumont, K., Yperman, J., Verbeken, E., Segers, P., Meuris, B., Vandenberghe, S., Flameng, W., Verdonck, P.R.: Design of a new pulsatile bioreactor for tissue engineered aortic heart valve formation. Artif. Organs **26**, 710–714 (2002)
116. Hildebrand, D.K., Wu, Z.J., Mayer, J.E., Jr., Sacks, M.S.: Design and hydrodynamic evaluation of a novel pulsatile bioreactor for biologically active heart valves. Ann. Biomed. Eng. **32**, 1039–1049 (2004)
117. Mol, A., Driessen, N.J., Rutten, M.C., Hoerstrup, S.P., Bouten, C.V., Baaijens, F.P.: Tissue engineering of human heart valve leaflets: a novel bioreactor for a strain-based conditioning approach. Ann. Biomed. Eng. **33**, 1778–1788 (2005)
118. Ruel, J., Lachance, G.: A new bioreactor for the development of tissue-engineered heart valves. Ann. Biomed. Eng. **37**, 674–681 (2009)
119. Rubbens, M.P., Mol, A., Boerboom, R.A., Bank, R.A., Baaijens, F.P., Bouten, C.V.: Intermittent straining accelerates the development of tissue properties in engineered heart valve tissue. Tissue Eng. Part A **15**, 999–1008 (2009)
120. Rubbens, M.P., Mol, A., van Marion, M.H., Hanemaaijer, R., Bank, R.A., Baaijens, F.P., Bouten, C.V.: Straining mode-dependent collagen remodeling in engineered cardiovascular tissue. Tissue Eng. Part A **15**, 841–849 (2009)
121. Alenghat, F.J., Ingber, D.E.: Mechanotransduction: all signals point to cytoskeleton, matrix, and integrins. Sci. STKE **2002**, e6 (2002)
122. Chiquet, M., Renedo, A.S., Huber, F., Fluck, M.: How do fibroblasts translate mechanical signals into changes in extracellular matrix production? Matrix Biol. **22**, 73–80 (2003)
123. Lehoux, S., Tedgui, A.: Cellular mechanics and gene expression in blood vessels. J. Biomech. **36**, 631–643 (2003)
124. Wang, N., Butler, J.P., Ingber, D.E.: Mechanotransduction across the cell surface and through the cytoskeleton. Science **260**, 1124–1127 (1993)
125. Mol, A., Bouten, C.V., Zund, G., Gunter, C.I., Visjager, J.F., Turina, M.I., Baaijens, F.P., Hoerstrup, S.P.: The relevance of large strains in functional tissue engineering of heart valves. Thorac. Cardiovasc. Surg. **51**, 78–83 (2003)
126. Schenke-Layland, K., Opitz, F., Gross, M., Doring, C., Halbhuber, K.J., Schirrmeister, F., Wahlers, T., Stock, U.A.: Complete dynamic repopulation of decellularized heart valves by application of defined physical signals-an in vitro study. Cardiovasc. Res. **60**, 497–509 (2003)
127. Carr-White, G.S., Afoke, A., Birks, E.J., Hughes, S., O'Halloran, A., Glennen, S., Edwards, S., Eastwood, M., Yacoub, M.H.: Aortic root characteristics of human pulmonary autografts. Circulation **102**, III15–III21 (2000)
128. Osman, L., Chester, A.H., Amrani, M., Yacoub, M.H., Smolenski, R.T.: A novel role of extracellular nucleotides in valve calcification—a potential target for atorvastatin. Circulation **114**, I566–I572 (2006)
129. Osman, L., Chester, A.H., Sarathchandra, P., Latif, N., Meng, W.F., Taylor, P.M., Yacoub, M.H.: A novel role of the sympatho-adrenergic system in regulating valve calcification. Circulation **116**, I282–I287 (2007)
130. Rajamannan, N.M., Subramaniam, M., Caira, F., Stock, S.R., Spelsberg, T.C.: Atorvastatin inhibits hypercholerolemia-induced calcification in the aortic valves via the Lrp5 receptor pathway. Circulation **112**, I229–I234 (2005)
131. Muller, A.M., Cronen, C., Kupferwasser, L.I., Oelert, H., Muller, K.M., Kirkpatrick, C.J.: Expression of endothelial cell adhesion molecules on heart valves: up-regulation in degeneration as well as acute endocarditis. J. Pathol. **191**, 54–60 (2000)
132. Otto, C.M., Kuusisto, J., Reichenbach, D.D., Gown, A.M., O'Brien, K.D.: Characterization of the early lesion of 'degenerative' valvular aortic stenosis. Histological and immunohistochemical studies. Circulation **90**, 844–853 (1994)

133. O'Brien, K.D., Kuusisto, J., Reichenbach, D.D., Ferguson, M., Giachelli, C., Alpers, C.E., Otto, C.M.: Osteopontin is expressed in human aortic valvular lesions. Circulation **92**, 2163–2168 (1995)
134. Rajamannan, N.M., Subramaniam, M., Rickard, D., Stock, S.R., Donovan, J., Springett, M., Orszulak, T., Fullerton, D.A., Tajik, A.J., Bonow, R.O., Spelsberg, T.: Human aortic valve calcification is associated with an osteoblast phenotype. Circulation **107**, 2181–2184 (2003)
135. Bryan, A.J., Angelini, G.D.: The biology of saphenous vein graft occlusion: etiology and strategies for prevention. Curr. Opin. Cardiol. **9**, 641–649 (1994)
136. Caira, F.C., Stock, S.R., Gleason, T.G., McGee, E.C., Huang, J., Bonow, R.O., Spelsberg, T.C., McCarthy, P.M., Rahimtoola, S.H., Rajamannan, N.M.: Human degenerative valve disease is associated with up-regulation of low-density lipoprotein receptor-related protein 5 receptor-mediated bone formation. J. Am. Coll. Cardiol. **47**, 1707–1712 (2006)
137. Mohler, E.R., III, Gannon, F., Reynolds, C., Zimmerman, R., Keane, M.G., Kaplan, F.S.: Bone formation and inflammation in cardiac valves. Circulation **103**, 1522–1528 (2001)
138. Hu, J.F., Gilmer, L., Hopkins, R., Wolfinbarger, L., Jr.: Effects of antibiotics on cellular viability in porcine heart valve tissue. Cardiovasc. Res. **23**, 960–964 (1989)
139. van Geldorp, M.W., Eric Jamieson, W.R., Kappetein, A.P., Ye, J., Fradet, G.J., Eijkemans, M.J., Grunkemeier, G.L., Bogers, A.J., Takkenberg, J.J.: Patient outcome after aortic valve replacement with a mechanical or biological prosthesis: weighing lifetime anticoagulant-related event risk against reoperation risk. J. Thorac. Cardiovasc. Surg. **137**, 881–886 (2009)

Author Index

A
Ahluwalia Arti, 29
Akhyari Payam, 49
Aubin Hug, 49

B
Barth Mareike, 49
Barbani Niccoletta, 187

C
Chester Adrian H., 243
Christman Karen L., 133
Cohen Smadar, 81
Covino Elvio, 215
Cristallini Caterina, 187

D
Di Nardo Paolo, 29
Doevendans Pieter A. F., 1
Donndorf Peter, 95
Dvir Tal, 81

F
Forte Giancarlo, 29

G
Gaetani Roberto, 1
Genovese Jorge A., 215
Giraud Marie Noëlle, 165
Giusti Paolo, 187

L
Leor Jonathan, 81
Lichtenberg Artur, 49

M
Minieri Marilena, 29
Messina Elisa, 1

R
Rainer Alberto, 215
Rosellini Elisabetta, 187

S
Singelyn Jennifer M., 133
Sluijter Joost P. G., 1
Spadaccio Cristiano, 215
Steinhoff Gustav, 95

T
Taylor Patricia M., 243
Tevaearai Hendrik, 165
Tirella Annalisa, 29

Y
Yacoub Magdi H., 243

Z
Zimmermann Wolfram-Hubertus, 111